Transforming Industry using Digital Twin Technology

Ashutosh Mishra • May El Barachi
Manoj Kumar
Editors

Transforming Industry using Digital Twin Technology

Editors
Ashutosh Mishra
Yonsei University
Incheon, Korea (Republic of)

May El Barachi
School of Computer Science
University of Wollongong in Dubai
Dubai, United Arab Emirates

Manoj Kumar
School of Computer Science
University of Wollongong in Dubai
Dubai, United Arab Emirates

ISBN 978-3-031-58522-7 ISBN 978-3-031-58523-4 (eBook)
https://doi.org/10.1007/978-3-031-58523-4

© The Editor(s) (if applicable) and The Author(s), under exclusive license to Springer Nature Switzerland AG 2024

This work is subject to copyright. All rights are solely and exclusively licensed by the Publisher, whether the whole or part of the material is concerned, specifically the rights of translation, reprinting, reuse of illustrations, recitation, broadcasting, reproduction on microfilms or in any other physical way, and transmission or information storage and retrieval, electronic adaptation, computer software, or by similar or dissimilar methodology now known or hereafter developed.

The use of general descriptive names, registered names, trademarks, service marks, etc. in this publication does not imply, even in the absence of a specific statement, that such names are exempt from the relevant protective laws and regulations and therefore free for general use.

The publisher, the authors and the editors are safe to assume that the advice and information in this book are believed to be true and accurate at the date of publication. Neither the publisher nor the authors or the editors give a warranty, expressed or implied, with respect to the material contained herein or for any errors or omissions that may have been made. The publisher remains neutral with regard to jurisdictional claims in published maps and institutional affiliations.

This Springer imprint is published by the registered company Springer Nature Switzerland AG
The registered company address is: Gewerbestrasse 11, 6330 Cham, Switzerland

Paper in this product is recyclable.

Contents

Digital Twins in Industry: Real-World Applications and Innovations.... 1
Shamik Tiwari and Amar Shukla

Artificial Intelligence in Digital Twins for Sustainable Future.......... 19
Pranati Rakshit, Nandini Saha, Shibam Nandi, and Pritha Gupta

The Place and Role of Digital Twin Applications: Directions
for Energy and Education Sector 45
Nurcan Kilinc-Ata and Ridvan Ata

The Role of Digital Trust in Enhancing Cyber Security Resilience 59
Praveen Kumar Malik

From Reactive to Proactive: Predicting and Optimizing Performance
for Competitive Advantage.. 69
Tapan Kumar Behera and Deep Manishkumar Dave

DT-AXYOLOV5: An Efficient Digital Twin–Assisted
Deep-Learning-Based Blockchain Framework for Patient Discomfort
Detection in Smart Healthcare System 95
J. Antony Vijay, C. D. Premkumar, and P. Revathi

Smart Factory Digital Twin for Performance Measurement,
Optimization, and Prediction...................................... 115
Suhas D. Joshi

Blockchain and Digital Twin 145
Durga Vinay Balla, Sravya Sri Kadiyala, and Nanda Kiran Kante

Personalize Learning Experience in Education Using Digital Twins
with Human-Centered Design and Pedagogy.......................... 165
A. Reethika and P. Kanaga Priya

Human Digital Twin Processes and their Future.................... 187
R. Hepziba Gnanamalar

Digital Twin Application in Various Sectors 219
M. Mythily, Beaulah David, and J. Antony Vijay

A Review of Digital Twin Applications in Various Sectors 239
P. Kanaga Priya and A. Reethika

Digital Twin-Enabled Solution for Smart City Applications 259
Pawan Whig, Balaram Yadav Kasula, Ashima Bhatnagar Bhatia,
Rahul Reddy Nadikattu, and Pavika Sharma

Combining Digital Twin Technology and Intelligent Transportation Systems for Smart Mobility 281
Ajay Dureja, Aman Dureja, Varun Kumar, and Sachin Sabharwal

Navigating the Digital Landscape: Challenges of BIM and Digital Twin Adaptation 297
Kamal Jaafar and Mohamed Awais

Index 325

Digital Twins in Industry: Real-World Applications and Innovations

Shamik Tiwari and Amar Shukla

Introduction

Digital twin (D-Twin) describes a virtual replica or simulation of a tangible entity, system, or process. It amalgamates data from numerous resources, such as sensors, simulations, and real-time analytics, to establish a digital counterpart mirroring the behavior and characteristics of its physical counterpart. A D-Twin is often made up of several parts that work together to produce a virtual copy of a real-world system or object. These elements consist of [1, 2]:

Physical Object: The entity or system that the D-Twin represents is known as the physical object. It could be a piece of machinery, infrastructure, a product, or even an entire factory. To gather data in real-time, the physical item is furnished with sensors, actuators, and other information gathering devices [3].

Data collection: To collect data from the actual object, sensors and other tools are employed. These sensors can measure a wide range of variables, including temperature, pressure, vibration, and others. Real-time measurements and archived data may both be present in the data gathered.

Data processing and integration: After data acquisition, the physical object is represented coherently by processing and integrating the collected data. Data from diverse sources must be cleaned, filtered, and aggregated to do this. Including extra data from outside sources, such as environmental or data about the supply chain, may also be a part of integration [4].

S. Tiwari (✉)
School of Computer Science & Engineering, IILM University, Gurugram, India

School of Computer Science, UPES, Dehradun, India
e-mail: shamik.tiwari@iilm.edu

A. Shukla
School of Computer Science, UPES, Dehradun, India

© The Author(s), under exclusive license to Springer Nature Switzerland AG 2024
A. Mishra et al. (eds.), *Transforming Industry using Digital Twin Technology*,
https://doi.org/10.1007/978-3-031-58523-4_1

Virtual model: Digital representation of the physical object is known as a virtual model. It is made with the use of simulation software, computer-aided design (CAD) software, or other modelling methods. The virtual model accurately depicts the physical object's form, structure, behavior, and functions.

Connectivity: The network architecture that permits interaction among the physical thing, sensors, data processing systems, and various other elements of the D-Twin ecosystem is referred to as connectivity. This connectivity can be cable or wireless and may make use of IoT, cloud computing, and edge computing technologies.

Analytics and AI: Data gathered from the physical object is processed using analytics and artificial intelligence techniques, then incorporated into the D-Twin. Predictive modelling, machine learning algorithms, and statistical analysis are some examples of these techniques. They aid in gaining insights, spotting patterns, spotting abnormalities, and making predictions about how the physical thing will behave and function.

Visualization and User Interface: Users may be able to interact with and view the virtual model and the accompanying data through the visualization component of the D-Twin. This can take the shape of augmented reality (AR) interfaces, dashboards, or 3D visualizations. The user interface gives people a way to keep an eye on and manage the actual item, evaluate data, and come to intelligent choices.

Feedback Loop: The D-Twin and the physical object interact in a feedback loop. The D-Twin is continuously updated with real-time data from the physical device, enabling continued monitoring, analysis, and adjusting. The D-Twin's views and forecasts can be employed to optimize the physical object's performance or to decide on upkeep, upgrades, or enhancements to the process.

Collectively, these elements form a dynamic and interactive D-Twin that offers a virtual version of a real-world system or item and facilitates observing, analysis, and decision-making. Figure 1 shows an example of D-Twin [5, 6].

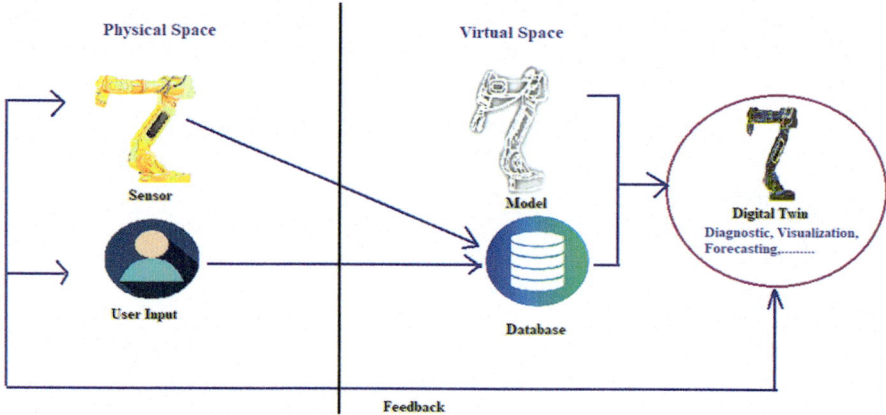

Fig. 1 Digital twin: An example

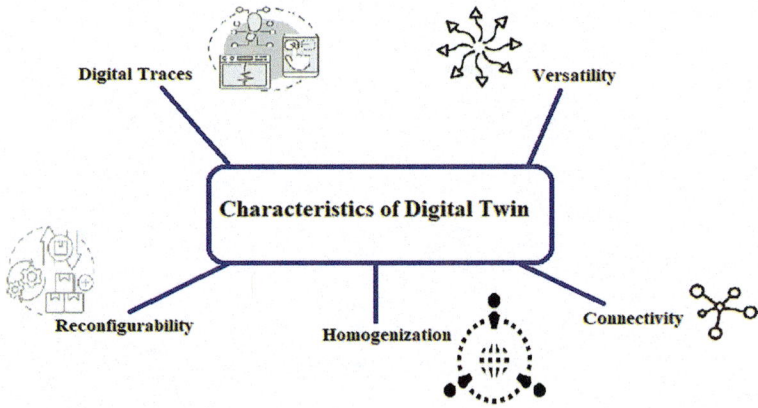

Fig. 2 Characteristics of digital twin

The main characteristics of D-Twins are connectivity, modularity, homogenization digital traces, and reconfigurability as provided in Fig. 2 [7, 8]. Connectivity: Connectivity is the foundation of a D-Twin. It makes it possible for the physical component and its digital counterpart to be connected. The sensors enable physical items to be connected so that data may be collected, processed, and shared via a variety of interconnection methods.

Homogenization: The homogeneity of data has consequences and is made possible by D-Twins. It makes it possible to separate information from its physical form.

Reconfigurability: Digital twins use sensors, AI methods, and statistical analysis to allow reconfigurability.

Digital Traces: Devices using D-Twins leave digital footprints. When a machine malfunctions, the traces can be used to identify the problem's underlying cause.

Versatility: The design and adaptation of products along with manufacturing units are examples of versatility. Manufacturers can modify machines and models because of the addition of flexibility to operational models.

Applications of Digital Twins

The adoption of D-Twin technologies is experiencing rapid growth, revolutionizing the operational landscape for businesses. Primarily applied in engineering and manufacturing, digital twins are instrumental in crafting accurate virtual representations and conducting simulations of operations. Here are a few examples of the diverse applications of D-Twins across various fields as shown in Fig. 3. Digital twins are flexible and adaptable. It enables them to be tailored to particular use cases and industry standards.

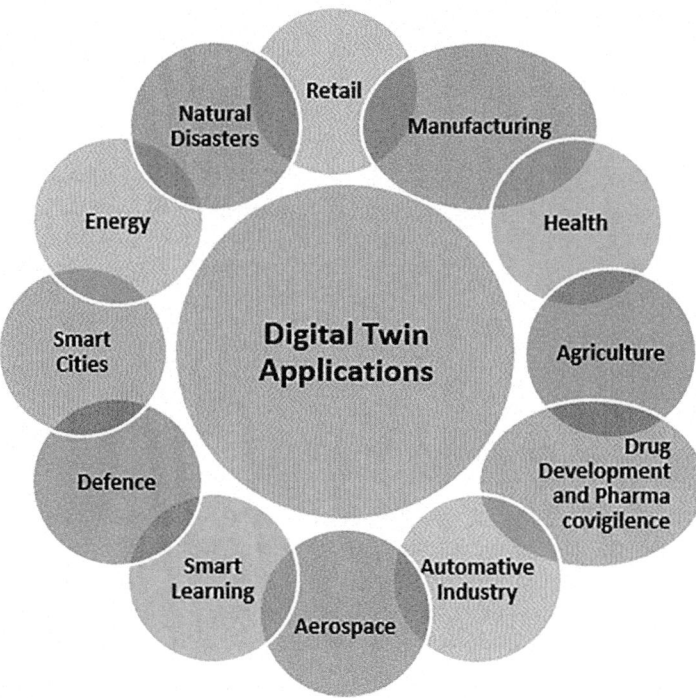

Fig. 3 Digital twin applications

Digital Twins in Manufacturing

In the manufacturing sector, digital twins find application in simulating diverse scenarios, enhancing production efficiency through the optimization of resources, and minimizing downtime. Crucial to the concept of smart manufacturing, data sourced from the plant, systems, supply chain, and equipment plays an integral role in leveraging the capabilities of digital twins. Manufacturers utilizing industry 4.0 apps employ the real-time data power of D-Twins to monitor and evaluate continuously shifting data in their manufacturing operations as shown in Fig. 4. Thanks to this technology [9], manufacturers now have the capability to test and validate a product even before it is introduced to the market. Digital twins aid engineers in identifying any potential process flaws before the product undergoes actual manufacturing by simulating the intended manufacturing procedure. The following few instances show how D-Twins can be used to meet various requirements for smart manufacturing [10, 11].

- Prototype testing and evaluation
- Shorten the time to market for products
- Enhance process and product performance
- Improve production efficiency

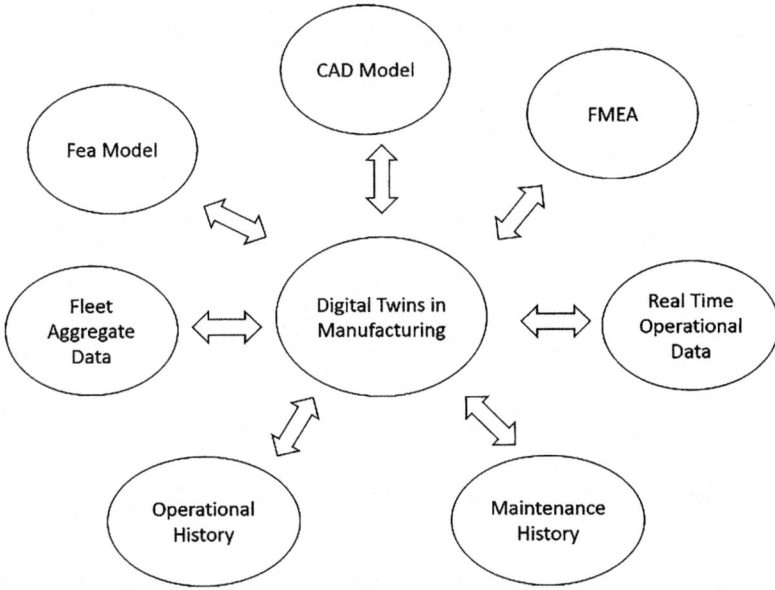

Fig. 4 Digital twins in manufacturing

- Equipment monitoring and preventive maintenance
- Turn on preventative maintenance
- Permit online commissioning
- For manufacturing organizations, D-Twins have several advantages, including:
- Reducing the duration of the manufacturing
- Predictive analysis results in little maintenance
- OEE (Overall Equipment Effectiveness) maximization
- Gaining fresh knowledge about the apparatus to improve performance
- The ability to conduct testing on a virtual duplicate without putting the real equipment at risk
- Evaluating a D-Twin with what-if questions
- Real-time asset management
- System optimization before deployment

Digital Twins in Automotive Industry

Digital twins can be used in the automotive industry to simulate complex systems like engines, transmissions, and suspensions. The D-Twin of the product encompasses the entire car, incorporating its software, engineering, electrical wiring, and physical behavior [12–14].

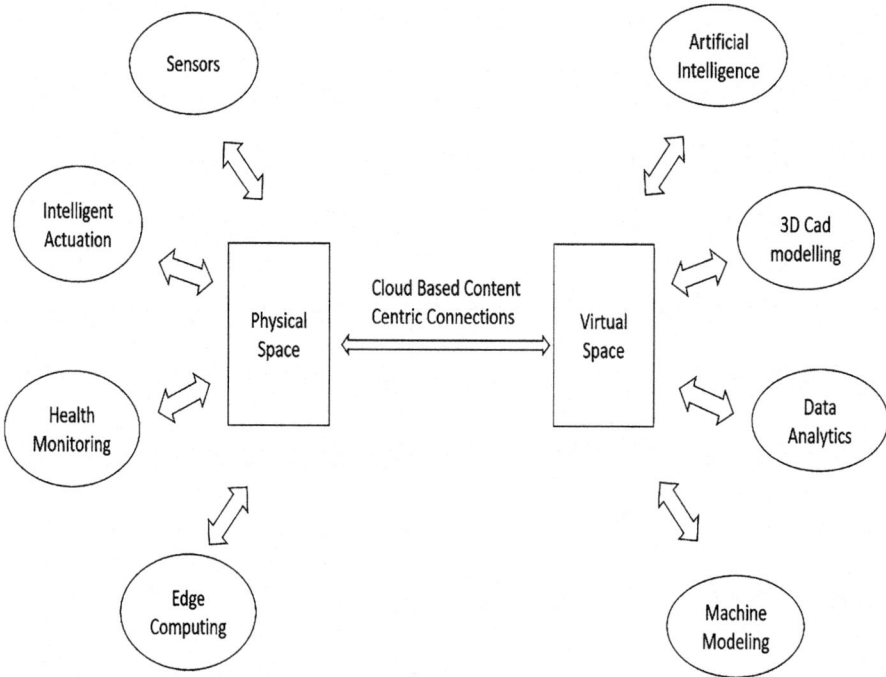

Fig. 5 Digital twins in automotive industry

This comprehensive representation enables the simulation and evaluation of each stage in the development process, facilitating the identification of issues and prediction of potential failures before the production of actual parts. Digital twins make it possible to observe some complex scenarios more effectively and can be used to determine specific circumstances that are needed for elaborate simulations as shown in Fig. 5. For instance, some situations are very challenging to replicate but can greatly improve sensors and programs [15].

Digital Twins in Aerospace

Digital twins can simulate and monitor the performance of airplanes or spacecraft in real-time, including the functionality of individual components. Digital twins offer a virtual depiction of systems like an airplane, a spacecraft, or even a semiconductor part inside a bigger system in the aerospace industry. Similar technology to Digital twin has been employed by the aerospace sector for some time. However, it has primarily been used to create digital models of military planes and commercial aircraft in airports to figure out how to enhance the equipment and achieve peak performance. Additionally, they have been effective in using the tracking capability

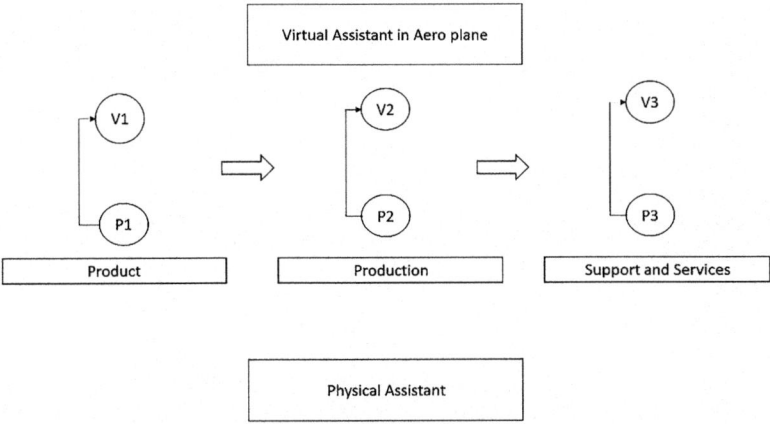

Fig. 6 Digital twins in health care

of D-Twins to follow airplanes around the globe to guarantee the effectiveness of their daily excursions. But as this technology advances and expands, more chances will start to emerge that will enable aircraft engineers to work more productively when it comes to carrying out new model tests, and when it applies to the business aspect of flying, passengers will ultimately have a more excellent and secure experience than ever earlier [16, 17].

To improve performance, the aviation sector still needs to fully implement D-Twin interfaces. When it pertains to a few of the most important components on which they must rely on such as aircraft monitoring, expansion, and examination, the air force and the armed forces, in contrast to the air force, are still unwilling to totally rely on digital twin as shown in Fig. 6. However, there has been a noticeable increase in the adoption of technology; as a result, it is going to be long before they build digital twin, and it will be ready for usage in the aviation sector [18].

Digital Twins in Healthcare

Digital twins can help doctors to test and diagnose medical conditions with greater accuracy, using real-time data from sensors and other devices. Futuristic precision healthcare is expected to include giving each patient an individualized diagnosis and course of medication, with simulation serving a growing part in healthcare. The development of D-Twin technology will make such customization feasible. The capacity to enhance the treatment of patients as well as study is one of the advantages of building a D-Twin in healthcare. For instance, by building a precise replica of a patient's brain, researchers can better understand diseases and how therapies affect human cells [19, 20].

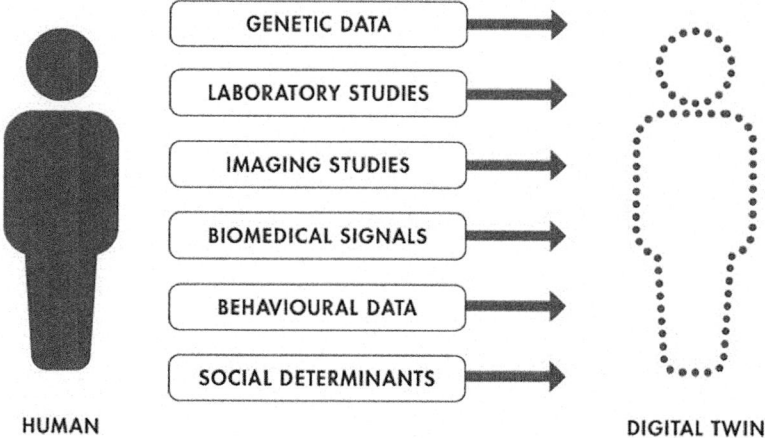

Fig. 7 Digital twin for healthcare [22]

A D-Twin is created when recreating a person or a patient employing recording of vital signs together with anatomical and physiological information. This data might originate from a variety of sources in the era of pervasive wearable technology and biological sensors. The virtual patient model can be updated with information from laboratory investigations and imaging scans performed during the patient's visit to a hospital as shown in Fig. 7. Additionally, a person's genetic, behavioral, and social variables could all be encoded in their D-Twin. A more comprehensive image of the medical condition of the individual is accessible to enhance decision-making when all these data are merged into one virtual depiction of the patient by D-Twin [21].

Digital Twins in Construction

Digital twins play a pivotal role in simulating construction projects, providing stakeholders with the ability to evaluate different outcomes and make well-informed decisions. Within the construction industry, a unified platform is utilized to generate digital twins by amalgamating various types of data. This integration includes 3D models, sensor data, and real-time performance data, enabling a comprehensive representation and analysis of construction projects as shown in Fig. 8. Through this platform, various scenarios, such as material selection, energy consumption, and service plans, can be simulated and optimized [23].

Furthermore, the use of digital twins can be instrumental in early detection of potential issues, resulting in decreased downtime and improved safety measures. An example of D-Twin application is modeling how extreme weather conditions might impact a structure or bridge. This allows engineers to anticipate potential issues and proactively implement protective measures to ensure the resilience and safety of the infrastructure [24].

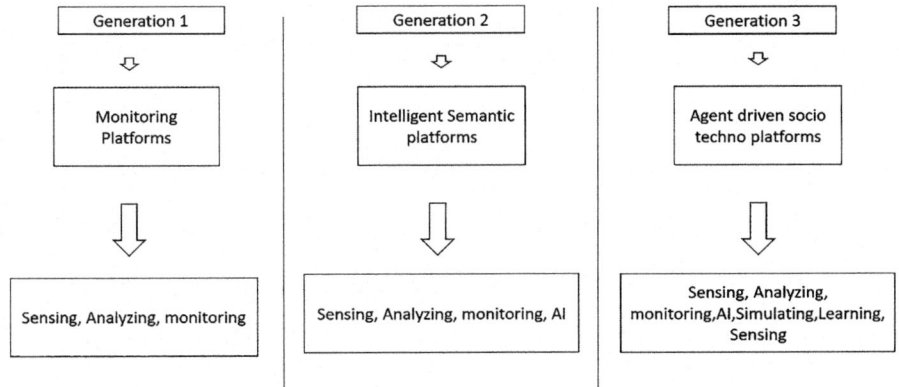

Fig. 8 Digital twin for construction

Digital Twins in Energy and Utilities

Digital twins can be used to predict energy consumption and optimize the use of resources, including renewable energy sources. The digitalization of the utilities sector creates potential for innovative, effective, and clever business practices, such as the application of D-Twin technologies. In fact, using a D-Twin for energy management is one of the finest ways to keep an eye on power usage and regulate energy sensors. The most important characteristics to search for in a D-Twin energy optimization solution are remote monitoring, automation capabilities, flexibility and scalability, and data and network security [25].

Planning and managing renewable energy infrastructure like solar farms or wind parks is made easier with the aid of D-Twins as shown in Fig. 9. They can model how many elements, including sunlight, wind patterns, and environmental variables, affect the efficiency of renewable energy resources. This data is useful for grid integration organizing, energy production prediction, and asset location [26].

Digital Twins in Smart Cities

Digital twins can be used to simulate how infrastructure and other systems work in smart cities, enabling city planners and administrators to optimize operations and increase efficiency. Urban planners as well as architects may visualize and simulate various urban development scenarios thanks to D-Twins. They can examine elements including building assignment, infrastructure needs, transit systems, and land usage as shown in Fig. 10. Decision-makers can make well-informed decisions on urban development, allocation of resources, and sustainability initiatives by modelling and analyzing these possible situations [27, 28].

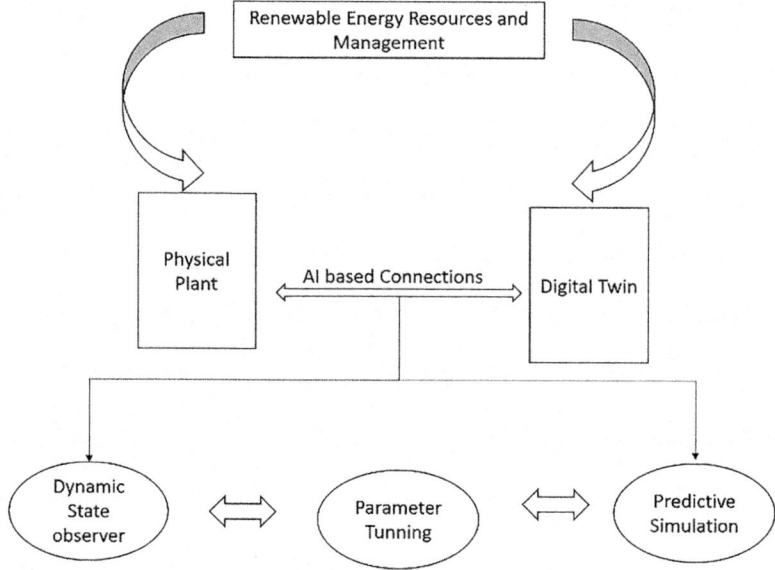

Fig. 9 Digital twins in energy resources and utilities

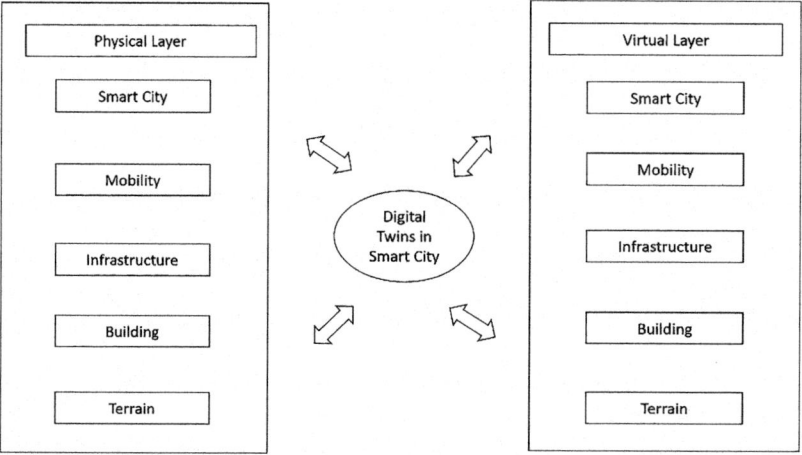

Fig. 10 Digital twins in smart cities

Cities' infrastructure, including roadways, connections, energy companies, and transportation networks, can be managed and maintained with the aid of D-Twins. Digital twins may track the health of infrastructure assets, forecast maintenance requirements, and optimize repairs by combining data from numerous sensors and internet of things (IoT) devices. This assures effective operations, reduces disruptions, and lengthens the infrastructure's entire lifecycle.

Digital Twins in Defense

Digital twins are employed in the defense sector to simulate and test a variety of technologies, including weapon systems, vehicles, and other defense-related equipment. This application helps mitigate the risk of human error and enhances overall performance by providing a virtual environment for rigorous testing and analysis. The efficiency and maintenance of military devices, such as automobiles, planes, or maritime vessels, can be tracked via D-Twins. The D-Twin can analyze the equipment's performance parameters, spot irregularities, and foretell required repairs by gathering data from sensors and other sources. This proactive strategy improves readiness for operation in general, decreases downtime, and optimizes maintenance plans [29, 30].

By recreating and visualizing the operational context, D-Twins can help with mission execution and evaluation. They can include information about the topography, the weather, and other pertinent variables to give commanders information and enhance their ability to make strategic choices. This makes it possible to more effectively allocate resources, evaluate risks, and carry out missions.

Digital Twins in Agriculture

Digital twins can facilitate farmers to improve crop yield and decrease waste by simulating different climate scenarios and optimizing resource use. In the world of agriculture, D-Twins are having a big impact on farming techniques and increasing total yields in agriculture. When referring to virtual reproductions or simulations of actual farming systems, such as crops, animals, and complete agricultural ecosystems, we employ the D-Twins in agriculture. Digital twins offer real-time monitoring, analysis, and forecasting capabilities for optimizing many parts of agricultural operations by utilizing data, sensors, and sophisticated analytics [31].

Digital twins can replicate agricultural development and growth while accounting for elements such as soil characteristics, weather patterns, and irrigation techniques.

Farmers may track the health, behavior, and well-being of livestock by using D-Twins to construct virtual duplicates of the animals. Body temperature, levels of activity, and feeding habits are just a few of the elements that sensors on animals may gather information on, which can then be included into the D-Twin as shown in Fig. 11.

By replicating the full agricultural activity, D-Twins assist precision farming practices. Based on soil and weather conditions, they can offer insights on the best planting strategies, seed choices, and fertilizer applications.

Digital twins can aid in the optimization of planting schedules, determination of the best time for harvesting, and management of potential hazards like outbreaks of

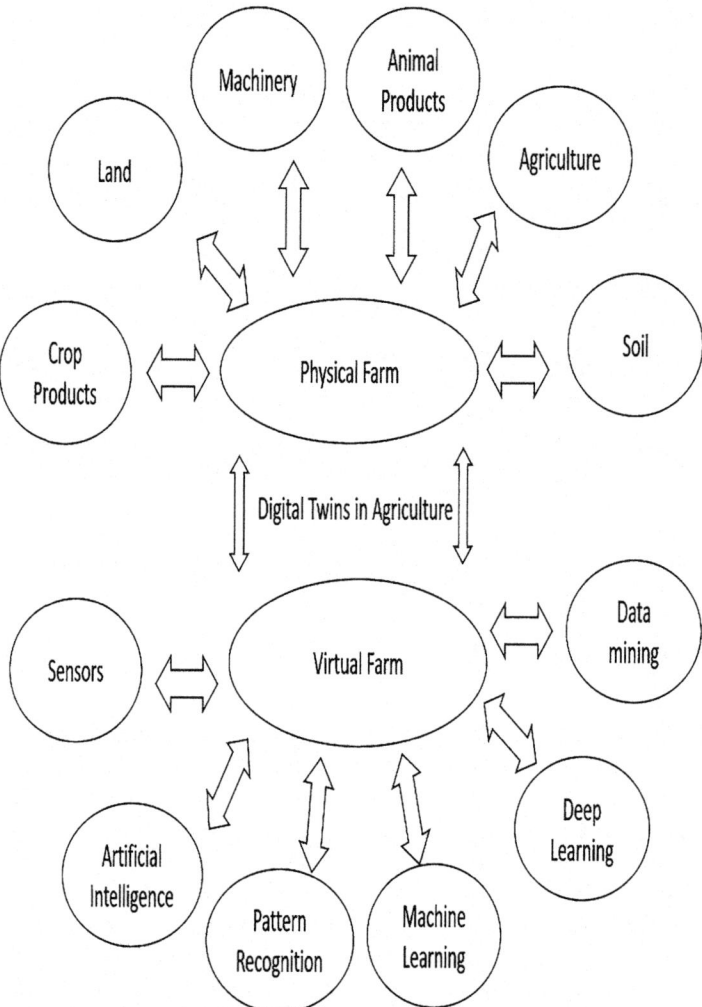

Fig. 11 Digital twins in agriculture

diseases or unfavorable weather occurrences by integrating data from numerous sources, like as weather forecasts, market pricing, and historical trends [32].

Digital twins can suggest exact irrigation plans, cut chemical inputs, and minimize environmental effect while preserving crop output by analyzing data on soil moisture levels, nutrients, and weather trends.

By simulating various layouts, infrastructural setups, and crop rotation techniques, D-Twins can help with the design and planning of agricultural operations. This aids farmers in making the most use of their land, evaluating the viability of new initiatives, and locating any possible obstacles or inconsistencies [33].

Digital Twins in Retail

Digital twins can provide retailers with an accurate insight into their supply chains, allowing them to optimize inventory management and logistics. Retailers can use these D-Twins as a potent tool to improve several aspects of their business, such as product design, managing inventory, consumer experience, as well as supply chain management.

During the design phase, retailers can make D-Twins of their products. Before producing the actual product, they can test and improve various features, materials, and combinations using these virtual models. It aids in lowering expenses, enhancing excellence, and speeding time to trade [34, 35].

Businesses can see their inventory quantities, locations, and circumstances in real time thanks to D-Twins. They may watch things as they move through the supply chain, keep an eye on stock levels, and improve replenishment procedures by combining data from sensors, RFID tags, and other sources. By doing this, you can manage your inventory effectively and avoid shortages and overstocking.

Digital Twins in Drug Development and Pharmacovigilance

In the context of drug development and pharmacovigilance, digital twins can be applied to various stages of the drug lifecycle, from preclinical studies to post-market surveillance.

In drug development, digital twins can help accelerate the discovery process by simulating the effects of potential drugs on virtual models of biological systems. By creating a virtual representation of a biological target, researchers can use computational methods to design drugs that are more likely to be effective and safe. Digital twins can also be used to optimize clinical trial design and predict drug efficacy and toxicity [36].

In the field of pharmacovigilance, digital twins play a crucial role in real-time drug safety monitoring. Through the creation of a virtual representation of a patient population, digital twins are instrumental in identifying potential adverse events and drug interactions before they manifest. Moreover, digital twins contribute to the identification of patient subpopulations at a heightened risk of adverse events, facilitating targeted monitoring and timely intervention measures [37].

Overall, digital twins have the potential to revolutionize drug development and pharmacovigilance by providing a more efficient and accurate way to predict drug behavior and optimize patient outcomes. However, there are still challenges to be addressed, such as data quality and privacy concerns before digital twins can be fully integrated into drug development and pharmacovigilance processes.

Simulate and Prepare for Natural Disasters Using Digital Twins

Digital twins are useful resources for planning and modelling natural disasters. Digital twins give organizations the ability to assess the possible effects of disasters, establish response strategies, and improve emergency preparedness by building digital copies of physical settings and infrastructure. Different kinds of natural disasters, including hurricanes, earthquakes, floods, and wildfires, can be simulated using D-Twins. Organizations can perform simulations to learn how the disaster might develop and impact their property, facilities, and neighborhoods by including pertinent data, such as historical climate trends, topographical data, and building designs. These simulations can be used to find weak points, assess evacuation options, and gauge the efficiency of different response tactics [38].

With the help of D-Twins, stakeholders can be trained and educated on catastrophe preparation and action. They can be utilized in scenario-based training sessions, giving staff members the chance to practice the duties they have in a lifelike virtual setting. This enhances planning, decision-making, and the efficiency of emergency responses. The rehabilitation and reconstruction process after a tragedy can be aided by D-Twins. Organizations can utilize D-Twins to assess destruction, plan rebuilding activities, and simulate the efficacy of various recovery techniques by preserving the as-is state of the affected regions prior to a disaster [39].

Digital Twins Enabled Smart Learning in Education

By providing intelligent learning experiences, D-Twins have the potential to revolutionize the education industry. Digital twins can improve teaching strategies and individualized training and offer immersive and engaging educational experiences by building virtual reproductions of actual learning environments [40].

The learning preferences, strengths, and shortcomings of pupils can be captured in virtual profiles made using D-Twins. Digital twins can offer personalized paths to learning and customizable material that cater to the needs of each student by analyzing data from numerous sources, including assessments, learning analytics, and educational assets. This individualized approach boosts involvement, allows students to learn at their own speed, and improves learning results. Students may perform virtual experiments as well as simulations in D-Twins that can mimic actual laboratories. This makes learning opportunities accessible that might not otherwise be possible due to limitations on time, money, or security. Students can investigate intricate scientific ideas, carry out investigations in a safe virtual setting, and track results in real-time, encouraging practical learning and analytical thinking [41].

Future of Digital Twins

Here are some potential trends and developments in the future of digital twins:

- Integration with IoT and Sensor Technologies: The synergy between digital twins and the Internet of Things continues to grow. As more devices and sensors are connected, digital twins can provide richer and more accurate representations of physical assets.
- Advancements in AI and Machine Learning: As artificial intelligence and machine learning technologies evolve; digital twins will likely benefit from more sophisticated analytics and predictive capabilities. This could enhance their ability to simulate and forecast the behavior of physical assets.
- Extended Use Cases: The application of digital twins is likely to expand into new industries and use cases. Beyond manufacturing and industrial settings, digital twins may find applications in healthcare, smart cities, logistics, and more.
- Blockchain for Security and Trust: Blockchain technology could be integrated to enhance the security and trustworthiness of digital twin data. This is especially crucial in industries where data integrity is paramount, such as healthcare and critical infrastructure.
- Decentralized Digital Twins: Decentralized and distributed architectures, such as blockchain, could be used to create decentralized digital twins. This approach may provide greater resilience, security, and scalability.
- Standardization and Interoperability: The development of industry standards and interoperability protocols will be essential for the widespread adoption of digital twins. This ensures that different systems and platforms can seamlessly work together.
- Human Digital Twins: The concept of creating digital twins for individuals (human digital twins) could gain traction, especially in healthcare and personalized medicine. This could aid in better understanding individual health and predicting potential health issues.
- Ethical and Privacy Considerations: As the use of digital twins becomes more prevalent, there will likely be increased attention to ethical and privacy considerations. Striking a balance between collecting valuable data and protecting individuals' privacy will be crucial.
- Edge Computing for Real-Time Processing: With the rise of edge computing, the processing and analysis of data from digital twins may move closer to the source, enabling real-time insights and reducing latency.
- Lifecycle Management: Digital twins may play a more significant role throughout the entire lifecycle of a product or asset, from design and manufacturing to operation, maintenance, and decommissioning.

Conclusion

The numerous ways that D-Twin technology is applied across various businesses and sectors are referred to as D-Twin applications. A D-Twin is a virtual version of a real-world system, process, or item that allows for surveillance, assessment, and modelling in real-time. Some important D-Twin applications are presented in this chapter. These are only a few instances of D-Twin applications. Technology has the power to transform several industries by delivering real-time insights, allowing predictive capabilities, and streamlining processes based on simulations of data. The use of D-Twins in the future has an opportunity to significantly progress numerous fields. With improvements in AI, IoT, simulation capabilities, and collaborative tools driving their wider use and providing significant advantages in terms of productivity, effectiveness, and development across sectors, the future of D-Twins appears bright.

References

1. Fuller, A., Fan, Z., Day, C., & Barlow, C. (2020). Digital twin: Enabling technologies, challenges and open research. *IEEE Access, 28*(8), 108952–108971.
2. Zheng, Y., Yang, S., & Cheng, H. (2019). An application framework of digital twin and its case study. *Journal of Ambient Intelligence and Humanized Computing, 13*(10), 1141–1153.
3. Liu, M., Fang, S., Dong, H., & Xu, C. (2021). Review of digital twin about concepts, technologies, and industrial applications. *Journal of Manufacturing Systems, 1*(58), 346–361.
4. Tao, F., Zhang, H., Liu, A., & Nee, A. Y. (2018). Digital twin in industry: State-of-the-art. *IEEE Transactions on Industrial Informatics, 15*(4), 2405–2415.
5. Haag, S., & Anderl, R. (2018). Digital twin–Proof of concept. *Manufacturing Letters, 1*(15), 64–66.
6. Wright, L., & Davidson, S. (2020). How to tell the difference between a model and a digital twin. *Advanced Modeling and Simulation in Engineering Sciences, 7*(1), 1–3.
7. Jones, D., Snider, C., Nassehi, A., Yon, J., & Hicks, B. (2020). Characterising the Digital Twin: A systematic literature review. *CIRP Journal of Manufacturing Science and Technology, 1*(29), 36–52.
8. VanDerHorn, E., & Mahadevan, S. (2021). Digital Twin: Generalization, characterization and implementation. *Decision Support Systems, 1*(145), 113524.
9. Kholopov, V. A., Antonov, S. V., Kurnasov, E. V., & Kashirskaya, E. N. (2019). Digital twins in manufacturing. *Russian Engineering Research, 39*, 1014–1020.
10. Kritzinger, W., Karner, M., Traar, G., Henjes, J., & Sihn, W. (2018). Digital Twin in manufacturing: A categorical literature review and classification. *Ifac-PapersOnline, 51*(11), 1016–1022.
11. Onaji, I., Tiwari, D., Soulatiantork, P., Song, B., & Tiwari, A. (2022). Digital twin in manufacturing: Conceptual framework and case studies. *International Journal of Computer Integrated Manufacturing, 35*(8), 831–858.
12. Piromalis, D., & Kantaros, A. (2022). Digital twins in the automotive industry: The road toward physical-digital convergence. *Applied System Innovation, 5*(4), 65.
13. Damjanovic-Behrendt, V. (2018). A digital twin-based privacy enhancement mechanism for the automotive industry. In *2018 International conference on intelligent systems (IS)* (pp. 272–279). IEEE.

14. Veledar, O., Damjanovic-Behrendt, V., & Macher, G. (2019, September 18–20). Digital twins for dependability improvement of autonomous driving. In *Systems, software and services process improvement: 26th European conference, EuroSPI 2019*, Edinburgh, UK, Proceedings 26 2019 (pp. 415–426). Springer International Publishing.
15. Biesinger, F., Kraß, B., & Weyrich, M. (2019). A survey on the necessity for a digital twin of production in the automotive industry. In *2019 23rd International conference on mechatronics technology (ICMT)* (pp. 1–8). IEEE.
16. Phanden, R. K., Sharma, P., & Dubey, A. (2021). A review on simulation in digital twin for aerospace, manufacturing and robotics. *Materials Today: Proceedings, 38*, 174–178.
17. Li, L., Aslam, S., Wileman, A., & Perinpanayagam, S. (2021). Digital twin in aerospace industry: A gentle introduction. *IEEE Access, 10*, 9543–9562.
18. Liu, S., Bao, J., Lu, Y., Li, J., Lu, S., & Sun, X. (2021). Digital twin modeling method based on biomimicry for machining aerospace components. *Journal of Manufacturing Systems, 58*, 180–195.
19. Ahmadi-Assalemi, G., Al-Khateeb, H., Maple, C., Epiphaniou, G., Alhaboby, Z. A., Alkaabi, S., & Alhaboby, D. (2020). Digital twins for precision healthcare. In *Cyber defence in the age of AI, smart societies and augmented humanity* (pp. 133–158). Springer.
20. Kaul, R., Ossai, C., Forkan, A. R., Jayaraman, P. P., Zelcer, J., Vaughan, S., & Wickramasinghe, N. (2023). The role of AI for developing digital twins in healthcare: The case of cancer care. *Wiley Interdisciplinary Reviews: Data Mining and Knowledge Discovery, 13*(1), e1480.
21. Angulo, C., Gonzalez-Abril, L., Raya, C., & Ortega, J. A. (2020). A proposal to evolving towards digital twins in healthcare. In *Bioinformatics and biomedical engineering: 8th international work-conference, IWBBIO 2020*, Granada, Spain, May 6–8, 2020, Proceedings (pp. 418–426). Springer International Publishing.
22. Suomi, V. *Next-generation digital twins in healthcare*. https://www.medtechintelligence.com/column/next-generation-digital-twins-in-healthcare/
23. Kor, M., Yitmen, I., & Alizadehsalehi, S. (2023). An investigation for integration of deep learning and digital twins towards Construction 4.0. *Smart and Sustainable Built Environment, 12*(3), 461–487.
24. Akanmu, A. A., Anumba, C. J., & Ogunseiju, O. O. (2021). Towards next generation cyber-physical systems and digital twins for construction. *Journal of Information Technology in Construction, 26*, 505–525.
25. Teng, S. Y., Touš, M., Leong, W. D., How, B. S., Lam, H. L., & Máša, V. (2021). Recent advances on industrial data-driven energy savings: Digital twins and infrastructures. *Renewable and Sustainable Energy Reviews, 135*, 110208.
26. Onile, A. E., Machlev, R., Petlenkov, E., Levron, Y., & Belikov, J. (2021). Uses of the digital twins concept for energy services, intelligent recommendation systems, and demand side management: A review. *Energy Reports, 7*, 997–1015.
27. Dembski, F., Wössner, U., Letzgus, M., Ruddat, M., & Yamu, C. (2020). Urban digital twins for smart cities and citizens: The case study of Herrenberg, Germany. *Sustainability, 12*(6), 2307.
28. Zheng, T., Liu, M., Puthal, D., Yi, P., Wu, Y., & He, X. (2022, May 24). Smart grid: Cyber attacks, critical defense approaches, and digital twin. arXiv preprint arXiv:2205.11783.
29. Wu, J., Yang, Y., Cheng, X. U., Zuo, H., & Cheng, Z. (2020). The development of digital twin technology review. In *2020 Chinese Automation Congress (CAC)* (pp. 4901–4906). IEEE.
30. Kraft, E. M. (2016). The air force digital thread/digital twin-life cycle integration and use of computational and experimental knowledge. In *54th AIAA aerospace sciences meeting* (p. 0897).
31. Pylianidis, C., Osinga, S., & Athanasiadis, I. N. (2021). Introducing digital twins to agriculture. *Computers and Electronics in Agriculture, 184*, 105942.
32. Purcell, W., & Neubauer, T. (2022). Digital twins in agriculture: A state-of-the-art review. *Smart Agricultural Technology, 3*, 100094.

33. Purcell, W., Neubauer, T., & Mallinger, K. (2023). Digital Twins in agriculture: Challenges and opportunities for environmental sustainability. *Current Opinion in Environmental Sustainability, 61*, 101252.
34. Kümpel, M., Mueller, C. A., & Beetz, M. (2021). Semantic digital twins for retail logistics. In *Dynamics in logistics: Twenty-five years of interdisciplinary logistics research in Bremen, Germany* (pp. 129–153). Springer International Publishing.
35. Maïzi, Y., & Bendavid, Y. (2021). Building a digital twin for IoT smart stores: A case in retail and apparel industry. *International Journal of Simulation and Process Modelling, 16*(2), 147–160.
36. Kumar, S. H., Talasila, D., Gowrav, M. P., & Gangadharappa, H. V. (2020). Adaptations of Pharma 4.0 from Industry 4.0. *Drug Invention Today, 14*(3), 405.
37. Paulick, K., Seidel, S., Lange, C., Kemmer, A., Cruz-Bournazou, M. N., Baier, A., & Haehn, D. (2022). Promoting sustainability through next-generation biologics drug development. *Sustainability, 14*(8), 4401.
38. Kaewunruen, S., Peng, S., & Phil-Ebosie, O. (2020). Digital twin aided sustainability and vulnerability audit for subway stations. *Sustainability, 12*(19), 7873.
39. Ham, Y., & Kim, J. (2020). Participatory sensing and digital twin city: Updating virtual city models for enhanced risk-informed decision-making. *Journal of Management in Engineering, 36*(3), 04020005.
40. Mashaly, M. (2021). Connecting the twins: A review on digital twin technology & its networking requirements. *Procedia Computer Science, 184*, 299–305.
41. Guc, F., Viola, J., & Chen, Y. (2021). Digital twins enabled remote laboratory learning experience for mechatronics education. In *2021 IEEE 1st international conference on digital twins and parallel intelligence (DTPI)* (pp. 242–245). IEEE.

Artificial Intelligence in Digital Twins for Sustainable Future

Pranati Rakshit, Nandini Saha, Shibam Nandi, and Pritha Gupta

Introduction

Artificial intelligence, often abbreviated as AI, represents a domain within computer science dedicated to constructing intelligent machines capable of emulating human-like cognitive functions. Its primary objective revolves around the development of algorithms and systems that empower computers to engage in reasoning, decision-making, and data-driven learning processes. This multifaceted field spans various subfields such as machine learning, natural language processing, computer vision, and robotics, each contributing to the overarching goal of creating intelligent systems. By harnessing AI technologies, researchers aim to bridge the gap between human intellect and machine computation, thereby revolutionizing industries, advancing scientific discovery, and augmenting human capabilities in diverse domains.

Deep learning, which uses neural networks with numerous layers to learn complicated patterns, machine learning, which allows computers to learn from data without explicit programming, and natural language processing, which enables computers to comprehend and produce human language, are some of the AI techniques. Robotics, computer vision, data mining, and pattern recognition are just a few of the industries where AI is used. With the potential to revolutionize technology and enhance our daily lives, it is changing sectors.

The concept of a digital twin revolves around crafting a virtual representation of a physical entity, enabling the display, analysis, and optimization of its performance in real-time. Through the integration of sensors and IoT devices, data is collected from the physical entity and transmitted to its digital counterpart, facilitating an accurate emulation of its behavior. This digital model serves as a dynamic reflection

---

P. Rakshit (✉) · N. Saha · S. Nandi · P. Gupta
Department of Computer Science and Engineering, JIS College of Engineering, Kalyani, Nadia, India

© The Author(s), under exclusive license to Springer Nature Switzerland AG 2024
A. Mishra et al. (eds.), *Transforming Industry using Digital Twin Technology*,
https://doi.org/10.1007/978-3-031-58523-4_2

of the physical entity, providing insights into its operation, facilitating predictive maintenance, and supporting decision-making processes across various industries.

To qualify as a digital twin (DT), synchronization between the virtual and physical attributes is essential at all times. In the context of a patient-centric Healthcare DT, this entails the creation of a virtual patient within a digital environment, requiring a significant dataset that accurately mirrors the patient's physical condition. This data convergence enables real-time monitoring, analysis, and simulation of the patient's health status, facilitating personalized healthcare interventions and improving overall patient outcomes. Merely possessing one aspect of patient data is insufficient for the Healthcare DT to form a comprehensive understanding of a patient who may encounter unpredictable health incidents. It necessitates access to all relevant information concerning the patient's medication, hospital records, and overall healthcare management [1–4]. Furthermore, it is important to acknowledge that the human body is intricate, influenced by various external factors such as environmental conditions, age, social activities, and more. Moreover, when delving into the causal factors of a disease or co-morbidity, individual variability becomes a critical consideration, as different factors can exert varying effects on different individuals [5]. Hence, the accumulation of comprehensive patient data from birth to death becomes paramount. This wealth of data enables the digital twin (DT) to analyze and accurately deduce the current state of the patient, as well as predict future threats based on real-time data insights [6]. This fundamental principle underpins the development of patient-centric healthcare DTs, offering invaluable benefits in scenarios such as remote patient monitoring. In such instances, the DT can provide all requisite data, enabling effective inquiry resolution.

Blockchain operates as a decentralized ledger system, where the ledger comprising data from all blocks is distributed across multiple peer nodes within the network. The replication process across a distributed network readily facilitates the attainment of immutability and non-repudiation [7]. Data stored within a blockchain, along with additional blocks, are exclusively appendable at the blockchain's terminus. These blocks are intricately connected to their antecedent blocks through the utilization of Cryptographic Hash Functions [8]. Once a block becomes part of the chain, it is inherently shielded from alteration or removal [9].

Therefore, blockchain incorporates a consensus mechanism that relies on prior consensus on rules and adheres to the principle of majority dominance [10]. This is why the blockchain system relies primarily on consensus algorithms, which guarantee consistency across distributed nodes [11].

Blockchain can be generally categorized into two types:

The concept of a Public Blockchain involves enabling unrestricted participation for creating, validating blocks, and recording data onto the blockchain ledger. This type of blockchain is often referred to as a permissionless blockchain due to its open access nature [12]. Within this blockchain framework, participants remain anonymous, and all transaction records are accessible to the public. While it has demonstrated notable strength in terms of security, it also exhibits limitations in propagation speed, leading to restricted transaction throughput and comparatively higher latency [13]. Prominent instances of public blockchain systems include Bitcoin [14] and Ethereum [15]. In the healthcare domain, notable examples of public blockchain systems encompass MedRec [16] and FHIRChain [17]. Conversely, a Private Blockchain functions as a permissioned system, permitting only a predefined set of entities to participate in the validation process. This strategy addresses concerns regarding energy consumption while upholding stringent security protocols [18]. The synchronization of blocks across the distributed network in a private blockchain is expedited due to the smaller participant pool engaged in block validation [19]. Private blockchain solutions are particularly crucial in scenarios where data privacy is imperative, such as in financial reports and health data management. Prominent private blockchain platforms include Hyperledger Fabric, Hyperledger Sawtooth, and Corda [20]. In the healthcare sector, various private blockchain-based systems have been developed [21–24]. Due to its remarkable versatility and practicality, it has captured the interest of the researchers. In this overview, we present a summary of several noteworthy recent research findings [25]. The article offers a construction case study on a hospital DT in China, which had already undergone the construction phase. The hospital twin had already been constructed prior to the authors' delineation of its development using the Continuous Lifecycle Integration method. During the construction process, numerous sensors were strategically installed to capture real-time data from the hospital, enabling centralized control of the entire system through the use of DT.

Alternatively, Liu and colleagues introduced a cloud-centric framework integrating healthcare Distributed Technologies (DT) [26]. The main motive of this project is to provide the medical services to the elders who are not able to take medical facilities due to their disease. The authors have designed a system that consists of four components: The chapter discusses various elements such as physical objects, virtual objects, cloud healthcare service platforms, and healthcare data. While the authors have covered significant aspects, they have not explicitly included any algorithms for predictive measures.

In this particular chapter [27], the authors utilize edge computing to create a healthcare Twin aimed at mitigating heart diseases by leveraging real-time data collected through IoT devices, primarily smartphones. The collected data undergoes data fusion transformation before being stored in a central data storage system. The twin comprises three main structures: data source, AI-inference engine, and multimodal interaction and smart service. The authors' primary focus is on training a Convolutional Neural Network (CNN), although challenges related to data storage and security remain unaddressed. Shamanna et al. [28] have presented a related study on Precision Nutrition applied to DT, introducing Twin Precision Nutrition (TPN). The research focuses on monitoring a specific group of type 2 diabetic patients aged 64 years, aiming to reduce HbA1c levels in their blood. The TPN platform utilizes data collected from body sensors and a mobile app to track and analyze various health signals, enabling personalized treatment for the patients. However, the authors have not detailed the specific mechanism through which they conducted the analysis despite utilizing real-time data.

In their publication, Barbiero and co-authors [29] introduced an innovative architecture that combines the functionalities of a generative model with a graph-based depiction of pathophysiological circumstances. Through the utilization of synthetic data and the enrichment of navigable states within the biological system, their framework facilitates the emulation of intricate clinical scenarios that would have otherwise presented difficulties for thorough examination.

The authors have adopted multiple data models to systematically gather information and have leveraged graph neural networks for deep learning to generate predictions regarding the patient's physiological state evolution.

Petrova et al. [30] have introduced a digital twin (DT) platform tailored for studying behavioral changes in patients suffering from cognitive disorders, with a specific emphasis on multiple sclerosis. The platform encompasses several key components, including data collection functionality specifically tailored for DT purposes. Additionally, an advanced analytical application is integrated within the platform, enabling data aggregation, enrichment, analysis, and visualization, facilitating the generation of novel insights and decision support. The authors have indicated that patient data will be sourced from Electronic Health Records (EHRs), open clinical datasets, social networks, and external applications. However, they have not outlined the necessary measures to ensure data integrity and confidentiality, which may present potential vulnerabilities.

In their publication [31], the authors have introduced a digital twin system for risk diagnosis, aiming to improve decision-making processes for liver diseases by incorporating explainable artificial intelligence. The authors have utilized their developed Random Forest (RDF) model along with the Local Interpretable Model-Agnostic Explanations (LIME) library, which is a state-of-the-art tool for explainable AI. Recognizing the sensitivity of healthcare matters, the authors have emphasized the importance of using explainable AI techniques. While the authors have provided comprehensive details about the algorithms employed, they have not addressed aspects related to storage facilities and security measures, which are crucial in healthcare systems.

In the realm of Healthcare DT, a noteworthy and recent contribution can be found in [32]. The authors present a framework that offers significant benefits for digital healthcare and healthcare operations. The framework revolves around an intelligent context-aware healthcare system primarily focused on diagnosing heart problems and detecting heart disease through the classification of ECG heart rhythms using DT. This system seamlessly integrates artificial intelligence (AI), data analytics, IoT, virtual and augmented reality, as well as digital and physical objects. By combining these technologies, the framework enables real-time data analysis and proactive problem resolution through continuous status monitoring and provides valuable insights for risk management, cost reduction, and predicting future opportunities. However, similar to the aforementioned works, the authors have not addressed the issue of safeguarding the stored data.

The authors have presented a visionary concept that explores the integration of multi-agent systems with DT in the healthcare field [33]. In this context, multi-agent systems refer to software agents capable of providing responses prior to taking actions. From a DT perspective, these agents serve as a framework for developing intelligent systems that combine AI and Distributed AI techniques, incorporating a certain level of autonomy on top of DTs to leverage their inherent features. However, the authors have not included any empirical process or analysis to substantiate their proposal.

The healthcare sector is witnessing significant advancements in DT, indicating substantial progress. However, a critical challenge lies in the gathering of detailed and complex data from the physical environment. Additionally, ensuring the secure storage of this vast amount of data and preserving its integrity and confidentiality remain a paramount concern for DT in the healthcare sector.

Digital Twin (DT)

A digital twin (DT) refers to the representation of a digital asset's anatomy in a digital environment, mirroring a physical phenomenon extracted from the real world. Serving as a complex system, it ensures consistency between the digital and physical realms while also acquiring cognitive insights about the physical environment. The primary role of a DT is to facilitate interaction between the physical and digital domains, making it a crucial component of its functionality. The rapid advancement of cutting-edge technologies such as Radio-Frequency Identification (RFID) and the Internet of Things (IoT) facilitates the convenient collection of data from various aspects of a physical phenomenon. Figure 1 illustrates the structural model and technical composition of a digital twin.

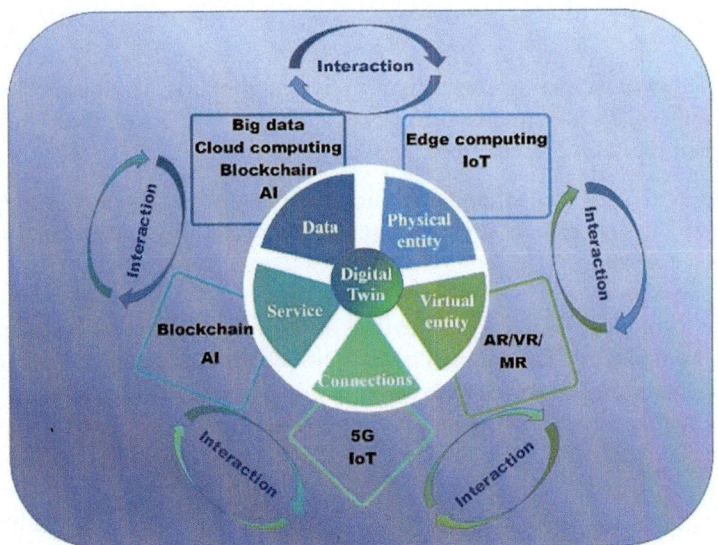

Fig. 1 Structural model and technical composition of digital twin [34]

Categorization

DT can be categorized into two types based on its intended purpose of utilization:

(a) Digital Twin for Developing a Product

The representative DT possesses all the information required to develop the physical product, even though the product itself is yet to be created. By leveraging prior knowledge, the representative DT can anticipate the workflow and behavior of the product through factors such as the current development status, work allocation, and product description. As an illustration, a DT can be introduced during the manufacturing stage of a hospital facility. This is categorized as part of Product Lifecycle Management within the healthcare industry.

(b) Digital Twin for an Individual Instance

This particular form of DT possesses knowledge regarding either a tangible product or a non-spatial phenomenon, and it has the capability to continuously synchronize the virtual representation with real-time data sourced from IoT devices within the physical environment [2]. Suppose an automobile has been constructed with sensors integrated into every crucial component. In such a scenario, the DT would continuously receive data from the vehicle, enabling it to evaluate the current condition of the car through vehicle health management [3]. Remote analysis can facilitate the determination of maintenance intervals for parts, product longevity, and other related factors, thus enabling the extraction of valuable insights regarding the specific instance.

Characteristics of Digital Twin

- *Connectivity*

- In virtual twin technology, we connect physical property and their virtual opposite numbers. We attach sensors to physical objects to decorate their connectivity with their digital representations.

- Data from the physical components are acquired and incorporated through these sensors. This integration allows the sensors to speak the accumulated information to a consumer.
- *Homogenization*
- Digital technology is also characterized by the homogenization of records from bodily components. This approach that a virtual representation just like the physical object may be created for the usage of the accrued statistics. This era can also enable statistics to be decoupled from physical artifacts [35].
- *Reprogrammable*

- A digital twin can enable the replicated bodily product to be reprogrammed. This may be used as the premise for developing new versions of the initial product.
- An actual example of this may be with the case of engines. A virtual twin for a present-day engine may be reprogrammed to enhance gas efficiency and productiveness.
- *Digital traces*
- This era includes virtual lines which can be left while creating a digital twin. Digital strains are utilized by engineers in the diagnosis of problems when machines destroy.

Revolutionization of Industries by Digital Twin

The digital twin era has the potential to convert industries by offering new insights into how bodily systems and approaches function [36]. Figure 2 depicts the application of digital twin in real estate. Here are some methods in which the digital twin era is getting used to revolutionize special industries:

- *Manufacturing*
- Digital twin era can be used to simulate the performance of producing structures and identify bottlenecks or inefficiencies. This can help producers optimize procedures, reduce waste, and enhance productivity.
- *Healthcare*

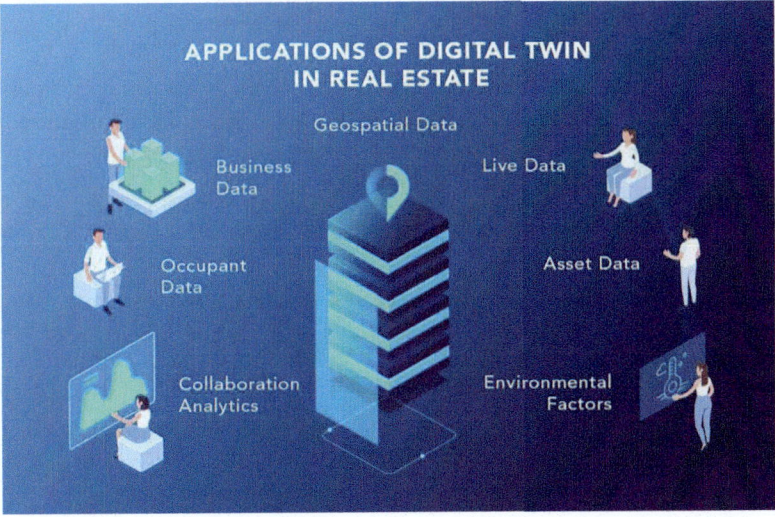

Fig. 2 Application of DT in real estate [37]

- In the healthcare enterprise, the digital twin generation may be used to create personalized models of sufferers' bodies. This can help medical doctors simulate distinctive treatments and expect their effectiveness earlier than they are applied to the affected person. Digital twins can also be used to reveal patients' vital signs in actual time, offering early warning signs and symptoms of capacity fitness issues [38].
- *Construction*
- Digital twin generation can be used to simulate constructing designs and predict how they may carry out in unique situations. This can help architects and engineers optimize constructing designs, lessen prices, and improve protection.
- *Transportation*
- Digital twin technology may be used to reveal the performance of cars in real-time, identify potential issues, and optimize gasoline efficiency. This can assist transportation agencies in reducing prices and enhancing safety.
- *Energy*
- Digital twin era may be used to reveal the overall performance of energy structures and identify possibilities for optimization. This can assist strengthen corporations reduce expenses, improve performance, and decrease carbon emissions.

AI and Digital Twin

Digital twins and AI integration is a rapidly expanding topic that has the potential to completely transform a variety of businesses. Digital twins can be used to enhance decision-making, optimize performance, and avoid issues by fusing the real-time data from physical objects or systems with the predictive capabilities of AI.

The creation and use of digital twin technology heavily relies on artificial intelligence (AI). AI improves the capabilities of digital twins, which are virtual representations of actual things, systems, or processes. It does this by enabling sophisticated analytic, automation, and decision-making.

The Area Where Digital Twins and AI Can Work Together

Machine Learning

Data from digital twins can be analyzed using machine learning algorithms to spot trends and patterns. The performance of the physical object or system can then be predicted using this knowledge.

Natural Language Processing

Text data from digital twins can be analyzed using natural language processing (NLP) [36, 38, 39]. Then, using this data, recommendations and insights can be drawn.

Computer Vision

Images and videos from digital twins may be analyzed using computer vision. Then, you may use this data to locate things, monitor motion, and spot irregularities.

Robotics

Robotics may be used to communicate with real-world systems or physical items. The digital twin can then be updated with this data to increase accuracy.

How Digital Twins and AI Can Work Together

Although it is still in its infancy, the combination of AI and digital twins has the potential to completely transform a wide range of sectors. The opportunities for integration with digital twins will only expand as AI technology advances. Figure 2 shows the structural model and technical composition of digital twin.

Data Analysis and Modelling

To analyze and understand data gathered from sensors, devices, and other sources in the physical system connected to the digital twin, AI techniques like machine learning and deep learning are used.

This data analysis aids in the development of precise models that faithfully capture the traits and behavior of the real-world equivalent.

Predictive Analytic

Using historical data and real-time inputs from the physical system, AI algorithms can be utilized to forecast future behavior and consequences.

Digital twins can foresee prospective problems, performance deterioration, or abnormalities by utilizing machine learning models, enabling proactive maintenance and optimization [39].

Simulation and Scenario Analysis

Using the digital twin, AI can simulate various scenarios and what-if assessments. Digital twins can aid in assessing the effects of various situations, modifications, or interventions in a secure virtual environment before putting them into practice in the real world by utilizing machine learning and computational models.

Control and Optimization

By utilizing real-time data from the digital twin, AI systems can optimist the performance of the physical system. These algorithms can make decisions and give the physical system control signals, which improves performance and efficiency.

For instance, digital twins powered by AI can optimist production processes in the manufacturing industry to maximize output and reduce energy use.

Making Decisions on One's Own

AI can enable autonomous decision-making within the digital twin in complicated systems. The digital twin can make judgement in real-time, adapt to changing circumstances, and react to dynamic events without human intervention by combining AI algorithms with its capabilities.

This is especially helpful for industrial automation, smart cities, and driverless cars.

Real-Time Monitoring and Analytics

The practice of continually gathering and analyzing data as it is created is known as real-time monitoring and analytics. This enables organizations to acquire quick insights and make wise decisions in real-time.

With this strategy, companies may react quickly to shifting circumstances, find anomalies, spot patterns, and improve continuing operations [40].

Cognitive Capabilities

Digital twins' usability and efficacy are greatly improved by cognitive capabilities driven by AI. A digital twin is a virtual replica of a real-world system, process, or item that enables businesses to imitate, track, and examine data from the actual world in a virtual setting.

Understanding context and interpreting large volumes of data gathered from sensors, devices, and other sources in real-time are possible using AI algorithms. AI may spot patterns, correlations, and abnormalities that human operators might not see right away by comprehending the context of the data.

Autonomous Decision-Making

The capacity of an AI system or an autonomous agent to decide and act without direct human interaction is referred to as autonomous decision-making. It entails using machine learning, AI algorithms, and other computational methods to analyze data, evaluate situations, and make decisions based on models, rules, or learnt patterns.

In order to find patterns, correlations, and trends, autonomous decision-making relies on data analysis, both historical and current. AI systems may get insights and make wise judgments based on the facts at hand by digesting massive amounts of data.

Data Fusion and Integration

The process of mixing and merging data from several sources to provide a cohesive and complete view of information is known as data fusion and integration. It seeks to do away with data silos, improve data quality, and provide users a comprehensive view of the data. An overview of data fusion and integration is provided below:

Data sources including databases, files, sensors, apps, APIs, streaming data, and more, and data fusion and integration require combining data from these sources. These sources might differ in their forms, organizational schemes, or degree of quality.

Human-Machine Collaboration

The term "human-machine collaboration" describes the fusion and cooperation of people and machines or artificial intelligence (AI) systems to work towards a shared objective. It integrates the distinct advantages and skills of both people and technology, resulting in synergies that improve production, judgement, and problem-solving.

While computers excel in data processing, pattern recognition, automation, and scalability, humans have cognitive abilities like creativity, critical thinking, empathy, and intuition. Combining these advantages allows human-machine collaboration to make greater use of each party's complementing assets and skills [41].

Continuous Learning and Improvement

The constant process of gaining new information, honed skills, and improved performance is known as continuous learning and improvement. It entails actively seeking feedback, thinking back on experiences, and using that knowledge to change, learn from errors, and get better results.

Receiving feedback on one's performance, whether from superiors, peers, or self-evaluation, is the first step in the continuous learning process. Feedback sheds light on one's skills, flaws, and potential improvement areas.

Innovation and Future Potential

The development and use of novel concepts, procedures, goods, or services that result in a significant improvement are referred to as innovations. In order to handle problems and grab opportunities, it entails the use of creative thinking, problem-solving, and developing technology.

By promoting productivity, competitiveness, and the establishment of new sectors, innovation is essential for encouraging economic growth. It encourages entrepreneurship, draws funding, and opens doors for employment. Innovation-focused countries and organizations are better equipped to adjust to shifting market dynamics and profit from new trends.

How AI and Digital Twin Can Evolve the Future World

The combination of AI and digital twin technology has the potential to revolutionize and shape the future world in various ways:

Improved Decision-Making

AI-powered digital twins can provide organizations with real-time insights, predictive analytics, and scenario modelling capabilities. This enables more informed decision-making and strategic planning, leading to optimized operations, resource allocation, and risk management.

Enhanced Efficiency and Productivity

By simulating and optimizing processes, systems, and infrastructure, digital twins powered by AI can identify inefficiencies, streamline operations, and improve productivity. They enable organizations to automate tasks, optimize workflows, and make data-driven improvements, resulting in increased efficiency and resource utilization.

Sustainable Development

AI-driven digital twins can contribute to sustainable development goals by enabling organizations to design and operate sustainable infrastructure, optimize energy consumption, and promote responsible production and consumption patterns. They facilitate the modelling and analysis of complex systems, helping identify sustainable practices and reduce environmental impact.

Accelerated Innovation

Digital twins, coupled with AI capabilities, provide a platform for rapid experimentation and innovation. By creating virtual replicas of physical systems, organizations can test new ideas, products, and technologies in a risk-free environment. AI algorithms can analyze vast amounts of data generated by digital twins, uncover patterns, and drive innovation through insights and optimization.

Improved Maintenance and Predictive Maintenance

AI-powered digital twins can monitor real-time data from physical assets, allowing organizations to predict maintenance needs and prevent costly breakdowns. By analyzing sensor data and historical patterns, AI algorithms can detect anomalies, predict failures, and recommend proactive maintenance actions, maximizing asset lifespan and minimizing downtime.

Personalized Experiences

AI and digital twins can enable personalized experiences across various domains. For example, in healthcare, digital twins can model patient data, enabling personalized treatment plans and remote monitoring. In retail, digital twins can create personalized shopping experiences based on customer preferences and behaviors, improving customer satisfaction and loyalty [42].

Smarter Cities and Infrastructure

AI-powered digital twins can transform cities and infrastructure management. By simulating and optimizing transportation systems, energy grids, and urban planning, digital twins can enhance efficiency, sustainability, and resilience. They enable intelligent traffic management, energy optimization, and urban development strategies, improving the overall quality of life in cities.

Remote Operations and Collaboration

Digital twins, combined with AI, facilitate remote operations and collaboration. They enable real-time monitoring, remote control, and remote diagnostics of physical assets and systems. This capability is particularly valuable in industries such as manufacturing, energy, and healthcare, where experts can remotely analyze data, provide guidance, and collaborate on problem-solving, reducing the need for physical presence.

Smart Manufacturing and Circular Economy

Highlight how AI-powered digital twins can optimize manufacturing processes, reduce waste, and promote the circular economy. Discuss how AI can monitor and analyze production lines, supply chains, and product lifecycles to identify opportunities for resource optimization, waste reduction, and material recycling.

Intelligent Transportation Systems

Discuss how AI and digital twins can improve transportation systems' efficiency, reduce emissions, and enhance mobility. Explain how AI algorithms can analyze traffic patterns, optimize route planning, manage logistics, and support the development of smart and connected infrastructure.

Urban Planning and Smart Cities

Explore how AI in digital twins can contribute to sustainable urban planning and the development of smart cities. Discuss how AI algorithms can analyze data from various sources, including IoT devices, to optimize urban infrastructure, improve traffic flow, reduce energy consumption, and enhance citizen services.

As AI and digital twin technology continue to advance, their integration and widespread adoption hold tremendous potential to drive innovation, enhance sustainability, and improve various aspects of our lives and industries, shaping a more efficient, interconnected, and intelligent future world.

AI in Decision Support Systems

AI plays a significant role in enhancing decision support systems (DSS) by providing advanced analytics, intelligent insights, and automated decision-making capabilities. Here's how AI is integrated into decision support systems.

Data Analysis and Pattern Recognition

AI algorithms can analyze vast amounts of data and identify patterns, trends, and correlations that may not be easily discernible to humans. By applying techniques such as machine learning and data mining, AI can extract valuable insights from data, enabling better decision-making.

Predictive Analytics

AI-powered predictive models can forecast future outcomes based on historical data. These models can assist decision support systems by providing probabilistic predictions, identifying potential risks or opportunities, and supporting scenario analysis to assess the potential impact of different decisions.

Automated Decision-Making

AI can automate routine or rule-based decisions within a decision support system. By encoding decision rules or using machine learning algorithms, AI can analyze data, evaluate different options, and recommend or even execute decisions without human intervention. This can significantly streamline decision-making processes and improve efficiency.

Natural Language Processing

AI techniques like natural language processing (NLP) enable decision support systems to understand and interpret unstructured data, such as text documents or customer feedback. NLP algorithms can extract relevant information, perform sentiment analysis, and summarize textual data, providing decision-makers with valuable insights.

Intelligent Recommendations

AI algorithms can generate personalized recommendations based on user preferences, historical data, or collaborative filtering techniques. These recommendations can assist decision-makers in identifying relevant information, alternative courses of action, or potential solutions, helping them make informed decisions.

Cognitive Reasoning

AI can simulate cognitive reasoning capabilities within decision support systems. By employing techniques like expert systems or knowledge-based systems, AI can mimic human expertise, reasoning, and decision-making processes. This enables the system to provide intelligent guidance and support to decision-makers.

Real-Time Monitoring and Alerts

AI can continuously monitor data streams and trigger alerts or notifications based on predefined conditions or anomalies. This real-time monitoring helps decision support systems stay updated with the latest information, identify emerging trends or issues promptly, and enable proactive decision-making.

Visualization and Interactive Interfaces

AI-driven decision support systems can leverage data visualization techniques to present complex information in a visually intuitive and interactive manner. Graphs, charts, and interactive dashboards enable decision-makers to explore data, uncover insights, and gain a deeper understanding of the factors influencing their decisions.

Continuous Learning and Improvement

AI enables decision support systems to learn from user feedback, data feedback, and changing conditions. By incorporating feedback loops and adaptive algorithms, AI can continuously improve decision-making capabilities, refine models, and adapt to evolving circumstances.

Steps in AI Integration

Integrating AI into a digital twin involves several steps to enable the twin to leverage AI capabilities effectively. Clearly articulate the goals and objectives of integrating AI into the digital twin. Identify the specific areas where AI can enhance the twin's capabilities, such as predictive analytics, anomaly detection, optimization, or automation.

Here are the key steps in AI integration for a digital twin:
- *Data collection and preprocessing*
- Gather relevant data from various sources that are necessary for AI analysis. This may include historical data, real-time sensor data, external data feeds, or data from other systems. Preprocess the data to ensure its quality, consistency, and compatibility with AI algorithms.
- *AI model development*
- Develop AI models that are tailored to the specific objectives of the digital twin. This may involve selecting appropriate algorithms, training the models using relevant data, and optimizing them for the desired outcomes. Common AI techniques used in digital twins include machine learning, deep learning, and statistical modelling.
- *Integration with the digital twin platform*
- Integrate the AI models into the digital twin platform or framework. This can involve developing APIs or connectors to enable seamless communication between the twin and the AI algorithms. Ensure that the AI models can receive input data from the digital twin and provide actionable outputs or recommendations.
- *Real-time data processing*
- Enable real-time data processing capabilities to support the AI integration. Implement mechanisms to receive and process streaming data from sensors, IoT devices, or other sources in real-time. This allows the AI models to continuously analyze and learn from the latest data, enabling dynamic decision-making and responsiveness.
- *Continuous learning and improvement*
- Implement mechanisms for continuous learning and improvement of the AI models within the digital twin. This can involve feedback loops, retraining the models with new data, or updating the models as new insights or algorithms become available. Continuous learning ensures that the AI capabilities of the twin stay up-to-date and adapt to changing conditions.
- *Visualization and insights*
- Develop visualizations and dashboards that present the AI-generated insights and recommendations in a user-friendly manner. This enables stakeholders to understand and interpret the AI-driven analytics provided by the digital twin, facilitating decision-making and action.
- *Validation and testing*

- Validate and test the integrated AI capabilities within the digital twin to ensure their accuracy, reliability, and performance. Conduct thorough testing scenarios to verify the effectiveness of the AI models and their alignment with the objectives of the digital twin. Address any issues or limitations identified during the validation process.
- *Deployment and monitoring*
- Deploy the integrated AI capabilities into the operational environment and monitor their performance continuously. Establish monitoring mechanisms to track the performance of the AI models, assess their impact on the digital twin's outcomes, and identify opportunities for further refinement or optimization.
- *Iterative improvement*
- Embrace an iterative improvement process for the AI integration. Continuously gather feedback, measure the impact of AI on the digital twin's performance, and refine the AI models and their integration based on the insights gained. This iterative approach ensures that the AI integration evolves and delivers increasing value over time.

By following these steps, organizations can successfully integrate AI capabilities into their digital twins, enabling advanced analytics, intelligent decision-making, and improved operational outcomes.

Data Security and Privacy Using AI

Data security and privacy are crucial considerations when utilizing AI in digital twin technology. Here are some key aspects to address for ensuring data security and privacy.

Data Encryption

Implement strong encryption techniques to safeguard data at rest and in transit. Encryption helps protect sensitive information from unauthorized access or interception.

Access Control

Employ robust access control mechanisms to restrict data access based on user roles and privileges. Only authorized individuals should have access to specific data and functionalities within the digital twin system.

User Authentication

Implement strong user authentication methods such as multi-factor authentication to verify the identity of users accessing the digital twin. This helps prevent unauthorized access and data breaches.

Anonymization and Pseudonymization

When dealing with sensitive or personally identifiable information, consider anonymizing or pseudonymizing the data. Anonymization removes identifying information, while pseudonymization replaces identifiable data with artificial identifiers, protecting the privacy of individuals.

Secure Data Storage

Ensure that the data storage infrastructure is secure, utilizing industry best practices for data protection. This includes using secure servers, employing firewalls, and implementing intrusion detection systems.

Regular Updates and Patching

Keep the digital twin system up to date with the latest security patches and updates. Regularly applying security fixes helps address vulnerabilities and minimize the risk of exploitation.

Audit Logs and Monitoring

Implement logging and monitoring mechanisms to track user activities and detect any suspicious behavior. This enables the identification of potential security incidents and facilitates prompt response.

Data Minimization

Collect and store only the necessary data within the digital twin system. Minimizing the amount of stored data reduces the potential impact in case of a security breach and helps protect individual privacy.

Data Transfer

Use secure protocols (e.g., HTTPS, VPNs) for data transfer between different components of the digital twin system. Encryption and secure channels ensure data integrity and confidentiality during transit.

Privacy Policies and User Consent

Clearly communicate privacy policies to users and obtain their informed consent regarding data collection, usage, and sharing practices. Transparency empowers users to make informed decisions and fosters trust.

Regular Security Assessments

Conduct periodic security assessments, penetration testing, and vulnerability scans to identify and address any potential weaknesses or vulnerabilities in the digital twin system.

Compliance with Regulations

Ensure compliance with applicable data protection regulations, such as the General Data Protection Regulation (GDPR) or other regional data protection laws, depending on the jurisdiction in which the digital twin is deployed.

By implementing these measures, organizations can enhance data security and privacy in AI-powered digital twin systems, safeguarding sensitive information and ensuring user trust.

Digital Twin for Healthcare

Two innovative technologies have emerged in the healthcare industry in recent years: blockchain and digital twins. From patient data management to medical research and supply chain logistics, both of these technologies have the potential to revolutionize healthcare. Let's take a look how blockchain and digital twin are being applied in healthcare and their benefits.

Blockchain technology is essentially a decentralized and transparent digital ledger that securely records and verifies transactions between multiple computers or nodes. A primary application of this technology in healthcare is to enhance data security, privacy, interoperability, and the management of medical records. Blockchain technology allows healthcare organizations to store and share patient data in an immutable and tamper-proof manner. In addition to maintaining the confidentiality of sensitive medical information, this technology also allows authorized parties such as healthcare providers and patients to securely access and update relevant data in real-time. It can also streamline the exchange of medical records between different healthcare institutions which helps in reducing duplication of tests and improving overall care coordination.

Also, blockchain can facilitate clinical research and trials by providing a trusted platform for data sharing and collaboration. Researchers can securely access anonymized patient data from various sources, allowing them to conduct studies on a larger scale and gain more comprehensive insights. This can clearly accelerate the development of new treatments and therapies to improve patient outcomes.

Digital twin technology involves creating a virtual replica or model of a physical object or system. In healthcare, these are used to monitor and simulate individual patients, medical devices, and entire healthcare facilities. By combining data from various sources such as electronic health records, wearables, and IoT devices, a digital twin can provide an overall view of a patient's health status in real-time which allows healthcare providers to make more accurate diagnoses, personalize treatment plans, and monitor patient progress remotely.

It can also improve the management of medical devices and equipment. By creating virtual replicas of these assets, healthcare organizations can track their performance, predict maintenance needs, and ensure optimal utilization which leads to cost savings, improved operational efficiency, and better patient care.

When combined, blockchain and digital twin technologies offer a powerful solution for healthcare. The secure and transparent nature of blockchain enhances the reliability and trustworthiness of the data used in digital twins. This ensures the integrity and privacy of patient information while providing a robust foundation for real-time monitoring and decision-making.

Sustainable Development Goals (SDGs)

Introduce the concept of SDGs and how they aim to address global challenges such as climate change, resource management, and social equality. Digital twin technology has the potential to contribute to several Sustainable Development Goals (SDGs) in the future. The contribution of digital twins to the SDGs will depend on how they are designed, implemented, and used. Organizations and policymakers need to ensure that digital twin projects are aligned with the SDGs and incorporate sustainability principles throughout their lifecycle. Additionally, considering ethical considerations, inclusivity, and addressing potential unintended consequences are crucial for maximizing the positive impact of digital twin technology on sustainable development.

Here are some examples of how digital twins can align with specific SDGs.

SDG: Industry, Innovation, and Infrastructure
Digital twins can optimize industrial processes, enhance productivity, and improve the efficiency of infrastructure. By simulating and analyzing various scenarios, digital twins can help design and operate sustainable infrastructure systems, reducing resource consumption and promoting innovation.

SDG: Sustainable Cities and Communities
Digital twins can support urban planning and development by simulating and optimizing the performance of cities. They can aid in designing smart and resilient infrastructure, optimizing energy consumption, managing transportation systems, and improving the overall quality of life in urban areas.

SDG: Responsible Consumption and Production
Digital twins can facilitate more sustainable production and consumption patterns. By modeling and analyzing production processes, supply chains, and product lifecycles, organizations can identify areas for efficiency improvements, waste reduction, and the adoption of circular economy principles.

SDG: Climate Action
Digital twins can help address climate change challenges by enabling the modeling and analysis of energy systems, optimizing energy usage, and supporting the integration of renewable energy sources. They can assist in developing climate resilience strategies, facilitating the monitoring of environmental parameters, and promoting sustainable resource management.

SDG: Life on Land
Digital twins can contribute to biodiversity conservation and land management. By simulating ecosystems and analyzing land use patterns, digital twins can support sustainable agriculture practices, land restoration efforts, and the monitoring of protected areas, enabling informed decision-making for the preservation of biodiversity.

SDG: Partnerships for the Goals
Digital twins can foster collaboration and partnerships among different stakeholders. They enable data sharing, visualization, and analysis, facilitating multidisciplinary collaboration for sustainable development initiatives. Digital twins can support public-private partnerships and enhance knowledge sharing for achieving the SDGs.

SDG: Good Health and Well-Being
Digital twins can be used to simulate the spread of diseases, develop new treatments, and improve the efficiency of healthcare systems. For example, digital twins have been used to simulate the spread of COVID-19, which has helped to inform public health interventions.

SDG: Affordable and Clean Energy
Digital twins can be used to optimize the efficiency of energy systems, reduce energy consumption, and develop new renewable energy sources. For example, digital twins have been used to optimize the operation of power grids, which has helped to reduce energy waste.

SDG: Zero Hunger
Digital twins can be used to improve agricultural productivity, reduce food waste, and ensure food security.

SDG: Clean Water and Sanitation
Digital twins can be used to improve the efficiency of water systems, reduce water pollution, and ensure access to safe water for all.

SDG: Decent Work and Economic Growth
Digital twins can be used to improve the efficiency of businesses, create new jobs, and promote economic growth.

SDG: Reduced Inequality
Digital twins can be used to improve access to education and healthcare, reduce poverty, and promote social inclusion.

Conclusion

The integration of Artificial Intelligence (AI) in digital twins holds immense promise for creating a sustainable future. By leveraging AI algorithms and techniques within digital twin frameworks, industries can achieve greater efficiency, optimization, and informed decision-making.

AI-enabled digital twins contribute to sustainable development by offering real-time monitoring, predictive maintenance, and optimization capabilities. They empower industries to monitor environmental factors, optimize energy consumption, and promote resource efficiency. Through data analysis and machine learning, AI can identify patterns, anomalies, and trends, enabling proactive measures to address environmental challenges.

Moreover, AI in digital twins supports sustainable practices across various sectors. It enables smart manufacturing and circular economy initiatives by reducing waste, optimizing processes, and promoting recycling. It enhances transportation systems, optimizing routes, reducing emissions, and improving mobility. AI also aids in urban planning, enabling the development of smart cities that prioritize energy efficiency, infrastructure optimization, and citizen services.

However, the integration of AI in digital twins for a sustainable future must be guided by ethical considerations. Ensuring fairness, transparency, accountability, and privacy protection is crucial to prevent biases, maintain public trust, and respect individual rights.

To realize the full potential of AI in digital twins, collaboration among governments, industries, research institutions, and communities is essential. Together, they can harness the power of AI to drive sustainable development, mitigate environmental impact, and create innovative solutions that address global challenges.

In summary, AI in digital twins has the potential to revolutionize industries and contribute significantly to a sustainable future. By leveraging AI's capabilities, we can optimize resource utilization, reduce environmental footprint, and foster long-term sustainability across sectors, ultimately shaping a better world for generations to come.

References

1. Fuller, A., Fan, Z., Day, C., et al. (2020). Digital twin: Enabling technologies, challenges and open research. *IEEE Access, 8*, 108952–108971.
2. Chowdhury, M. J. M., Ferdous, M. S., Biswas, K., Chowdhury, N., & Muthukkumarasamy, V. (2020). A survey on blockchain-based platforms for IoT use-cases. *Knowledge Engineering Review, 35*, 1–24.
3. Ezhilarasu, C. M., Skaf, Z., & Jennions, I. K. (2019). Understanding the role of a digital twin in integrated vehicle health management (IVHM). In *Proceedings of the IEEE international conference on systems, man and cybernetics (SMC)* (pp. 1484–1491).
4. Ahmadi-Assalemi, G., Al-Khateeb, H., Maple, C., Epiphaniou, G., Alhaboby, Z. A., Alkaabi, S., et al. (2020). Digital twins for precision healthcare. In *Cyber defence in the age of AI smart societies and augmented humanity* (pp. 133–158). Springer.
5. D'Auria, E., Abrahams, M., Zuccotti, G., & Venter, C. (2019). Personalized nutrition approach in food allergy: Is it prime time yet? *Nutrients, 11*(2), 359.
6. Jouan, P., & Hallot, P. (2020). Digital twin: Research framework to support preventive conservation policies. *ISPRS International Journal of Geo-Information, 9*(4), 228.
7. Chowdhury, M. J. M., Ferdous, M. S., Biswas, K., Chowdhury, N., Kayes, A. S. M., Alazab, M., et al. (2019). A comparative analysis of distributed ledger technology platforms. *IEEE Access, 7*, 167930–167943.
8. Dagher, G. G., Mohler, J., Milojkovic, M., & Marella, P. B. (2018). Ancile: Privacy-preserving framework for access control and interoperability of electronic health records using blockchain technology. *Sustainable Cities and Society, 39*, 283–297.
9. Crosby, M., Pattanayak, P., Verma, S., & Kalyanaraman, V. (2016). Blockchain technology: Beyond bitcoin. *Apply Innovation, 2*(6), 71.
10. Mingxiao, D., Xiaofeng, M., Zhe, Z., Xiangwei, W., & Qijun, C. (2017, October). A review on consensus algorithm of blockchain. In *Proceedings of the IEEE international conference on systems, man, and cybernetics (SMC)* (pp. 2567–2572).
11. Ferdous, M. S., Chowdhury, M. J. M., Hoque, M. A., & Colman, A. (2020). Blockchain consensus algorithms: A survey. *arXiv:2001.07091*.
12. Alhadhrami, Z., Alghfeli, S., Alghfeli, M., Abedlla, J. A., & Shuaib, K. (2017, November). Introducing blockchains for healthcare. In *Proceedings of the international conference on electrical and computing technologies and applications (ICECTA)* (pp. 1–4).
13. Zheng, Z., Xie, S., Dai, H., Chen, X., & Wang, H. (2017, June). An overview of blockchain technology: Architecture consensus and future trends. In *Proceedings of the IEEE international congress on big data (BigData congress)* (pp. 557–564).
14. Nakamot, S. (2008). Bitcoin: A peer-to-peer electronic cash system. *Decentralized Business Review*, 21260.

15. Buterin, V., et al. (2014). *A next-generation smart contract and decentralized application platform* (Vol. 3, No. 37). Zug, Switzerland.
16. Azaria, A., Ekblaw, A., Vieira, T., & Lippman, A. (2016, August). MedRec: Using blockchain for medical data access and permission management. In *Proceedings of the 2nd international conference on open and big data (OBD)* (pp. 25–30).
17. Zhang, P., White, J., Schmidt, D. C., Lenz, G., & Rosenbloom, S. T. (2018). FHIRChain: Applying blockchain to securely and scalably share clinical data. *Computational and Structural Biotechnology Journal, 16*, 267–278.
18. Alom, I., Eshita, R. M., Harun, A. I., Ferdous, M. S., Shuhan, M. K. B., Chowdhury, M. J. M., et al. (2021. May). Dynamic management of identity federations using blockchain. In *Proceedings of the IEEE international conference on Blockchain and cryptocurrency (ICBC)* (pp. 1–9).
19. Mohan, C. (2019, June). State of public and private blockchains: Myths and reality. In *Proceedings of the 2019 international conference on Management of Data* (pp. 404–411).
20. *Corda.* (2022, April). [Online] Available https://www.r3.com/reports/corda-technical-whitepaper/
21. Xia, Q. I., Sifah, E. B., Asamoah, K. O., Gao, J., Du, X., & Guizani, M. (2017). MeDShare: Trust-less medical data sharing among cloud service providers via blockchain. *IEEE Access, 5*, 14757–14767.
22. Grieves, M., & Vickers, J. (2017). Digital twin: Mitigating unpredictable undesirable emergent behavior in complex systems. In *Transdisciplinary perspectives on complex systems* (pp. 85–113). Springer.
23. Zhuang, C., Liu, J., & Xiong, H. (2018). Digital twin-based smart production management and control framework for the complex product assembly shop-floor. *International Journal of Advanced Manufacturing Technology, 96*(1), 1149–1163.
24. Shostack, A. (2014). *Threat modeling: Designing for security*. Wiley.
25. Peng, Y., Zhang, M., Yu, F., Xu, J., & Gao, S. (2020). Digital twin hospital buildings: An exemplary case study through continuous lifecycle integration. *Advances in Civil Engineering, 2020*, 1–13.
26. Liu, Y., Zhang, L., Yang, Y., Zhou, L., Ren, L., Wang, F., et al. (2019). A novel cloud-based framework for the elderly healthcare services using digital twin. *IEEE Access, 7*, 49088–49101.
27. Martinez-Velazquez, R., Gamez, R., & El Saddik, A. (2019, June). Cardio twin: A digital twin of the human heart running on the edge. In *Proceedings of the IEEE international symposium on medical measurements and applications (MeMeA)* (pp. 1–6).
28. Shamanna, P., Saboo, B., Damodharan, S., Mohammed, J., Mohamed, M., Poon, T., et al. (2020). Reducing HbA1c in type 2 diabetes using digital twin technology-enabled precision nutrition: A retrospective analysis. *Diabetes Therapy, 11*(11), 2703–2714.
29. Barbiero, P., Torné, R. V., & Lió, P. (2021). Graph representation forecasting of patient's medical conditions: Toward a digital twin. *Frontiers in Genetics, 12*, 652907.
30. Petrova-Antonova, D., Spasov, I., Krasteva, I., Manova, I., & Ilieva, S. (2020). A digital twin platform for diagnostics and rehabilitation of multiple sclerosis. In *Proceedings of the international conference on computational science and its applications* (pp. 503–518).
31. Rao, D. J., & Mane, S. (2019). Digital twin approach to clinical DSS with explainable AI. *arXiv:1910.13520*.
32. Elayan, H., Aloqaily, M., & Guizani, M. (2021). Digital twin for intelligent context-aware IoT healthcare systems. *IEEE Internet of Things Journal, 8*(23), 16749–16757.
33. Croatti, A., Gabellini, M., Montagna, S., & Ricci, A. (2020). On the integration of agents and digital twins in healthcare. *Journal of Medical Systems, 44*(9), 1–8.
34. https://www.researchgate.net/figure/Structural-model-and-technical-composition-of-digital-twin_fig1_361987128
35. Hyerledger Fabric. (2022, April). [Online]. Available https://www.hyperledger.org/use/fabric
36. Rakshit, P., Sarkar, P., Ghosh, D., Roy, S., Talukder, S., & Chakraborty, P. S. (2022). Sentiment analysis of twitter data using deep learning. Advances in Communication, Devices and Networking: Proceedings of ICCDN 2021, pp. 495–501. Singapore: Springer Nature Singapore.

37. https://softengi.com/blog/use-cases-and-applications-of-digital-twin/
38. Rakshit, P., Gupta, S., & Das, T. (2022). Sentiment analysis to find sentence polarity on Tweet Data. Machine Learning in Information and Communication Technology: Proceedings of ICICT 2021, SMIT, pp. 197–202. Singapore: Springer Nature Singapore. (pp. 51–58).
39. Rakshit, P., & Sarkar, A. (2024). A supervised deep learning-based sentiment analysis by the implementation of Word2Vec and GloVe Embedding techniques. *Multimedia Tools and Applications*, 1–34.
40. McConaghy, T., Marques, R., Müller, A., Jonghe, D. D., McConaghy, T., McMullen, G., et al. (2022, May). *BigchainDB 2.0 the blockchain database*. [Online]. Available https://www.bigchaindb.com/whitepaper/bigchaindb-whitepaper.pdf
41. Benet, J. (2014). IPFS–content addressed versioned P2P file system. *arXiv:1407.3561*.
42. https://ieeexplore.ieee.org/document/9771180/references#references

The Place and Role of Digital Twin Applications: Directions for Energy and Education Sector

Nurcan Kilinc-Ata and Ridvan Ata

Introduction

The EIA anticipates a roughly 50% rise in world energy demand by 2050 [7]. Furthermore, fossil fuel-based CO_2 emissions, which will account for nearly 91% of emissions in 2022, make up the majority of all global emissions in recent years [13]. Therefore, it is essential to look for more efficient ways of operating for changes in energy resources and costs due to the rising demand in the global energy sector and the negative consequences that energy resources have on the environment and air. To effectively increase efficiency in the energy sector, techniques and advances in technology are thus crucial. To enable the quick transformation of power systems and enhance operational flexibility, the application of DT in the energy sector is one of these technologies [27].

DT has been used for the past 20 years in a variety of industries, albeit its use in the energy sector is still limited [22]. Education [23], manufacturing [15], technology and industrial applications [17], uses in the production industry [5], agriculture [21], and several other industries are instances of these diverse sectors that are at the implementation phase of DT.

Regarding education, as aforementioned above, the capacity of DT to depict real-world objects by connecting them with real-world data is what gives it its true strength. Using real-world data in 3D visualization, simulation, modeling, and prediction, DT enables the collaborative and immersive design, development, testing,

---

N. Kilinc-Ata (✉)
College of Economics and Management, Al-Qasimia University, Sharjah, UAE

Institute for Statistical Studies and Economics of Knowledge, National Research University Higher School of Economics, Moscow, Russia

R. Ata
Faculty of Education, Department of Computer Education and Instructional Technology, Mugla Sitki Kocman University, Mugla, Türkiye

© The Author(s), under exclusive license to Springer Nature Switzerland AG 2024
A. Mishra et al. (eds.), *Transforming Industry using Digital Twin Technology*,
https://doi.org/10.1007/978-3-031-58523-4_3

deployment, and operation of intricate and comprehensive scenarios. This lets the users engage with dynamic contents that effectively simulate real-world conditions and visualize the outcomes with augmented, mixed, and virtual reality concepts. Balla et al. [1] indicate that the main purposes of using DT technologies include simulation, monitoring, control, design, optimizations, maintenance, validation, prediction, and customization. With DT deployments in education, the ultimate immersive teaching and learning experiences can be possible. For instance, learners can nonetheless conduct risky, complicated, or high-budget tasks easily. Besides, abstract concepts can be learned with a high level of engagement by using DT technologies in AR or VR simulated learning experiences. Although papers on DTs frequently concentrate on manufacturing and commercial applications, this book chapter also focuses on the integration of DT technologies into higher education.

The crucial main of the present chapter is to fill the gap in the literature because to the best knowledge of the authors, no other investigation of DT applications in the energy and education sectors has been as comprehensive and rigorous. The following sections describe the existing applications of DT investigation in the energy and education sectors as well as the primary contributions of the current study.

The following sections comprise the remainder of the chapter: Background facts on DT are presented in the second section. In particular, the definition, characteristics, and trend of DT are discussed. The third section of the chapter examines DT applications in the energy area and explains why DT is significant for this industry. Applications of DT in the sphere of education are provided in the fourth section of the chapter. The importance of DT in the sectors of energy and education is underlined after the chapter.

The Background of Digital Twins

The concept of the DT, its components, function, and trends are thoroughly covered in the sub-titles of this section before the applications of DT in the sectors of energy and education.

Characteristics of the Concept of Digital Twins

DTs are technologies (both artificial and/or physical) or models based on computers that mimic, imitate, replicate, or "twin" the existence of a physical entity. This physical entity could be an individual, an item, a process, or a characteristic related to a person [2]. Shahzad et al. [26] state that DT describes the notion as an exact duplicate of an actual procedure displayed beside the procedure and typically matching what happens in reality precisely and instantaneously.

Kritzinger et al. [16] classified the DT concept as a virtual model, digital shadows, or DT based on the degree of data integration. Within the virtual model, there is no automated data interchange between the physical and the digital model. This implies that any modifications made to the physical model do not influence the virtual model after it is created. On the other hand, a digital shadow is a one-way data transfer that occurs between the digital and physical representations. Changes to the physical model impact the digital model, not the other way around. However, if there is data flow and complete integration in both ways between the current physical model and the digital model, it is known as the DT idea. The digital model is influenced as the physical model is changed and vice versa.

DT has many characteristics and the following are some of DT's notable characteristics.

- DT needs to have access to effective high-dimensional data (de)coding and analysis techniques in addition to data fusion techniques to handle data. This would allow DT to combine various data sources and generate results that are more reliable, accurate, and relevant [30].
- DT uses statistical, pattern recognition, and unsupervised/supervised learning technologies to characterize the input data coming from the physical twin and/or the Internet of Things surroundings. Evaluating the data makes it possible to recognize changes as well as identify important patterns and trends [2].
- The DT may self-adapt and self-parameterize, allowing it to mimic the physical twin during its entire existence [35].
- The DT employs artificial intelligence to foresee future conditions as well as significant shifts (like failures) in the lifecycle of a product [3].
- The DT makes decisions that are pertinent to its future by using the output of descriptive and forecasting methods as input into analytics that are prescriptive by using a complicated collection of objectives, needs, and restrictions (represented by the historic and stationary data) to computationally regulate a set of highly valuable alternatives [33].
- The DT also offers modeling and simulation tools for expressing the physical twin's current state as well as other "what-if" scenarios naturally and realistically [2].

The Development Trend of Digital Twins

DT is a flexible technology that can be effectively utilized in many different industries, but organizations are still hesitant to use it. This is because a DT is not a product that directly generates income, but rather a technology meant to cut costs and maximize efficiency, making it challenging to calculate the Return on Investment (ROI) [20].

The size of the global DT market was estimated at USD 11.12 billion in 2022, and the industry is expected to develop at a compound annual growth rate (CAGR) of 37.5% between 2023 and 2030. DT enables the building of predictive models and aids in success evaluation before launching actual prototypes. The worldwide epidemic has spurred the use of DT technology in applications in several non-manufacturing sectors, notably healthcare, real estate, telecommunications, and sales, boosting the market's level of competition. Combining technological advances

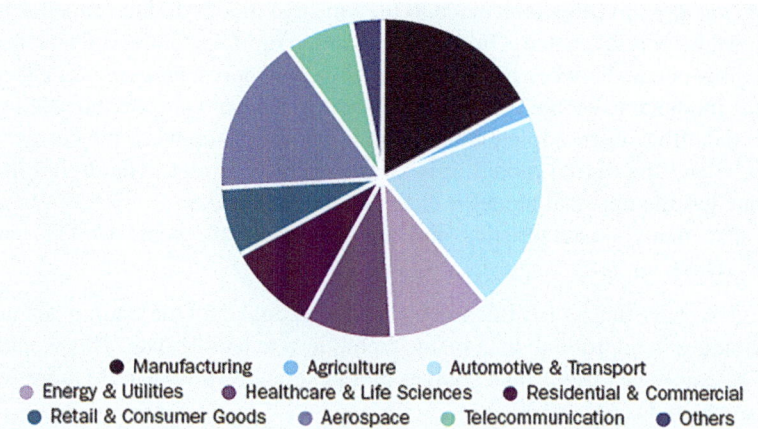

Fig. 1 The market for digital twins worldwide in 2022. (Source: GVR Report [12])

from DT with cutting-edge innovations like cloud computing, machine learning, and the Internet of Things is anticipated to further drive market growth [12]. The DT market in 2022 is shown in the figure below.

According to Fig. 1, the industry with the biggest revenue share in 2022 was automotive and transportation, which contributed more than 20% of total revenue. The growth in the automobile industry is followed by the telecommunications, housing, and healthcare sectors. In addition, the market for DT solutions, which produce 3D models for impending smart city projects, is predicted to see faster growth due to their growing application in the residential and commercial sectors.

The expected value of the global electrical DT market in terms of revenue was 0.8 billion dollars in 2021, and it is anticipated to expand to 1.3 billion dollars by 2026, increasing at a CAGR of 12.2% over that time. The key element driving the market is the increasing demand for decentralizing distributed energy resources via electrical DT and integrating variable RE with the grid. The trend in the global electrical DT market is shown in Fig. 2.

Digital Twins Applications in the Energy Sector

DTs are virtual, frequently real-time reproductions of the actual grid assets used in the energy sector. Energy companies use DT to enhance planning and requirements, operational effectiveness, and staff training. The usage of DT in the energy industry also allows testing events to be conducted anywhere and anytime. While it will not completely replace on-site testing, it considerably reduces physical testing, reduces costs, and contributes to improving the quality of the protection system [4]. The DT's main objectives in the energy sector include controlling market

Fig. 2 Global Electrical DT Market Trend. (Source: Market Research Report [18])

dynamics, cost savings, revenue growth, downtime reduction, and improved operations [34] (Table 1).

Why Is the Digital Twin Important for the Energy Industry?

The DT is utilized in the energy sector to improve efficiency and lower costs. A digital replication of a genuine asset, including an offshore oil and gas system, electricity plant, turbine for wind energy, or solar power plant, could be produced due to this technology [27]. DT also keeps an eye on, analyzes, and improves energy systems, such as the mix, cost, and efficiency of the energy used [11]. The DT offers information and perspectives in real time that can be utilized to identify, treat, and forecast operational issues in energy systems. Energy firms can enhance operational efficiency and prevent costly downtime by monitoring mirror data to identify issues early and make informed decisions [14]. Energy firms can model and evaluate modifications using a DT before taking any action. In the long run, this helps them decrease risk, improve safety, increase productivity, and save money by enabling them to make informed and cost-effective decisions. Energy firms are also able to comprehend how market conditions and environmental changes impact their ability to produce energy through the use of DT analysis [36].

Recently, the concept of a DT has gained momentum in several sectors, including oil and gas due to its capacity to integrate simulation, machine learning, and operational data [31]. The gas and oil industry is a challenging and dangerous workplace, and for this industry to produce safely and affordably, processes must be optimized [9]. Additionally, DT can determine the most effective approaches to

Table 1 Outline of digital twin in the energy sector

Application type	Description
DT applications in the energy generation area	
The electrical power sector, encompassing electricity production and delivery	Using DT to operate flexible power plants Emerging technologies' commercial potential in the decentralized energy sector DT for cloud-based and Internet of Things-based energy cyber-physical systems Grid control using Digital Dynamic Mirror (DDM) Networked Microgrids DT Resistance against cyber attacks Analyzing DT to maintain Empty Cell
Renewable energy (RE) sector	Wind farm predictive maintenance RE system for power generation
Nuclear energy sector	A semi-autonomous management and control system can be used by advanced reactors to maintain the peak fuel temperature below the allowable limit
Batteries, energy storage, and the Internet of Vehicles (IoV)	Analyzing the energy use, charging, and waiting time statistics for electric vehicles DT framework should be used in a cloud-computing environment to calculate battery module.
Planning sustainable projects	Utilized by a real-world system of energy
DT applications in energy saving	
Generation engineering	Production automation; energy conservation via virtual calibration Creating energy-flexibility solutions with limitations for procedures and goods for greenhouse farmers
Monitoring devices, equipment, tools, and compressors	High lift aircraft system Cyber-physical system (CPS) for controlling and monitoring compressors The amount of energy used by cutting instruments
Manufacturing	Energy use in the shop floor for refurbishing Energy management and green manufacturing Industrial robots and individuals, measurements of human behaviors, and kinetic energy Improves the use of energy and resources for the grinding wheel
Information technology facilities and the construction sector	IT systems cooling and server fan controller modification Creating IoT solutions for properties using business cases and industry practices
Pumping systems	To adapt operating settings to accomplish improved efficiency in reaction to changes in downstream conditions, imagine placing pumps in a series or a parallel configuration
Ventilation system	Design of the mine ventilation system of control

Source: Sleiti et al. [27]

produce, transmit, and distribute energy by facilitating energy optimization and modeling of various scenarios [28]. Figure 3 below shows an example of a solar plant using IoT machine learning (ML) and DT technologies to enhance photovoltaic plant performance while streamlining maintenance tasks.

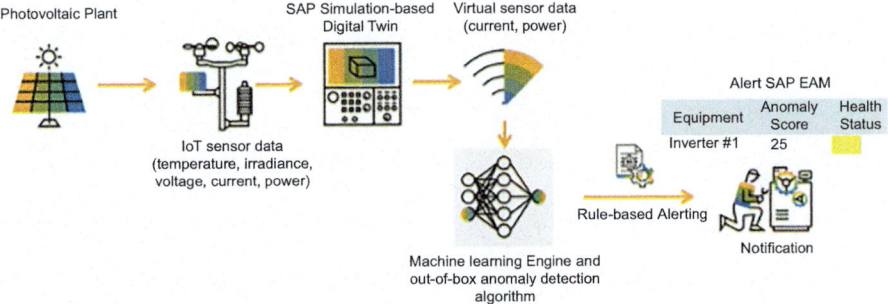

Fig. 3 IOT, ML, and DT technology-based solar plant. (Source: Gouda [11])

Digital Twins Applications in Education

As was pointed out previously, due to developments in VR, AR, artificial intelligence (AI), and IoT, DT technology has entered into various industries including manufacturing, engineering, and health care. In the meantime, DT has begun to penetrate education. A notable example of DT applications is the study carried out by Eriksson et al. [8] in which industrial-like, digital automation labs are virtually built with real data integration highlighting the spread of distance education and restriction of accessing physical labs due to factors such as pandemics or disasters. It has been reported that the DT model successfully simulated the behaviors of the real system and completed the manufacturing cycles without any problems. In support of this, Sepasgozar [25] must introduce cutting-edge tools to architecture, engineering, and construction students to explain the challenging operations of drilling, piling, and boring as well as an excavator DT. The purpose of these virtual apps is to offer extensive visits so that students can investigate and learn in a virtual setting about construction topics and heavy equipment developing mixed-reality modules. As a part of the module, an excavator DT that allows for two-way communication between an excavator's physical entity and itself in real time is developed for students to operate and learn different movements of the digger. Another good example of this is the project developed by Roy [24] using the Blender/UPBGE platform with Arduino microcontroller and Python script. For instance, Roy developed a sliding gate programming the behavior of the digital model with Python script and linked the PC to an Arduino board to simultaneously execute the same program on the real system. The sliding gate has been designed within UPBGE which allows users to read data for visualization in 3D scenes and interact with 3D objects to make real physical action (Fig. 4). More details about the project can be found at https://gitlab.com/phroy.

As one of the implementations of DT technology, it is possible to develop imitations of items that mimic the actions of their real counterpart to assess how well they work or behave in circumstances and make necessary adjustments. As an example of this, the following paragraph suggests the scheme and development of a DT

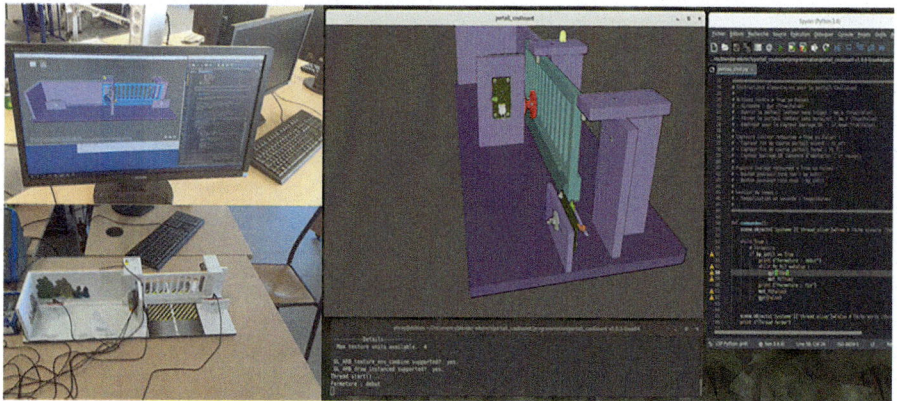

Fig. 4 Sliding gate design. (Source: Roy [24])

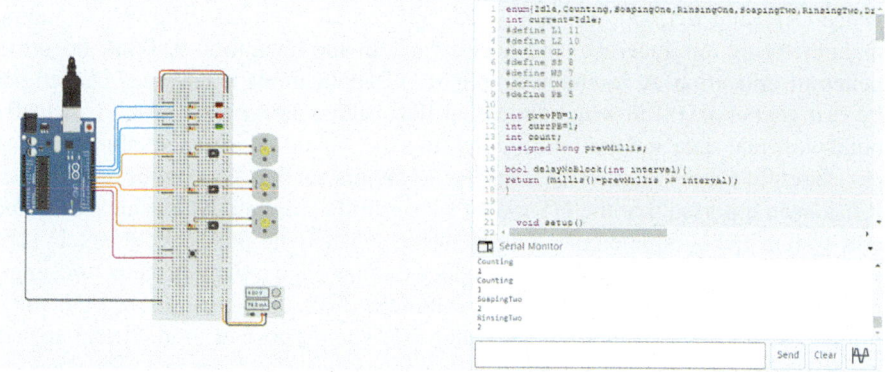

Fig. 5 Virtual module on Tinkercad

inspired by the study of Moise et al. [19] in which a virtual washing machine is simulated using Arduino on Tinkercad program which is an online free 3D modeling application for education and training. Students can use this platform to test their algorithms in this way.

Here, the following procedures are used to build the design for a virtual dishwasher. In the first stage, the virtual module that includes Arduino Uno R3, 2 Red LED, 1 Green LED, 6 kohm Resistors, 1 Pushbutton, 3DC Motors, 3 NPN Transistors (BJT), 15,5 Power Supply is created. The prototype is designed using Tinkercad (Fig. 5). In this example, the motor in the washing machine is a DC motor, which is used for soaping, rinsing, and drying controlled by transistors. When the pushbutton is pressed, the soaping module is ensured with a red LED that lights up, the rinsing unit is postulated with a red LED that lights up after a short while, and the drying module is provided with a green LED that lights up. LEDs are used as feedback for the functionality of the module.

When a dishwasher is turned on, the machine goes into an idle state until the door is closed; at that point, the dishwasher loads the dirty dishes and closes. After the door is shut, the dishwasher enters the counting mode where it can be programmed and waits for the user to select the program. Pushing the button causes the dishwasher to start the washing cycle after the user has finished selecting their chosen settings. (The user is supposed to create the script. An example of the script can be accessed on Tinkercad gallery circuit search.) For the actual product test, after user's design and test their algorithm in the virtual platform, a server that has multiple mini dishwashers associated with it can use the same algorithm. Here a video camera can be used for every dishwasher to give feedback to users. To push the physical buttons, the users can create a user interface that can be well-ordered distantly. Thanks to the video camera, checking and troubleshooting can be carried out for feedback. The proposal of DT prototyping design can be seen in Fig. 6.

Overall, Turner [32] indicates the most common applications of DT in education as follows: (i) Virtual Classrooms: DT enables the creation of virtual learning environments, where students can access educational resources, collaborate with peers, and engage in projects from any location. (ii) Virtual Labs: DT provides simulated lab experiences, allowing students to conduct experiments and analyze data within a virtual setting, bypassing physical limitations and potential risks. (iii) Personalized Learning: DT facilitates customized learning experiences tailored to individual students. By leveraging DT, students gain access to resources that match their specific

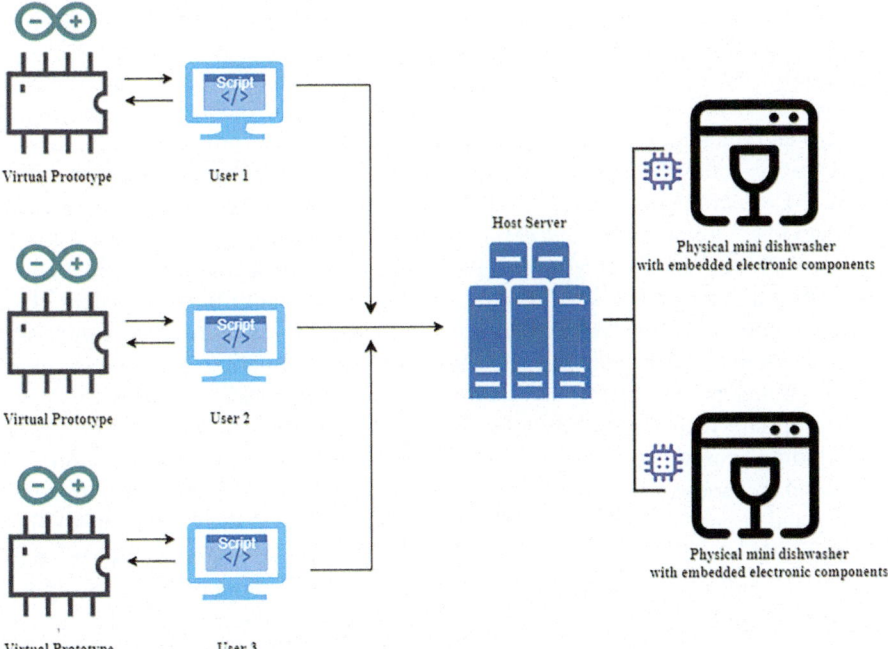

Fig. 6 DT prototyping design

needs and learning styles, resulting in enhanced learning progress. (iv) Assessment and Evaluation: DT can aid in a more accurate assessment of student progress. Educators can utilize DT to evaluate students' comprehension of concepts and monitor their advancement over time. (v) Gamification: DT can create captivating and immersive learning experiences. Educators harness gaming elements within DT, such as rewards and challenges, to motivate students and foster motivation for learning.

Why Is the Digital Twin Important for the Education Sector?

The use of DT technologies in education also seems to be promising. Significant benefits of DT in education are highlighted by Turner [32] as follows:

Enhanced Student Engagement: DT has the potential to create immersive learning experiences, utilizing VR and AR technologies. Through more engaging interactions with items and others, pupils discover their digital surroundings [25].

Improved Learning Outcomes: DT can contribute to the creation of effective and stimulating educational opportunities. By creating opportunities for students and modeling real-world situations to interact with the environment, DT facilitates a deeper understanding of complex concepts and enhances skill development [23].

Personalized Learning Experiences: DT allows for the customization of educational opportunities adapted to each pupil's demands and preferences. Through tailored resources and content, students can access materials that align with their specific learning styles, leading to accelerated progress [10].

Enhanced Assessment: DT offers a more accurate means of assessing students' progress. Educators can utilize DT to measure students' comprehension of concepts and track their advancement over time with greater precision [8].

Furthermore, some researchers state various principles when developing and utilizing DT effectively in education [6, 29, 32]. These principles are (i) Understand Your Goals: Before creating a DT, it is crucial to comprehend the intended objectives. What outcomes do you seek to achieve? What data is necessary to gather? Identifying these aspects helps in developing an efficient DT. (ii) Create a Plan: Developing a DT involves a multifaceted process, necessitating the creation of a comprehensive plan. This plan should encompass a timeline, budget allocation, and strategies for testing and evaluating the DT's performance. (iii) Select Appropriate Technology: Various technologies can be employed to construct DT, including sensors, simulation software, and algorithms for machine learning. It is vital to choose the most suitable technology that ensures the DT's effectiveness and dependability. (iv) Harness Data: Data serves as the fundamental building block of a DT; thus, leveraging data effectively is crucial. Acquiring data from the physical object and utilizing it to construct accurate models and simulations is paramount. (v) Conduct Testing and Evaluation: Once the DT is developed, it is imperative to conduct thorough testing and evaluation procedures. This ensures the DT's accuracy and reliability in its intended application.

Conclusion

Two important results were determined in the study on DT energy applications in the energy area. First, there's an obvious upward tendency in the application of DT technology across a range of energy-related sectors. Second, DT technologies are being developed for the energy industry to reduce energy consumption or increase energy efficiency. The energy sector's implementation of DT technology is crucial and necessary for a transition to sustainable energy. Ultimately, the DT is assisting the energy, utilities, oil, and gas sectors in enhancing reliability, efficiency, and safety factors that are crucial in a sector that provides services to millions of consumers and companies. However, even though DT innovations are still currently in their beginnings, they offer several chances for multinational expansion, company growth, and entrepreneurial activity in the energy sector. These opportunities will help accelerate the energy transition and promote the growth of sustainable energy in the years ahead. By simulating and analyzing the performance of energy systems and components, as well as diagnosing issues efficiently in the virtual world, the DT may speed up innovation, foster consensus, and reduce time and costs.

From the viewpoint of the educational sector, the use of DT technology as a tool to enhance the learning environment and give students more varied and immersive experiences is noted based on relevant studies and activities. In the instructional design process, it can be stated that using DT technology is a beneficial tool for integrating fragmented information, making predictions, providing accurate assessments, and giving rapid feedback. In this regard, it can be said that DT technology, similar to other areas, is a powerful tool/system that can be employed in education. In the research conducted on DT technology in education, it is observed that various models have been proposed, but there has not been enough focus on the implementation level. Researchers need to conduct studies evaluating the functionality of previously established models or create their models to assess their advantages and disadvantages. In this regard, it can be said that conducting experimental- or design-based investigations on the usage of DT applications in schooling would be beneficial for the teaching and learning process.

References

1. Balla, M., Haffner, O., Kučera, E., & Cigánek, J. (2023). Educational case studies: Creating a digital twin of the production line in TIA portal, unity, and game4automation framework. *Sensors, 23*(10), 4977.
2. Barricelli, B. R., Casiraghi, E., & Fogli, D. (2019). A survey on digital twin: Definitions, characteristics, applications, and design implications. *IEEE Access, 7*, 167653–167671.
3. Boje, C., Guerriero, A., Kubicki, S., & Rezgui, Y. (2020). Towards a semantic construction digital twin: Directions for future research. *Automation in Construction, 114*, 103179.
4. Bonetti, A. (2022). *How digital twins are used in the energy sector.* https://www.iec.ch/blog/how-digital-twins-are-used-energy-sector. 16 May 2023.

5. Cimino, C., Negri, E., & Fumagalli, L. (2019). Review of digital twin applications in manufacturing. *Computers in Industry, 113*, 103130.
6. Deniz, S., Müller, U. C., Steiner, I., & Sergi, T. (2022). Online (remote) teaching for laboratory-based courses using "digital twins" of the experiments. *Journal of Engineering for Gas Turbines and Power, 144*(5), 051016.
7. EIA. (2020). *EIA projects a nearly 50% increase in world energy usage by 2050, led by growth in Asia*. https://www.eia.gov/todayinenergy/detail.php?id=42342#:~:text=World%20industrial%20sector%20energy%20use,British%20thermal%20units%20(Btu). 11 May 2023.
8. Eriksson, K., Alsaleh, A., Behzad Far, S., & Stjern, D. (2022). Applying digital twin technology in higher education: An automation line case study. *Advances in Transdisciplinary Engineering, 21*, 461–472.
9. Fataliyev, T. K., & Mehdiyev, S. A. (2018). Analysis and new approaches to the solution of problems of operation of oil and gas complex as a cyber-physical system. *International Journal of Information Technology and Computer Science, 10*(11), 67–76.
10. Furini, M., Gaggi, O., Mirri, S., Montangero, M., Pelle, E., Poggi, F., & Prandi, C. (2022). Digital twins and artificial intelligence: As pillars of personalized learning models. *Communications of the ACM, 65*(4), 98–104.
11. Gouda, H. (2023). *Why digital twin for the energy industry?* https://blogs.sap.com/2023/03/25/why-digital-twin-for-the-energy-industry/. 16 May 2023.
12. GVR Report. (2023). *Digital twin market size, share & trends analysis report by end-use (manufacturing, agriculture), by solution (component, process, system), by region, and segment forecasts, 2023–2030*. https://www.grandviewresearch.com/industry-analysis/digital-twin-market. 15 May 2023.
13. Hausfather, Z., & Friedlingstein, P. (2022). *Analysis: Global CO_2 emissions from fossil fuels hit a record high in 2022*. https://www.carbonbrief.org/analysis-global-co2-emissions-from-fossil-fuels-hit-record-high-in-2022/. 11 May 2023.
14. Jafari, M., Kavousi-Fard, A., Chen, T., & Karimi, M. (2023). A review on digital twin technology in smart grid, transportation system, and smart city: Challenges and future. *IEEE Access, 11*, 17471.
15. Jones, D., Snider, C., Nassehi, A., Yon, J., & Hicks, B. (2020). Characterizing the digital twin: A systematic literature review. *CIRP Journal of Manufacturing Science and Technology, 29*, 36–52.
16. Kritzinger, W., Karner, M., Traar, G., Henjes, J., & Sihn, W. (2018). Digital twin in manufacturing: A categorical literature review and classification. In *Proceedings of the 16th IFAC symposium on information control problems in manufacturing (INCOM 2018)* (Vol. 51, pp. 1016–1022). Bergamo, Italy, 11–13 June 2018.
17. Liu, M., Fang, S., Dong, H., & Xu, C. (2021). Review of digital twin about concepts, technologies, and industrial applications. *Journal of Manufacturing Systems, 58*, 346–361.
18. Market Research Report. (2021). *Electrical digital twin market*. https://www.marketsandmarkets.com/Market-Reports/electrical-digital-twin-market-192874390.html. 16 May 2023.
19. Moise, V. M., Elisei, N., Dumitraşcu, A., Niculescu, A. M., & Pavel, D. M. (2021, May). Design of a virtual embedded system for mini washing machine. In *2021 44th international spring seminar on electronics technology (ISSE)* (pp. 1–4). IEEE.
20. Mihai, S., Yaqoob, M., Hung, D. V., Davis, W., Towakel, P., Raza, M., et al. (2022). Digital twins: A survey on enabling technologies, challenges, trends, and prospects. *IEEE Communications Surveys & Tutorials, 24*(4), 2255–2291.
21. Nasirahmadi, A., & Hensel, O. (2022). Toward the next generation of digitalization in agriculture based on the digital twin paradigm. *Sensors, 22*(2), 498.
22. Negri, E., Fumagalli, L., Cimino, C., & Macchi, M. (2019). FMU-supported simulation for CPS digital twin. *Procedia Manufacturing, 28*, 201–206.
23. Razzaq, S., Shah, B., Iqbal, F., Ilyas, M., Maqbool, F., & Rocha, A. (2022). DeepClassRooms: A deep learning-based digital twin framework for on-campus classrooms. *Neural Computing and Applications, 35*, 1–10.

24. Roy, P. (2023). *Blender/UPBGE + Python + Arduino = Digital twin*. Retrieved from https://blenderartists.org/t/blender-upbge-python-arduino-digital-twin/1456983
25. Sepasgozar, S. M. (2020). Digital twin and web-based virtual gaming technologies for online education: A case of construction management and engineering. *Applied Sciences, 10*(13), 4678.
26. Shahzad, M., Shafiq, M. T., Douglas, D., & Kassem, M. (2022). Digital twins in built environments: An investigation of the characteristics, applications, and challenges. *Buildings, 12*(2), 120.
27. Sleiti, A. K., Kapat, J. S., & Vesely, L. (2022). Digital twin in the energy industry: Proposed robust digital twin for power plant and other complex capital-intensive large engineering systems. *Energy Reports, 8*, 3704–3726.
28. Strielkowski, W., Rausser, G., & Kuzmin, E. (2022). Digital revolution in the energy sector: Effects of using digital twin technology. In *Digital transformation in industry: Digital twins and new business models* (pp. 43–55). Springer.
29. Tagliabue, L. C., Cecconi, F. R., Maltese, S., Rinaldi, S., Ciribini, A. L. C., & Flammini, A. (2021). Leveraging digital twin for sustainability assessment of an educational building. *Sustainability, 13*(2), 480.
30. Tao, F., Cheng, J., Qi, Q., Zhang, M., Zhang, H., & Sui, F. (2018a). Digital twin-driven product design, manufacturing, and service with big data. *The International Journal of Advanced Manufacturing Technology, 94*, 3563–3576.
31. Tao, F., Zhang, H., Liu, A., & Nee, A. Y. (2018b). Digital twin in industry: State-of-the-art. *IEEE Transactions on Industrial Informatics, 15*(4), 2405–2415.
32. Turner, C. (2023). *Unleashing the potential of digital twins in education—A revolution in learning*. Retrieved from https://www.linkedin.com/pulse/unleashing-potential-digital-twins-education-learning-turner/?trk=pulse-article_more-articles_related-content-card
33. VanDerHorn, E., & Mahadevan, S. (2021). Digital twin: Generalization, characterization, and implementation. *Decision Support Systems, 145*, 113524.
34. Wanasinghe, T. R., Wroblewski, L., Petersen, B. K., Gosine, R. G., James, L. A., De Silva, O., et al. (2020). Digital twin for the oil and gas industry: Overview, research trends, opportunities, and challenges. *IEEE Access, 8*, 104175–104197.
35. Yitmen, I., Alizadehsalehi, S., Akıner, İ., & Akıner, M. E. (2021). An adapted model of cognitive digital twins for building lifecycle management. *Applied Sciences, 11*(9), 4276.
36. Yu, W., Patros, P., Young, B., Klinac, E., & Walmsley, T. G. (2022). Energy digital twin technology for industrial energy management: Classification, challenges, and future. *Renewable and Sustainable Energy Reviews, 161*, 112407.

The Role of Digital Trust in Enhancing Cyber Security Resilience

Praveen Kumar Malik

Introduction

Digital trust and cyber-security are two important concepts in the modern world. Digital trust is the confidence that people have in digital systems and services. It is based on the assurance that the systems and services are secure, reliable, and trustworthy [1, 2]. Cyber security is a critical component of digital trust, as it helps to ensure that digital systems and services are secure and reliable. Digital trust and cybersecurity are essential for the safe and secure use of digital assets. Without trust, users may be hesitant to use digital services or applications, as they may not feel secure in their data or transactions. Cybersecurity is also important, as it helps protect networks, systems, and programs from malicious attacks. Without proper security measures in place, digital assets can be vulnerable to attack, leading to data loss or theft. Digital trust and cyber security are essential for the protection of individuals, businesses, and governments in the digital age. Digital trust is the confidence that individuals and organizations have in the security, privacy, and reliability of digital systems and services [3–5]. Cyber security is the practice of protecting networks, systems, and programs from digital attacks. The importance of digital trust and cyber security cannot be overstated. Without trust and security, individuals, businesses, and governments are vulnerable to malicious actors who can exploit weaknesses in digital systems and services to steal data, disrupt operations, and cause financial losses. Cyber security is essential for protecting the privacy of individuals and organizations, as well as for ensuring the integrity of digital systems and services. Additionally, digital trust and cyber security are important for protecting national security and critical infrastructure, as well as for promoting economic growth and innovation [3].

P. K. Malik (✉)
School of Electronics and Electrical Engineering, Lovely Professional University, Phagwara, Punjab, India

Improving Digital Trust and Cyber Security

There are several ways to improve digital trust and cybersecurity. One way is to use strong authentication methods, such as two-factor authentication or biometric authentication. This helps ensure that only authorized users can access digital assets. Additionally, organizations should use encryption to protect data in transit and at rest. Finally, organizations should implement regular security audits and vulnerability assessments to identify potential weaknesses in their systems [6]. Some of the ways to improve it are as follows:

1. *Implement strong authentication:* This can be done through two-factor authentication, biometric authentication, or other methods.
2. *Implement encryption:* This can be done through a variety of methods, such as SSL/TLS, AES, and PGP.
3. *Implement access control:* This can be done through role-based access control, user authentication, and other methods.
4. *Implement security policies:* Security policies are documents that outline the rules and regulations for using a system or network. These policies should be regularly updated and enforced to ensure that users are following the proper security protocols.
5. *Educate users:* Educating users on the importance of digital security is essential for improving digital trust and cyber security. This can include providing users with information on how to use the system securely, as well as the consequences of not following security protocols.
6. *Monitor activity:* Monitoring user activity on a system or network can help identify potential security threats. This can be done through log monitoring, network monitoring, and other methods.
7. *Implement vulnerability management:* It is the process of identifying, assessing, and mitigating security vulnerabilities in a system or network. This can be done through regular vulnerability scans and patch management (Fig. 1).

Thrust Area in Digital Trust and Cyber Security

Cybersecurity Risk Management

It involves the evaluation of threats and vulnerabilities, the implementation of security controls, and the monitoring of the effectiveness of those controls. The goal of cyber security risk management is to reduce the likelihood and impact of a security incident and to ensure that the organization is prepared to respond in the event of an attack [7].

The process of cyber security risk management typically includes the following steps:

Improving Digital Trust and Cyber security

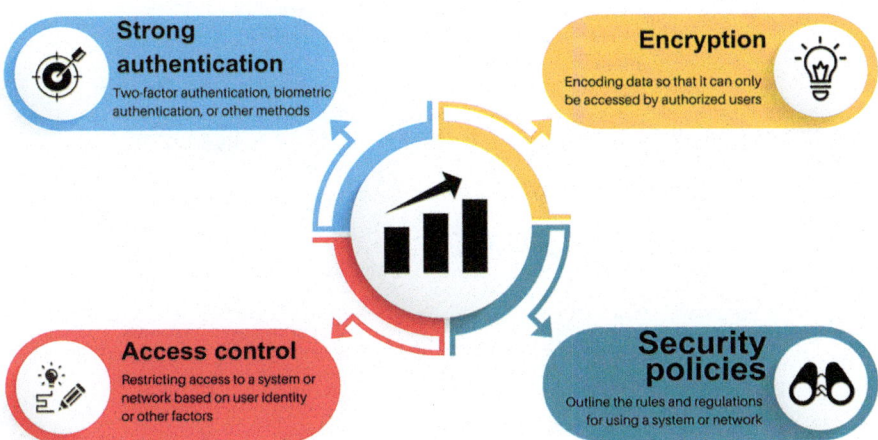

Fig. 1 Digital trust and cyber security

1. Identifying and assessing risks: This involves identifying potential threats and vulnerabilities and assessing their potential impact on the organization.
2. Implementing security controls: This involves implementing technical, administrative, and physical controls to mitigate the identified risks.
3. Monitoring and reviewing: This involves monitoring the effectiveness of the security controls and regularly reviewing and updating them as needed.
4. Responding to incidents: This involves having a plan in place to respond to security incidents and ensuring that the organization is prepared to respond quickly and effectively.

Cyber Threat Intelligence

Cyber threat intelligence (CTI) is a type of intelligence that focuses on potential threats and risks to an organization's digital assets and infrastructure. It is used to identify, analyze, and respond to cyber threats in a timely and effective manner. CTI can be used to detect and prevent malicious activity, such as malware, phishing, and other cyber-attacks. It can also be used to assess an organization's risk posture and develop strategies to mitigate those risks. CTI can also be used to inform an organization's security policies and procedures [8] (Fig. 2).

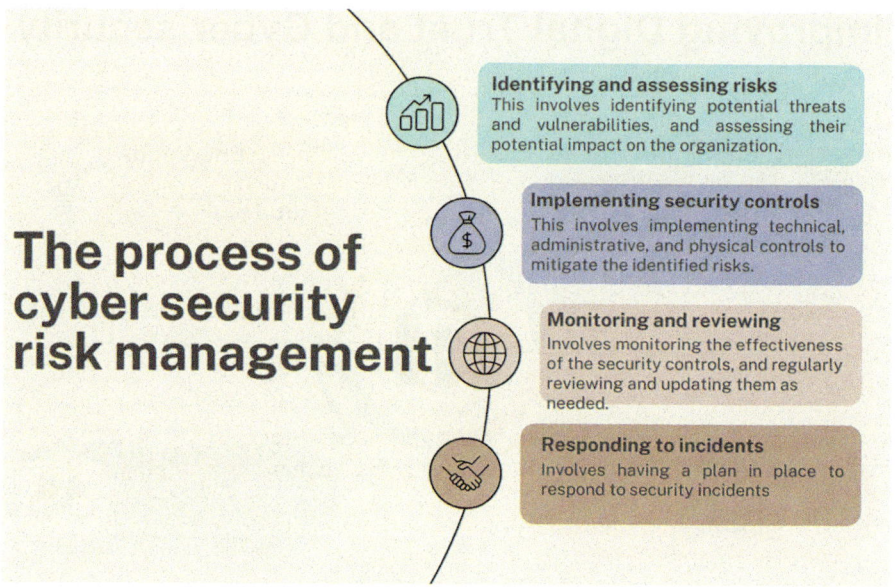

Fig. 2 Process of cyber security risk management

Identity and Access Management

A set of guidelines and tools called Identity and Access Management (IAM) is used to control digital identities and the access privileges that go along with them. It is employed to manage who has access to the data, apps, and other resources within an organization. IAM is a vital part of any company's security plan since it makes sure that sensitive data is only accessible to those who are authorized. Authentication, access control, and identity management are the three primary parts of identity and access management (IAM). The process of establishing, controlling, and upholding digital identities is known as identity management. Assigning roles and permissions, controlling access privileges, and creating user accounts are all included in this. The process of allowing or refusing access to resources in accordance with a user's identification and role is known as access control. The process of confirming a user's identity before allowing them to access a resource is called authentication. IAM is a crucial component of any organization's security plan since it makes sure that critical data is only accessible to those who are authorized. It also aids in lowering the possibility of future security issues and data breaches [9, 10].

Data Protection and Privacy

Protection of personal information and privacy refers to a collection of rules, regulations, and standards that govern the manner in which data is gathered, stored, utilized, and distributed. The purpose of this system is to safeguard the privacy of persons and the personal information they possess as well as to guarantee that data is utilized in a responsible and ethical manner. The rules that govern data protection and privacy differ from country to country, but in general, they mandate that enterprises must acquire, keep, and utilize personal data in a responsible and secure manner. Additionally, they mandate that organizations extend certain rights to individuals, such as the ability to access and update their personal data, which is a need for businesses. The protection of personal information from unauthorized access, use, and disclosure is another responsibility that organizations have to fulfill. They must make certain that they have sufficient security measures in place. Encryption, access control, and data minimization are some examples of safeguards that may fall under this category. The education of individuals of their rights and responsibilities in relation to their personal data is another component of data protection and privacy. Organizations have a responsibility to provide individuals with information that is both clear and concise regarding the utilization of their data and the means by which they can exercise their rights [11, 12].

Network Security

Network security that is a practice of preventing unauthorized access, misuse, and modification can be defined as follows:

1. *Firewalls:* When it comes to network security, a firewall is a device that monitors and controls network traffic coming into and going out of the network depending on certain security criteria that have been defined. The implementation of firewalls might take the form of either hardware, software, or a combination of the two.
2. *Intrusion Detection Systems (IDS):* An IDS is a system that monitors network traffic and attempts to detect malicious activity. It can be used to detect malicious activities such as port scans, denial of service attacks, and malicious code.
3. *Encryption:* The process of converting data into a format that cannot be read by individuals who are not allowed to do so is known as digital encryption. For the purpose of protecting data while it is in transit, such as when it is transmitted over the Internet, as well as data while it is at rest, such as when it is saved on a hard drive, encryption is utilized.
4. *Access Control Lists (ACLs):* An ACL is a set of rules that specify which users or systems are allowed to access certain network resources. ACLs can be used to control access to files, directories, and other network resources.
5. *Virtual Private Networks (VPNs):* A VPN is a secure connection between two or more computers over the Internet. It is used to protect data in transit by encrypting the data and authenticating the sender and receiver.
6. *Network Segmentation:* Network segmentation is the process of dividing a network into smaller, more secure segments. This can be done using firewalls, VLANs, and other network (Fig. 3).

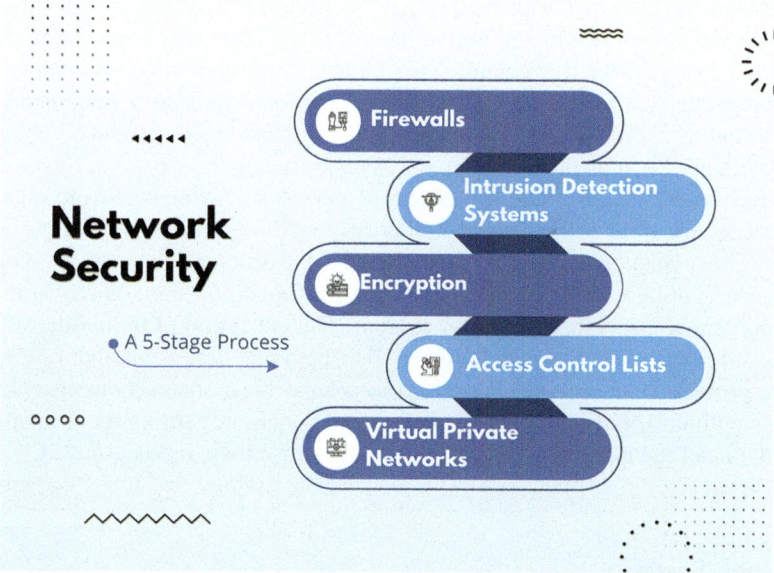

Fig. 3 Network security which is a practice of preventing unauthorized access, misuse, modification

Cloud Security

On the other hand, cloud security refers to the collection of rules, methods, and controls that are utilized to safeguard data, applications, and the infrastructure that is linked with cloud computing. In the larger context of computer security, it is a sub-domain of information security, network security, and cybersecurity in general. Providing protection for data, apps, and the infrastructure that is linked with cloud computing is the goal of cloud security. It encompasses a wide range of controls, including as identity and access management, data security, threat detection and response, data privacy, compliance, and a number of other controls. Additionally, cloud security encompasses the safeguarding of data when it is in motion, such as when it is being transported between cloud providers and their respective customers. The security of cloud computing is a responsibility that is shared between cloud providers and their consumers. It is the responsibility of the cloud provider to ensure the safety of their infrastructure, while it is the responsibility of the customer to ensure the safety of their data and applications. Therefore, it is the responsibility of the customers to ensure the safety of their data and applications, while the cloud service providers are responsible for ensuring the safety of their infrastructure [13, 14].

Cryptography

In the presence of third parties, cryptography refers to the practice and study of methods that are used to ensure the confidentiality of communication. In the presence of adversaries, it is a subfield of mathematics that focuses on the development of methods for ensuring the confidentiality of communication. There are a variety of applications that make use of cryptography, including online banking, electronic commerce, and secure communication over the Internet. Additionally, it is utilized to safeguard information that is kept on mobile devices and desktops. The protection of data against illegal access, modification, or disclosure is accomplished through the use of cryptography. In addition to this, it can be utilized to guarantee the genuineness and completeness of the data. The usage of cryptography is a method that can be employed to safeguard data from eavesdropping, manipulation, and other types of attacks. The use of cryptography can also be employed to safeguard data against being stolen or misappropriated [15].

Cybercrime Prevention

Cybercrime prevention in India is a growing concern due to the increasing number of cybercrimes being committed in the country. The Government of India has taken several steps to address this issue, including the establishment of the National Cyber Security Coordination Centre (NCSC) in 2018. This center is responsible for coordinating the efforts of various government agencies and private organizations to prevent, detect, and respond to cybercrimes. The government has also enacted the Information Technology Act, 2000, which provides for the legal framework for the regulation of cybercrimes in India. The Act also provides for the establishment of the Cyber Appellate Tribunal, which is responsible for hearing appeals against orders passed by the Controller of Certifying Authorities. The government has also launched several initiatives to raise awareness about cybercrime prevention, including the National Cyber Safety and Security Standards (NCSSS). This initiative provides guidance on how to protect oneself from cybercrimes and provides a platform for the exchange of information between various stakeholders. The government has also set up the National Cyber Crime Reporting Portal (NCCRP) to enable citizens to report cybercrimes. The portal also provides information about cybercrime prevention and the steps to be taken in case of a cybercrime. In addition, the government has taken steps to strengthen the cyber security infrastructure in the country. This includes the establishment of the CERT-In (Computer Emergency Response Team-India), which is responsible for responding to cyber security incidents and providing technical assistance to organizations. The government has also launched the National Critical Information Infrastructure Protection Centre (NCIIPC) to protect critical information infrastructure in the country.

Conclusion

Digital trust and cyber security are essential for the safe and secure use of digital assets. Without trust, users may be hesitant to use digital services or applications, as they may not feel secure in their data or transactions. Cyber security is also important, as it helps protect networks, systems, and programs from malicious attacks. Organizations should use strong authentication methods, encryption, and regular security audits to improve digital trust and cyber security.

References

1. Abbass, H. A., Leu, G., & Merrick, K. (2016). A review of theoretical and practical challenges of trusted autonomy in big data. *IEEE Access, 4*, 2808–2830. https://doi.org/10.1109/ACCESS.2016.2571058
2. Syed Mustapha, S. M. F. D., & Gupta, P. (2024). DBSCAN inspired task scheduling algorithm for cloud infrastructure. *Internet of Things and Cyber-Physical Systems, 4*, 32–39. ISSN 2667-3452. https://doi.org/10.1016/j.iotcps.2023.07.001
3. Fachrunnisa, O., & Hussain, F. K. (2013). A methodology for maintaining trust in industrial digital ecosystems. *IEEE Transactions on Industrial Electronics, 60*(3), 1042–1058. https://doi.org/10.1109/TIE.2011.2173890
4. Wang, C., Cai, Z., Seo, D., & Li, Y. (2023). TMETA: trust management for the cold start of IoT services with digital-twin-aided blockchain. *IEEE Internet of Things Journal, 10*(24), 21337–21348. https://doi.org/10.1109/JIOT.2023.3285108
5. Binnar, P., Bhirud, S., & Kazi, F. (2024). Security analysis of cyber physical system using digital forensic incident response. *Cyber Security and Applications, 2*, 100034. ISSN 2772-9184. https://doi.org/10.1016/j.csa.2023.100034
6. Wang, H., Kang, X., Li, T., Lei, Z., Chu, C.-K., & Wang, H. (2023). An overview of trust standards for communication networks and future digital world. *IEEE Access, 11*, 42991–42998. https://doi.org/10.1109/ACCESS.2023.3270042
7. Admass, W. S., Munaye, Y. Y., & Diro, A. A. (2024). Cyber security: State of the art, challenges and future directions. *Cyber Security and Applications, 2*, 100031. ISSN 2772-9184. https://doi.org/10.1016/j.csa.2023.100031
8. Rejeb, A., Rejeb, K., Appolloni, A., Jagtap, S., Iranmanesh, M., Alghamdi, S., Alhasawi, Y., & Kayikci, Y. (2024). Unleashing the power of internet of things and blockchain: A comprehensive analysis and future directions. *Internet of Things and Cyber-Physical Systems, Volume, 4*, 1–18. ISSN 2667-3452. https://doi.org/10.1016/j.iotcps.2023.06.003
9. Sun, N., et al. (2022). Defining security requirements with the common criteria: Applications, adoptions, and challenges. *IEEE Access, 10*, 44756–44777. https://doi.org/10.1109/ACCESS.2022.3168716
10. Marican, M. N. Y., Razak, S. A., Selamat, A., & Othman, S. H. (2023). Cyber security maturity assessment framework for technology startups: A systematic literature review. *IEEE Access, 11*, 5442–5452. https://doi.org/10.1109/ACCESS.2022.3229766
11. Kavallieratos, G., Diamantopoulou, V., & Katsikas, S. K. (2020). Shipping 4.0: Security requirements for the cyber-enabled ship. *IEEE Transactions on Industrial Informatics, 16*(10), 6617–6625. https://doi.org/10.1109/TII.2020.2976840
12. Son, J., Choi, J., & Yoon, H. (2019). New complementary points of cyber security schemes for critical digital assets at nuclear power plants. *IEEE Access, 7*, 78379–78390. https://doi.org/10.1109/ACCESS.2019.2922335

13. Karie, N. M., Sahri, N. M., Yang, W., Valli, C., & Kebande, V. R. (2021). A review of security standards and frameworks for IoT-based smart environments. *IEEE Access, 9*, 121975–121995. https://doi.org/10.1109/ACCESS.2021.3109886
14. Sun, N., Li, C.-T., Chan, H., Islam, M. Z., Islam, M. R., & Armstrong, W. (2022). How do organizations seek cyber assurance? Investigations on the adoption of the common criteria and beyond. *IEEE Access, 10*, 71749–71763. https://doi.org/10.1109/ACCESS.2022.3187211
15. Sutradhar, S., Karforma, S., Bose, R., Roy, S., Djebali, S., & Bhattacharyya, D. (2024). Enhancing identity and access management using Hyperledger Fabric and OAuth 2.0: A block chain-based approach for security and scalability for healthcare industry. *Internet of Things and Cyber-Physical Systems, 4*, 49–67. ISSN 2667-3452. https://doi.org/10.1016/j.iotcps.2023.07.004

From Reactive to Proactive: Predicting and Optimizing Performance for Competitive Advantage

Tapan Kumar Behera and Deep Manishkumar Dave

Introduction

A large-scale software solution created to support intricate business activities and processes within an organization is referred to as an enterprise application, also known as enterprise software or a business application. These programs were created especially to cater to the various requirements of businesses, including data management, task automation, collaboration, and effective decision-making. Enterprise applications, such as those designed for human resources, finance, supply chain management, enterprise resource planning (ERP), and customer relationship management, typically offer a broad spectrum of features and can be tailored to suit various industries and corporate needs. They can be integrated with other systems within the enterprise's technological ecosystem and are characterized by their scalability, resilience, security, and ability to do so [5].

Performance testing is a crucial part of creating and deploying software applications. It is essential for ensuring that a program provides a seamless user experience and satisfies the appropriate performance standards. Organizations can find and fix performance bottlenecks that might impede user happiness by doing various performance tests on the application. Performance testing aids in the optimization of the application's performance and offers users a seamless and responsive experience, regardless of the issue—slow response times, high latency, or excessive resource usage [60]. One of the primary objectives of performance testing is to assess the capability and scalability of an application. It is critical to assess the application's capacity to meet rising demands as user loads and data volumes rise. Performance testing assists organizations in properly planning for future expansion and ensuring that the application can expand to accommodate the anticipated user base and data

---

T. K. Behera (✉) · D. M. Dave (✉)
Independent Researcher, Boston, MA, USA

© The Author(s), under exclusive license to Springer Nature Switzerland AG 2024
A. Mishra et al. (eds.), *Transforming Industry using Digital Twin Technology*,
https://doi.org/10.1007/978-3-031-58523-4_5

growth by identifying performance restrictions, scalability concerns, and capacity constraints.

Another crucial component of an application's performance is reliability and stability. Evaluation of an application's performance under various loads and stress situations is aided by performance testing. Organizations may address these concerns and improve the stability of their programs by recognizing potential problems like memory leaks, performance degradation, or system crashes. As a result, downtime is reduced, overall reliability is increased, and users can rely on the program to function consistently. Performance testing also has the important advantage of optimizing resource allocation. Organizations can choose the best distribution of servers, databases, and network bandwidth by evaluating the application's resource usage and load balancing capabilities. Performance testing identifies resource-intensive elements or bottlenecks that could affect the performance of the program. Organizations can improve resource utilization, boost system performance, and save infrastructure costs by optimizing resource allocation [1].

Applications frequently rely on interconnections with other systems, databases, or APIs in complex software ecosystems. Performance testing is essential for guaranteeing seamless interoperability and integration between various components and external systems. Organizations can spot any performance-related problems or communication bottlenecks that can occur during interactions by thoroughly testing the performance of integrated systems. This helps them to create a seamless user experience throughout the whole software ecosystem and optimize the system's overall performance. Performance testing also enables businesses to confirm that an application meets with specified performance SLAs [31, 32]. The performance goals and metrics that the application must fulfill are frequently stated in service level agreements. Organizations may make sure they are fulfilling their contractual responsibilities and achieving the expected performance levels by evaluating the application's performance against these predetermined criteria.

Performance testing has substantial effects on customer satisfaction and a company's entire reputation in addition to its technical advantages. Performance problems in software programs can result in irate users, unfavorable evaluations, and harm to a company's reputation. By engaging in thorough performance testing, companies have the opportunity to identify and address performance issues in the early stages of the development cycle [3]. The application will work as expected, provide a pleasant user experience, and eventually increase client satisfaction and loyalty thanks to this proactive strategy [28].

What's the Problem of Launching Software Without Performance Testing?

Launching software without first doing performance testing can have major ramifications and dangers for both the software supplier and the product's users. One significant problem is the possibility of a bad user experience. Without performance testing, the program may have poor reaction times, crashes, or be unresponsive, causing user annoyance and discontent. This can lead to unfavorable evaluations, customer loss, and harm to the software's reputation. Another issue is the occurrence of bottlenecks in performance. Performance restrictions and bottlenecks may go undetected if extensive testing is not performed. When confronted with concurrent users, heavy data loads, or complicated tasks, these bottlenecks might cause the program to suffer. As a result, under production-level utilization, performance suffers, response times increase, and system failures are possible [48].

Scalability issues are also a source of worry. It is impossible to tell if the program can efficiently scale to meet increasing user demands or support increased data quantities without doing performance testing. This can result in decreased performance, longer reaction times, and an inability to handle the expected workload efficiently [8]. As a result, the software's capacity to develop with the company and adapt to changing business demands is limited. Another significant element influenced by a lack of performance testing is resource consumption. The software's resource consumption habits may remain unknown without thorough testing. As a result, hardware resources like as CPU, memory, and network bandwidth may be allocated inefficiently. Inefficient resource use not only has an impact on performance, but it also raises infrastructure costs and demands unscheduled hardware upgrades [19].

The lack of performance testing increases the likelihood of system instability and downtime. Under some situations, the program may experience crashes, instability, or even full system failure if performance issues are not identified and addressed in advance. This interrupts corporate operations, results in financial losses, and undermines user trust in the program. Another consequence of deploying software without doing performance testing is failure to meet SLA criteria. Service level agreements sometimes include performance criteria that must be met by the program [44]. By ignoring performance testing, there is a greater chance of failing to achieve these SLA criteria, resulting in contractual violations and significant legal ramifications. Furthermore, the costs of releasing software without doing performance testing might be enormous. To detect and repair post-launch performance issues, extra support and maintenance activities are required. Addressing performance issues in a live environment takes time, resources, and money. Proper performance testing before to launch aids in the identification and correction of faults early in the development cycle, eliminating the need for reactive and costly post-launch changes [5].

The Digital Twin Technology

Digital twin technology is a concept within the realm of Industry 4.0 and involves generating a digital counterpart that mirrors a physical asset, system, or process. Its utility extends across various sectors, encompassing healthcare, automotive, aerospace, and more, as it aids in enhancing real-world operations [15] In recent years, digital twins have gained significant momentum as virtual replicas of physical systems generated through data, sensors, and algorithms. These replicas find utility in a range of applications, from predicting maintenance requirements to simulating real-time scenarios. Although the concept of digital twins has been in existence for some time, recent progress in artificial intelligence (AI) and the Internet of Things (IoT) has made it increasingly accessible and relevant across a wide range of industries [60]. This chapter aims to explore the world of digital twin technology and its impact on software performance testing.

At the core of digital twin technology is its ability to generate virtual duplicates of physical objects, processes, or systems using real-time data and advanced analytics. This empowers the simulation, analysis, and enhancement of the corresponding physical entities throughout their lifecycles, spanning design, development, operation, and maintenance [21]. These virtual replicas come into existence by gathering data from sensors and various sources integrated within or around the physical entities. The gathered data is then utilized by software programs to construct the digital twins. Such virtual representations enable the simulation of different scenarios and the testing of various modifications without impacting the actual physical counterparts [39, 40].

Digital twin technology offers several key advantages. To begin with, it offers immediate insights into the performance of physical objects or systems, making it easier to spot potential problems early and enabling proactive maintenance and optimization [14]. For instance, in manufacturing, digital twins can optimize production processes, minimize downtime, and improve product quality. Secondly, digital twins enable remote monitoring and control, empowering operators to oversee and manage physical objects or systems from anywhere in the world using computers or mobile devices [52]. This capability proves particularly valuable in industries like aerospace, where remote monitoring and control contribute to cost reduction and safety enhancement [3].

The History of Digital Twin

The concept of digital twin technology has gained significant popularity in recent times. It was Dr. Michael Grieves, a professor at the University of Michigan, who originally coined the term "digital twin" back in 2002. However, the Apollo 13 mission by NASA in 1970 is where the idea of digital twins first emerged. An oxygen tank exploded during the trip, endangering the crew's lives and seriously damaging the spacecraft. The astronauts had to be brought back safely; therefore, the engineers on the ground had to act rapidly. They experimented with several situations using a physical replica of the spaceship in an effort to identify the issue. From there on, this idea only grew stronger where today we see new dimensions in areas such as improved efficiency through better system design and performance [49].

Origins in Computer-Aided Design (CAD) and Simulation

CAD and simulation are where the concept of the digital twin first emerged. These early applications established the groundwork for the concept of generating virtual representations of actual assets by using computer models and simulations to mimic and study the behavior of physical items or systems [9].

Conceptualization of Digital Twins

In 2002, Dr. Michael Grieves, hailing from the University of Michigan, introduced the term "digital twin." Grieves aimed to bridge the divide between the physical and digital realms, fostering better comprehension and decision-making. He defined a digital twin as a virtual representation of a physical product that integrates real-time data to monitor and enhance its performance [28].

Application in Manufacturing and Product Development

Early on, the manufacturing sector saw substantial applications for digital twin technology. It was used to improve design, testing, and manufacturing processes by building virtual representations of products, parts, or production systems [12]. Before going into physical manufacturing, manufacturers may test, evaluate, and optimize the performance of their goods thanks to digital twins.

IoT and Data Analytics Integration

The development of the Internet of Things (IoT) and improvements in data analytics increased the potential of digital twins. Real-time data from physical assets may now be gathered thanks to the proliferation of sensors and linked devices and fed into digital twin models, enabling continuous evaluation of the performance and condition of the physical asset [59]. Up until the early 2010s, the technology was mostly restricted to industrial uses, but advances in sensor technology and cloud computing opened it up to other applications. Today, digital twins find applications across various industries, including manufacturing, healthcare, transportation, and construction [1].

Artificial Intelligence (AI) Advancements Include the Following

The inclusion of AI technologies, such as deep learning, machine learning, and predictive analytics, increased the power of digital twins [41]. Digital twins can now learn from historic data, forecast future actions, and improve performance based on intricate patterns and scenarios thanks to AI algorithms. This advancement made it possible to create digital twin models that were more complex and dynamic and could offer suggestions and insights that could be put into practice [43].

Expansion into Different Industries

As digital twin technology developed, other industries began to use it outside of manufacturing. Digital twins have been adopted by sectors including infrastructure, energy, transportation, agriculture, healthcare, and software performance to improve decision-making, streamline operations, and simplify maintenance procedures [20, 26, 44, 48].

Digital Twin Types

Digital twin models may be used to represent anything, from discrete parts to whole systems. Although every sort of digital twin performs the same basic task—virtually simulating a real-world item or system—their objectives and range drastically differ from one another [33]. There are four main categories of digital twins (Fig. 1):

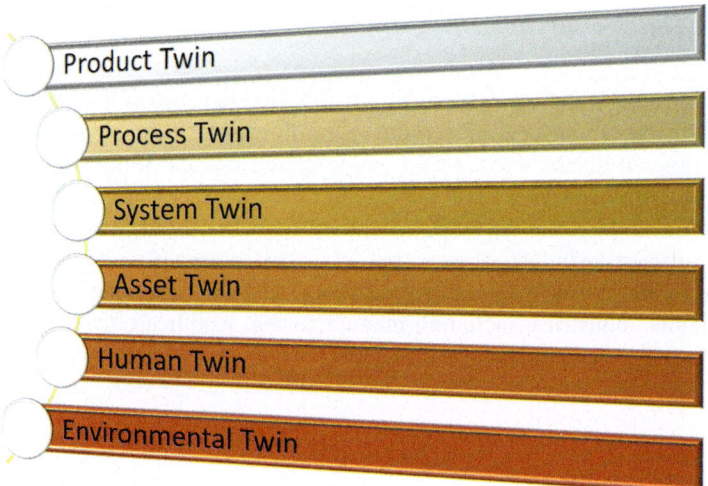

Fig. 1 Digital Twin types

Product Twins

Product twins are digital representations of actual goods or assets that incorporate all of the product's precise geometrical, functional, and behavioral details [31, 32]. Product twins are utilized at every stage of the lifespan of a product, from design and development to production, usage, and maintenance. They make it possible to test, analyze, and optimize product performance, quality, and dependability virtually [56, 57]. Digital twins of automobiles, appliances, structures, or consumer gadgets are a few examples.

Process Twins

They record the order of operations, dependencies, and performance metrics of a process with an emphasis on recreating and emulating certain industrial processes or workflows [53]. Process twins are frequently employed in the manufacturing, supply chain management, healthcare, and other sectors. They allow for the optimization, analysis, and improvement of process efficiency, productivity, and resource utilization [19].

System Twins

Complex systems or networks made up of interrelated parts or resources are represented by system twins. They mimic the behaviors and interactions of the system's many components. System twins make it possible to analyze, improve, and coordinate system-level performance. They are utilized in industries including infrastructure management, smart cities, energy, and transportation. Examples include digital twins of the infrastructure in smart cities, power grids, and transportation networks [31, 32].

Asset Twins

These twins concentrate on specific pieces of machinery, equipment, or other tangible property. They record the asset's precise features, performance information, and maintenance history [46]. Real-time monitoring, proactive maintenance, and performance enhancement of assets are made possible by asset twins [42]. They are frequently used in sectors including manufacturing, energy, and aerospace.

Human Twins

Human twins are a special type of digital twin that may represent a person or a group of individuals by capturing information about their physical characteristics, actions, interests, and health issues [45]. Human twins are digital representations of patients for personalized treatment planning, digital avatars for virtual reality experiences, or digital models used in ergonomic design [25]. Human twins can serve as models for studying and analyzing human factors in diverse fields like healthcare, personalized medicine, and fitness.

Environmental Twins

Environmental twins imitate and keep track of real-world ecosystems or geographic regions, collecting information on the weather, pollution levels, topography, and ecological aspects [27]. Urban planning, environmental monitoring, and disaster management all employ environmental twins. Digital twins of cities, forests, water bodies, or disaster-prone regions are just a few examples of how they make it possible to analyze, anticipate, and mitigate the effects of the environment [38].

How Digital Twin Helps Solve the Problems of Software Performance

Digital twin is a valuable tool for effectively addressing issues related to software performance. Organizations can monitor and assess performance in real-time by leveraging digital twins, as they facilitate the creation of a virtual representation of the software system. Continuous data collection and analysis aid in locating performance bottlenecks, patterns of resource consumption, and potential performance-related problems [56, 57]. With the help of this proactive strategy, companies can actively manage and optimize performance while making decisions that will improve user experience and system effectiveness. Performance issues are greatly reduced by using digital twins' predictive analytics and simulation capabilities. Digital Twins can forecast future performance trends and spot potential problems before they arise by utilizing historical data and cutting-edge analytical approaches [3]. Organizations can evaluate the effects of changes on performance and scalability through scenario simulations and load testing. To meet the demands of the anticipated workload, this enables companies to take preemptive measures, plan resource allocation, and enhance system performance.

The proactive issue detection and mitigation is made possible by digital twins. Organizations can spot performance irregularities and deviations in real-time by constantly monitoring the software's Digital Twin. Early identification of performance problems enables businesses to take prompt corrective action, reducing the impact on end users and halting further deterioration. With the help of digital twins, enterprises may proactively solve performance issues and guarantee the best possible software performance. Digital Twins are also excellent at allocating resources and maximizing performance [50]. Digital twins assist organizations in optimizing resource allocation and fine-tuning system setups by offering insights into resource use, dependencies, and performance hotspots. Organizations can find chances for performance improvement, make changes, and improve resource efficiency by utilizing the data and analytics capabilities of Digital Twins. By doing this, software systems are guaranteed to work at their very best, providing a smooth user experience.

Digital twins are also useful for designing and validating scalability. Prior to deployment, businesses can evaluate the software system's scalability by modeling various scenarios and load circumstances. With the use of digital twins, enterprises can virtually recreate the software environment to discover potential bottlenecks and make sure the system can accommodate rising demand without degrading performance. Organizations can successfully prepare for scalability and improve the performance of the software under a variety of circumstances with the aid of this proactive approach [9]. Additionally, during the whole program lifecycle, Digital Twins support ongoing performance enhancement. Organizations can find areas for improvement, adjust, and track the effects of optimizations by evaluating real-time performance data. Performance is maintained as a top priority thanks to this iterative method, which also makes it possible to make continuous improvements to keep performance levels high [18].

Digital twins provide further benefits such as real-time control and feedback. They enable autonomous adjustments, self-healing mechanisms, and adaptive optimization through interaction with monitoring and control systems, all based on real-time performance data. This dynamic feedback loop makes sure that software systems can adapt to shifting circumstances and continuously maintain peak performance levels. A complete solution to problems with software performance is provided by digital twins [10]. Digital twins enable enterprises to proactive manage, optimize, and improve software performance through real-time monitoring and analysis, predictive capabilities, optimization, scalability planning, and continuous improvement. Organizations can create high-performing software solutions, guarantee a seamless user experience, and meet their performance goals by utilizing the advantages of digital twins.

Digital Twin Performance Testing for Software Application

Digital twins provide further benefits such as real-time control and feedback. They enable autonomous adjustments, self-healing mechanisms, and adaptive optimization through interaction with monitoring and control systems, all based on real-time performance data [16].

There are several ways that may be employed to undertake digital twin performance testing. Among the most frequent approaches are:

- *Load Testing:* Load testing is simulating a high number of users or transactions to the digital twin to evaluate its scalability and performance under load.
- *Stress Testing:* Stress testing is exposing the digital twin to harsh circumstances to uncover any possible bottlenecks or malfunctions.
- *Fault Injection:* Fault injection is the process of introducing simulated defects into a digital twin to verify its robustness and fault tolerance.
- *Performance regression testing:* Performance regression testing is evaluating the digital twin's performance over time to find any performance regressions.

The approaches employed will be determined by the application and the digital twin's needs. To guarantee that the results are reliable and consistent, all digital twin performance testing should be performed in a controlled setting.

Here are a few benefits of conducting performance testing with digital twins:

- *Improved product quality:* Digital twin performance testing can assist to enhance product quality by detecting and fixing performance concerns early in the development process.
- *Reduced risk:* Digital twin performance testing can assist to lessen the risk of performance issues emerging in the real world by mimicking the product's performance in a controlled environment.
- *Improved decision-making:* Because digital twin performance testing provides insights into product performance, it can aid in decision-making about product design, development, and deployment.

Here are some pointers on how to do digital twin performance testing:

- *Begin early:* The sooner in the development process performance testing is undertaken, the easier it will be to discover and fix performance issues.
- *Use a number of approaches:* For all sorts of performance testing, no single approach is adequate. Use a range of ways to guarantee that all areas of the performance of the digital twin are reviewed.
- *Test in a controlled environment:* The findings of performance testing will be accurate only if they are conducted in a controlled setting. Use a representative test environment where digital twin can be deployed.
- *Repeat the tests:* Run the tests again and again to find any performance regressions.

In the development process of any software program utilizing a digital twin, performance testing with digital twins is a vital component. By following the suggestions above, you can verify that your digital twin has been extensively tested and that its performance satisfies your needs [17].

Challenges in Testing Digital Twins

Evaluating digital twins presents several distinct challenges that must be resolved to ensure precise and reliable outcomes. Here are some of the key issues encountered in testing digital twins.

- *Data Integration:* To produce an accurate picture of the physical system, digital twins depend on the integration of numerous data sources and sensors. However, integrating and syncing data from several sources can be complicated and difficult. To assure that the digital twin receives pertinent data at the appropriate moment and that the data is both accurate and consistent, rigorous data management and integration methods are essential [29].
- *Model Accuracy:* The digital twin model's correctness is critical for accurate testing. The test results may not reflect real-world performance if the model does not adequately capture the behavior and features of the actual system [16]. Creating exact and verified models that capture all key characteristics of the physical system may be difficult, necessitating subject expertise as well as significant testing and validation.
- *Scalability:* Digital twins are frequently used to depict complicated systems with many interrelated components. Scalability testing for such systems can be difficult since it entails replicating large-scale situations with a huge number of concurrent users, devices, or processes [50]. Creating and organizing the appropriate test data, coordinating the test environment, and evaluating the findings on a large scale may all be difficult.
- *Real-Time Simulation:* To effectively replicate the dynamic behavior of the physical system, many digital twins require real-time simulation and synchronization. It can be technically difficult to ensure that the simulation is synced with the actual system in real-time, especially when working with high-frequency data or systems with precise timing requirements. Real-time simulation necessitates the use of resilient and high-performance computer resources [3].
- *Testing Automation:* Testing digital twins sometimes entails complicated test scenarios that, to be efficient and successful, require automation. Nonetheless, due to the intricate and interconnected nature of system components, the need for real-time simulation, and the amalgamation of multiple data sources, automating the testing of digital twins can pose challenges [11]. It takes careful planning, tool selection, and scripting to create successful and maintainable automated testing frameworks for digital twins.
- *Security and Privacy:* Digital twins, particularly those linked to the Internet of Things (IoT) or cloud infrastructure, are vulnerable to security risks and data breaches. To uncover vulnerabilities, analyze risks, and confirm the efficiency of security solutions, testing the security and privacy elements of digital twins necessitates expertise and techniques [4]. It is critical to safeguard that sensitive data is secured and the digital twin is secure throughout its existence.
- *Test Environment Replication:* To effectively test digital twins, it is imperative to replicate the production environment as closely as possible. Hardware, software,

network settings, and other external requirements are all replicated. It can be difficult to create and manage such test environments, especially when working with complicated systems and large-scale deployments [23].

To address these issues, a mix of technical competence, domain knowledge, and effective testing methodologies is required. To overcome these obstacles and assure the reliability and efficacy of digital twin testing, organizations must invest in qualified employees, testing tools, and infrastructure.

AI-Based Testing Software to Test Load Performance Using Digital Twin

There are various AI-based testing software tools available in the market that can be employed to test load performance using a digital twin. These tools employ artificial intelligence and machine learning techniques to improve the testing process and provide more accurate results. Here are some examples:

- *Appvance IQ:* Appvance IQ is an AI-powered testing platform with load testing features. It uses machine learning techniques to develop and conduct load tests automatically based on user behavior patterns. The application can simulate thousands of virtual users and give detailed performance data.
- *Apica LoadTest:* Apica LoadTest is a cloud-based load testing solution that incorporates AI-driven analytics. It analyzes test data and identifies performance bottlenecks using machine learning methods. The program can mimic heavy user loads while also providing real-time monitoring and reporting.
- *LoadRunner:* Micro Focus's LoadRunner is a well-known load testing tool with AI-driven features. It uses machine learning techniques to produce load test scripts automatically and change load levels based on system performance. The tool gives information on performance metrics and helps in identifying any vulnerabilities.
- *NeoLoad:* Neotys' NeoLoad is an AI-powered load testing tool. It analyzes real-time data during load testing using machine learning and automatically adapts the test scenarios to maximize performance. For load testing, the program includes comprehensive analytics and reporting options.
- *BlazeMeter:* BlazeMeter is a load testing platform that uses AI-driven testing capabilities and is now part of the Broadcom family. Machine learning techniques are used to improve test performance, provide realistic load situations, and give intelligent reporting. The program allows you to test various protocols and technologies.
- *LoadUI:* SmartBear's LoadUI is an open-source load testing tool for simulating large user loads and assessing system performance. It has an easy-to-use UI and allows you to create complicated load testing scenarios.

- *Gatling:* It is an open-source load testing tool aimed for developers. It allows for the scripting and execution of load tests through the use of a domain-specific language (DSL). Gatling offers real-time data and statistics to help you analyze system performance.
- *Locust:* It is an open-source load testing tool that enables for the creation of load testing scenarios in Python. It provides a scalable and adaptable platform for creating user load and tracking performance indicators.
- *Artillery:* Artillery is an open-source load testing tool that focuses on simplicity and versatility. It allows you to create load test scenarios in YAML or JavaScript and gives real-time data to track system performance.
- *Flood Element:* Flood Element is a cloud-based load testing tool that simulates user behavior using actual web browsers. It allows for the writing and execution of load tests in JavaScript, as well as extensive performance insights.

These AI-based testing software tools offer various features and capabilities to effectively test load performance using a digital twin. Organizations can choose the tool that best fits their requirements and preferences.

Case Study

The critical importance of performance prediction and optimization is highlighted in this case study for retail applications. Specifically, in countries such as the USA, the surge in customer activity during Thanksgiving Day presents a substantial challenge. As the festivities commence, millions of users simultaneously engage with the application, executing operations such as logging in, conducting item searches, adding products to their carts, applying coupons, and completing the checkout process. The resulting spike in user traffic places an immense burden on the application's infrastructure, including the applications themselves, APIs, and database servers. The strain imposed on these components can lead to system overloads, causing server crashes and resulting in a complete website outage. Consequently, customers are unable to complete their purchases, leading to significant revenue loss and a substantial decline in trust towards the vendor [6, 7].

To mitigate these issues, the study emphasizes the importance of performance prediction and optimization. By accurately forecasting the anticipated surge in user activity, businesses can proactively prepare their infrastructure to handle the increased load. This involves conducting thorough assessments of the application's performance thresholds, scalability, and resource allocation. Additionally, employing optimization techniques such as load balancing, caching, and database optimization can help alleviate the strain on the system during peak periods.

The findings of this case study underscore the critical need for businesses to invest in performance prediction and optimization strategies. By effectively

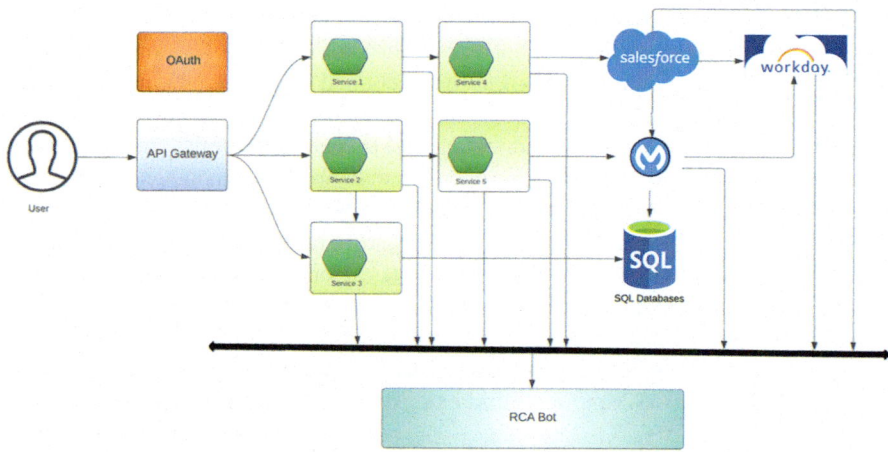

Fig. 2 Enterprise architecture of a retail application

managing the surge in user activity during peak shopping periods, companies can ensure uninterrupted service, minimize revenue loss, and uphold customer trust.

Figure 2 illustrates the flow of requests when a user logs in and performs a search for an item within the retail application. This process involves traversing multiple services to fulfill the user's request. By conducting performance tests, we can identify the specific services that contribute to longer processing times for requests. Optimizing the identified services is essential, as it leads to an enhancement in the overall response time of the application. Performance testing enables the measurement and analysis of various performance metrics, including response time, throughput, and resource utilization, at various stages of the user's interaction with the application. By collecting and analyzing this data, we can pinpoint the services that exhibit slower response times and require optimization.

Once the performance bottlenecks are identified, optimization techniques can be applied to enhance the efficiency of the corresponding services. This may involve fine-tuning the service's code, optimizing database queries, improving network communication, or employing caching mechanisms. By optimizing the identified services, we can reduce their processing time, ultimately improving the overall response time of the application. As shown in the Fig. 2, the service is optimized and on each iteration changes there is a performance test performed to check the improvement of the service. This test performed on the API's response time versus the incremental user loads per second. By leveraging performance testing and optimization strategies, businesses can enhance the user experience by ensuring faster response times and smoother interactions with the application. This, in turn, leads to increased customer satisfaction and improved overall performance of the retail application (Fig. 3).

Fig. 3 API Load testing on search Item Service

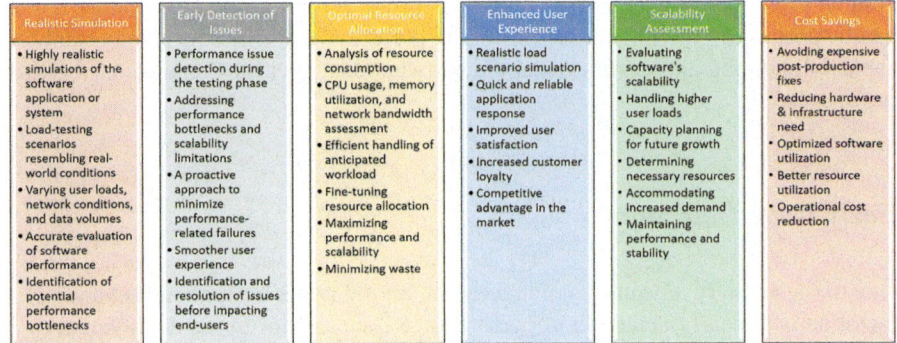

Fig. 4 Advantages of load testing and optimization using Digital Twins

Advantages of Load Testing and Optimization Using Digital Twin Technology

Leveraging Digital Twin methodologies for load testing and optimization provides several benefits that contribute to increased software performance and overall efficiency (Fig. 4).

Realistic Simulation

The use of Digital Twin technology for load testing and optimization has a significant benefit in terms of realistic simulation. Organizations can use Digital Twins to generate virtual clones of their software program or system that closely resemble real-world situations. This implies they may replicate a variety of performance-impacting factors such as user loads, network conditions, and data quantities [58]. Organizations may examine how the program works in different settings and find possible performance bottlenecks by properly duplicating these conditions. They can, for example, simulate heavy user loads during peak use periods to see if the program can manage the predicted number of concurrent users without degrading performance. They can also simulate network congestion or capacity variations to see how the program reacts under different network circumstances.

This authentic simulation provides crucial insights into the behavior and performance of the software within a controlled environment. It empowers enterprises to identify any performance limitations or vulnerabilities before deploying the software in a production environment [39, 40]. As a result, they can make educated recommendations about which optimizations and upgrades should be applied to increase the software's performance. Furthermore, realistic simulation allows businesses to foresee and plan for future scaling needs. By simulating user load and data volume growth, you can analyze how your program will grow and whether it can handle projected growth without sacrificing performance. This data is important for capacity planning and ensuring that the software can meet the demands of a growing user base.

Early Issue Detection

One of the primary advantages of embracing digital twins for load testing and optimization is the early detection of performance issues. Prior to the deployment of a software program or system in a production environment, digital twins provide an exceptional opportunity to assess the system's behavior and performance under diverse load conditions. [37]. Organizations may evaluate the digital twin under a variety of stress settings that reflect real-world usage scenarios. They can spot any potential performance bottlenecks, scalability challenges, or other problems by carefully monitoring the performance indicators and examining the data produced by the digital twin.

Early identification of these problems is essential because it enables companies to take proactive measures to fix them before they have an impact on end users. Organizations may considerably reduce the chance of failures or poor performance in the live production environment by identifying and fixing performance-related issues during the testing phase. For instance, if the digital twin reveals that the software application's response time significantly increases or begins to exhibit performance degradation at a certain load level, organizations can look into the underlying causes and take the necessary steps to improve the software's performance. This could entail refining wasteful code segments, adjusting resource allocation, or adding caching systems to speed up response times [2].

Organizations can guarantee a smoother user experience and prevent detrimental effects on productivity, client happiness, and brand reputation by taking these issues early on. A more efficient and successful software deployment in the production environment results from early identification of performance concerns, which also helps businesses to make data-driven decisions about capacity planning, infrastructure expenditures, and system architectural changes. Additionally, Digital Twins enable enterprises to iterate and improve their software development processes by enabling the early diagnosis of performance concerns. They may learn from the performance traits displayed by the digital twin and acquire insights into the behavior of the program by collecting performance data during the testing period [17]. With the use of this iterative method, enterprises may fine-tune their software to offer maximum performance and reliability.

Optimal Resource Allocation

Load testing using Digital Twins provides useful insights into the resource use patterns of a software program or system, allowing enterprises to optimize resource allocation and improve performance. The digital twin collects data on numerous resource consumption parameters, such as CPU use, memory utilization, and network bandwidth, during the stress testing procedure [31, 32]. Testers can acquire a thorough grasp of how the software application uses resources as the demand grows by studying this data under various load circumstances. Organizations can detect areas of inefficiency or over utilization that may impair performance or limit scalability by regularly monitoring resource usage trends.

For instance, the digital twin can show that the software application uses a lot of CPU power when it is heavily loaded, which could be a sign of a bottleneck. The code or configuration causing the excessive CPU utilization can be examined by testers, and improvements can be made to increase effectiveness. Algorithms may need to be optimized, pointless computations may need to be eliminated, or parallel processing strategies may need to be used to share the effort over several cores. Similar to this, memory usage patterns can be examined to find any possible memory leaks or excessive memory usage that may be affecting performance. Organizations can assure appropriate memory usage and prevent performance degradation due to excessive memory consumption by optimizing memory management tactics, such as allocating unused memory for freeing up space or implementing more effective data structures.

Another significant issue that may be investigated with the digital twin is network bandwidth use. Organizations can discover network bottlenecks or capacity restrictions by studying network traffic patterns under different load levels [36]. This data can assist them in making educated decisions regarding network infrastructure changes, caching mechanisms, and data compression strategies to maximize network use and overall performance. Organizations may enhance performance, reduce resource waste, and increase scalability by optimizing resource allocation based on insights from Digital Twins. The software program or system can efficiently manage the predicted demand and give optimal performance under diverse situations by fine-tuning resource consumption.

Enhanced User Experience

Load testing and optimization utilizing Digital Twins are critical in improving the user experience of software applications by assuring speedy and dependable response even under high load situations.

Realistic load scenarios may be generated while doing load testing with Digital Twins to accurately match real-world use patterns and user interactions. This covers scenarios with many users, concurrent transactions, and changing network circumstances [34]. Organizations may precisely evaluate how the software application functions under varying degrees of stress by duplicating these conditions in the digital twin environment. Optimizing software performance based on load testing findings allows companies to solve performance bottlenecks, scalability limits, and other issues that may have a negative influence on the user experience. Organizations may guarantee that the application responds promptly to user requests, executes transactions efficiently, and delivers a smooth user experience by proactively recognizing and correcting these issues [35].

There are various advantages to improving user experience through load testing and optimization with Digital Twins. For starters, enhanced responsiveness and dependability result in increased consumer satisfaction. Users are more likely to have a good opinion of the program and a greater degree of satisfaction when they engage with a software application that regularly performs effectively, reacts quickly, and handles their activities without delays or problems [56, 57]. Second, a better user experience leads to greater client loyalty. Users who are satisfied with the program are more likely to stay engaged with it, use it again, and promote it to others. This results in better user retention, improved consumer loyalty, and more chances for enterprises to create long-term connections with their user base.

A better user experience also gives you a competitive advantage in the market. In today's digital market, when customers have a plethora of software application options, having a superior user experience may set an organization's product or service apart from rivals. Companies that engage in load testing and optimization using Digital Twins to ensure optimal performance and a smooth user experience get a competitive advantage by attracting and maintaining consumers who appreciate dependability.

Scalability Assessment

The assessment of software applications' or systems' scalability is a crucial component of load testing and optimization, and digital twin technology offers a potent instrument for doing so. Organizations can simulate scenarios where user load steadily increases by applying increasing loads to the virtual replica of the software application or system using digital twins. This makes it possible for testers to watch and gauge how the program scales and functions as the workload increases. [18].

Various performance parameters are tracked and examined throughout the scalability assessment. Response times, throughput, resource use, and system stability are a few of them. The software application's handling of the added load, ability to maintain acceptable response times, and management of system resources like CPU, memory, and network bandwidth can all be evaluated by testers. For capacity planning and resource allocation, the insights obtained from scalability evaluation utilizing digital twins are helpful. Organizations may effectively estimate the resources needed to support future growth and rising user demands by understanding how the software scales under various load circumstances [51]. This entails proactively resolving any potential bottlenecks or constraints that might prevent scalability.

Organizations can enhance their hardware and software architecture through scalability evaluation using digital twins to ensure lag-free performance as user numbers rise. It aids in the early detection of possible scalability concerns, enabling for required adjustments to be made to accommodate future expansion and prevent performance deterioration or system failures. Additionally, scalability assessments help informing decisions on infrastructure improvements or scaling tactics. Organizations can determine the thresholds at which performance starts to deteriorate or resource utilization turns inefficient by carefully monitoring the behavior of the digital twin under increasing load [30]. To ensure scalability, this data directs the choice of the ideal configuration, which may include increasing the number of servers, modifying server capacity, or using cloud-based solutions.

Cost Savings

One of the most important advantages of load testing and optimization utilizing Digital Twins is cost savings. Organizations can gain considerable cost savings in a variety of areas by detecting and fixing performance issues early in the development cycle. First off, through load testing and optimization, performance issues can be found and fixed before they become costly post-production fixes. By replicating real-world load scenarios and meticulously analyzing the behavior and performance of the digital twin, organizations can preemptively identify bottlenecks, scalability limitations, and other performance-related issues [47]. Early attention to these problems enables developers to make the required modifications, such as code optimization, enhanced algorithms, or improved system configurations, before the software is released in the production environment. This not only lowers the need for expensive emergency situations but also saves time and effort.

Furthermore, load testing and optimization might result in increased resource efficiency. Organizations can optimize the allocation of hardware and infrastructure resources by examining the resource consumption patterns of the software application or system under various load levels. This optimization makes sure that the software is set up to effectively handle the projected workload without over- or under-allocating resources [22]. As a result, businesses might save money by forgoing needless investments in hardware, servers, or other infrastructure resources that might otherwise be needed to make up for subpar performance or insufficient scalability. In addition, improved software performance gained through load testing and optimization translates into operational cost reductions. When a software program or system performs well, it uses fewer resources to achieve the necessary levels of performance. This lowers the operating expenses of scaling up hardware, infrastructure, or cloud resources. Organizations may improve their operating budgets and achieve greater cost efficiency by improving resource use and avoiding waste.

Continuous Optimization

A significant benefit of employing Digital Twins for software performance testing is continuous tuning. It enables enterprises to continually analyze and enhance the performance of their software applications or systems throughout the lifespan of the application or system. By replicating real-world load scenarios and meticulously analyzing the behavior and performance of the digital twin, organizations can pre-emptively identify bottlenecks, scalability limitations, and other performance-related issues [55]. Organizations may measure important performance parameters such as response times, throughput, resource usage, and error rates over time by collecting data from the digital twin. This data enables them to detect any performance decrease caused by a variety of causes such as increasing user load, changes in the software environment, or application upgrades. Organizations may use continuous monitoring to discover any performance issues or bottlenecks that may develop, even after the product has been installed. Organizations may resolve performance issues and enhance software performance by evaluating data and identifying the main causes of performance decline [13].

Iterative cycles of testing, analysis, and improvement are used in continuous optimization. Enhancing software performance within organizations is achievable by implementing specific adjustments guided by the data obtained through the monitoring of the digital twin. This might include optimizing code, fine-tuning algorithms, altering system setups, or better resource allocation. Companies may guarantee that their software applications or systems stay performant and efficient over time by using continuous optimization. Continuous optimization enables proactive discovery and resolution of performance issues when new features are added, user loads grow, or the software environment develops [54]. This methodology assists businesses in maintaining optimal software performance, avoiding performance deterioration, and providing a high-quality user experience. Furthermore, ongoing optimization is consistent with agile development and DevOps techniques [10]. It helps businesses to constantly enhance the performance of their software as part of their ongoing development and operational processes [24]. This iterative strategy develops a culture of continuous development and allows firms to adjust swiftly to changing business needs or consumer expectations.

Conclusion

In this chapter, we examined the significance of performance testing in software applications prior to market release. We have stressed the importance of ensuring that software runs properly under a variety of scenarios to fulfill consumer expectations and corporate objectives. Organizations may improve the overall user experience by doing extensive performance testing to discover and fix possible bottlenecks, scalability concerns, and performance shortcomings. We explored the potential utility of digital twins in the realm of software testing. Digital twins are virtual counterparts of software systems that facilitate authentic simulations, monitoring, and performance analysis. We discussed various types of software testing that can be conducted using digital twins.

To demonstrate the implementation of performance testing with the aid of digital twins, we delved into a practical scenario featuring a retail application. By creating a virtual replica of the application, simulating real-world scenarios, generating test data, and scrutinizing performance metrics, we can unearth and rectify performance issues. The benefits of performance testing and optimization are significant. Organizations may increase customer satisfaction, customer retention, and revenue loss due to bad user experiences or system outages by ensuring that the software application runs properly under varying situations. Furthermore, performance testing assists firms in gaining insights into resource use, scalability needs, and areas for improvement, allowing them to enhance the performance of the software system and promote commercial success. Performance testing is an essential component of software development and deployment. Organizations may improve their testing capabilities, replicate real-world events, and optimize the performance of software applications by adopting Digital Twin technologies. This results in higher customer happiness, more income, and a competitive advantage in the market.

Key Term and Definitions

Digital Twin	A digital twin is a virtual portrayal of a tangible entity. This simulation relies on real-time data sourced from sensors affixed to the entity to replicate its behavior and oversee operations throughout its entire lifecycle.
Enterprise Application	Enterprise applications (EAs) are expansive software platforms crafted for use within business or government settings. An EA constitutes a multifaceted, scalable, component-based, widely distributed, and mission-critical system. EA software encompasses an array of business applications and organizational modeling tools meticulously crafted to deliver exceptional functionality. Enterprise architecture is employed in the development of these EAs.

Application Performance	The objective of enhancing application performance is to elevate the effectiveness, accessibility, and customer satisfaction associated with an organization's vital applications. So, it measures application performance, alerts administrators when performance baselines aren't met, provides visibility into root causes of performance issues, and automatically solves many performance issues before they affect users or the business.

References

1. Sasikumar, A., Vairavasundaram, S., Kotecha, K. V. I., Ravi, L., Selvachandran, G., & Abraham, A. (2023). Blockchain-based trust mechanism for digital twin empowered industrial internet of things. *Future Generation Computer Systems, 141*, 16–27. https://doi.org/10.1016/j.future.2022.11.002
2. Abdoune, F., Nouiri, M., Cardin, O., & Castagna, P. (2022). An enhanced methodology of fault detection and diagnosis based on digital twin. *IFAC-PapersOnLine, 55*(19), 43–48. https://doi.org/10.1016/j.ifacol.2022.09.181
3. Attaran, M., & Celik, B. G. (2023). Digital twin: Benefits, use cases, challenges, and opportunities. *Decision Analytics Journal, 6*, 100165. https://doi.org/10.1016/j.dajour.2023.100165
4. Banaeian Far, S., & Imani Rad, A. (2022). Applying digital twins in metaverse: User interface, security and privacy challenges. *Journal of Metaverse, 2*(1), 8–15. Retrieved from https://dergipark.org.tr/en/pub/jmv/issue/67967/1072189
5. Behera, T. K. (2023a). Architecture principles for enterprise software and mobile application development. In *Designing and developing innovative mobile applications* (pp. 1–20). IGI Global.
6. Behera, T. K. (2023b, June 09). *Enhancing search engine efficiency with elasticsearch aliases*. DZone. https://dzone.com/articles/enhancing-search-engine-efficiency-with-elasticsea
7. Behera, T. K., & Panda, B. S. (2023). Master data management using machine learning techniques: MDM Bot. *TechRxiv*. Preprint. https://doi.org/10.36227/techrxiv.21818040.v1
8. Behera, T. K., & Tripathi, K. (2022). Root cause analysis bot using machine learning techniques. *TechRxiv*. Preprint. https://doi.org/10.36227/techrxiv.21588159.v3
9. Bellalouna, F. (2021). Case study for design optimization using the digital twin approach. *Procedia CIRP, 100*, 595–600. https://doi.org/10.1016/j.procir.2021.05.129
10. Bomström, H., Kelanti, M., Annanperä, E., Liukkunen, K., Kilamo, T., Sievi-Korte, O., & Systä, K. (2023). Information needs and presentation in Agile Software Development. *Information and Software Technology, 162*, 107265. https://doi.org/10.1016/j.infsof.2023.107265
11. Boyes, H., & Watson, T. (2022). Digital twins: An analysis framework and open issues. *Computers in Industry, 143*, 103763. https://doi.org/10.1016/j.compind.2022.103763
12. Can, O., & Turkmen, A. (2023). Digital twin and manufacturing. *Digital Twin Driven Intelligent Systems and Emerging Metaverse*, 175–194. https://doi.org/10.1007/978-981-99-0252-1_8
13. Cao, Y., Tang, X., Gaidai, O., & Wang, F. (2022). Digital twin real time monitoring method of turbine blade performance based on numerical simulation. *Ocean Engineering, 263*, 112347. https://doi.org/10.1016/j.oceaneng.2022.112347
14. Chen, Y. (2022). Research on collaborative innovation of key common technologies in new energy vehicle industry based on digital twin technology. *Energy Reports, 8*, 15399–15407. https://doi.org/10.1016/j.egyr.2022.11.120

15. Cimino, C., Negri, E., & Fumagalli, L. (2019). Review of digital twin applications in manufacturing. *Computers in Industry, 113*, 103130. https://doi.org/10.1016/j.compind.2019.103130
16. da Silva Mendonça, R., de Oliveira Lins, S., de Bessa, I. V., de Carvalho Ayres, F. A., de Medeiros, R. L., & de Lucena, V. F. (2022). Digital twin applications: A survey of recent advances and challenges. *PRO, 10*(4), 744. https://doi.org/10.3390/pr10040744
17. Darvishi, H., Ciuonzo, D., Eide, E. R., & Rossi, P. S. (2021). Sensor-fault detection, isolation and accommodation for digital twins via modular data-driven architecture. *IEEE Sensors Journal, 21*(4), 4827–4838. https://doi.org/10.1109/jsen.2020.3029459
18. Dittmann, S., Zhang, P., Glodde, A., & Dietrich, F. (2021). Towards a scalable implementation of digital twins—A generic method to acquire Shopfloor data. *Procedia CIRP, 96*, 157–162. https://doi.org/10.1016/j.procir.2021.01.069
19. Evangeline, P., & Anandhakumar, A. (2020). Digital twin technology for "smart manufacturing". *Advances in Computers*, 35–49. https://doi.org/10.1016/bs.adcom.2019.10.009
20. Feng, H., Lv, H., & Lv, Z. (2023). Resilience towarded digital twins to improve the adaptability of transportation systems. *Transportation Research Part A: Policy and Practice, 173*, 103686. https://doi.org/10.1016/j.tra.2023.103686
21. Fuller, A., Fan, Z., Day, C., & Barlow, C. (2020). Digital twin: Enabling technologies, challenges and open research. *IEEE Access, 8*, 108952–108971. https://doi.org/10.1109/access.2020.2998358
22. Ge, Z., Song, Z., Ding, S. X., & Huang, B. (2017). Data mining and analytics in the process industry: The role of machine learning. *IEEE Access, 5*, 20590–20616. https://doi.org/10.1109/access.2017.2756872
23. Gómez-Abajo, P., Cañizares, P. C., Núñez, A., Guerra, E., & de Lara, J. (2023). Automated engineering of domain-specific metamorphic testing environments. *Information and Software Technology, 157*, 107164. https://doi.org/10.1016/j.infsof.2023.107164
24. Grande, R., Vizcaíno, A., & García, F. O. (2023). Is it worth adopting DevOps practices in global software engineering? Possible challenges and benefits. *Computer Standards & Interfaces, 87*, 103767. https://doi.org/10.1016/j.csi.2023.103767
25. Haleem, A., Javaid, M., Pratap Singh, R., & Suman, R. (2023). Exploring the revolution in healthcare systems through the applications of digital twin technology. *Biomedical Technology, 4*, 28–38. https://doi.org/10.1016/j.bmt.2023.02.001
26. Hassani, H., Huang, X., & MacFeely, S. (2022). Impactful digital twin in the healthcare revolution. *Big Data and Cognitive Computing, 6*(3), 83. https://doi.org/10.3390/bdcc6030083
27. Hoffmann, J., Bauer, P., Sandu, I., Wedi, N., Geenen, T., & Thiemert, D. (2023). Destination earth—A digital twin in support of climate services. *Climate Services, 30*, 100394. https://doi.org/10.1016/j.cliser.2023.100394
28. Hu, W., Zhang, T., Deng, X., Liu, Z., & Tan, J. (2021). Digital twin: A state-of-the-art review of its enabling technologies, applications and challenges. *Journal of Intelligent Manufacturing and Special Equipment, 2*(1), 1–34. https://doi.org/10.1108/jimse-12-2020-010
29. Huikkola, T., Kohtamäki, M., Rabetino, R., Makkonen, H., & Holtkamp, P. (2022). Overcoming the challenges of smart solution development: Co-alignment of processes, routines, and practices to manage product, service, and software integration. *Technovation, 118*, 102382. https://doi.org/10.1016/j.technovation.2021.102382
30. Jia, W., Wang, W., & Zhang, Z. (2022). From simple digital twin to complex digital twin part I: A novel modeling method for multi-scale and multi-scenario digital twin. *Advanced Engineering Informatics, 53*, 101706. https://doi.org/10.1016/j.aei.2022.101706
31. Jiang, L., Su, S., Pei, X., Chu, C., Yuan, Y., & Wang, K. (2023b). Product-part level digital twin modeling method for digital thread framework. *Computers & Industrial Engineering, 179*, 109168. https://doi.org/10.1016/j.cie.2023.109168
32. Jiang, Y., Li, M., Wu, W., Wu, X., Zhang, X., Huang, X., Zhong, R. Y., & Huang, G. G. Q. (2023a). Multi-domain ubiquitous digital twin model for information management of complex infrastructure systems. *Advanced Engineering Informatics, 56*, 101951. https://doi.org/10.1016/j.aei.2023.101951

33. Julien, N., & Martin, E. (2021). How to characterize a digital twin: A usage-driven classification. *IFAC-PapersOnLine, 54*(1), 894–899. https://doi.org/10.1016/j.ifacol.2021.08.106
34. Kalantari, S., Pourjabar, S., Xu, T. B., & Kan, J. (2022). Developing and user-testing a "digital twins" prototyping tool for architectural design. *Automation in Construction, 135*, 104140. https://doi.org/10.1016/j.autcon.2022.104140
35. Kherbache, M., Maimour, M., & Rondeau, E. (2022). Digital twin network for the IIOT using eclipse ditto and hono. *IFAC-PapersOnLine, 55*(8), 37–42. https://doi.org/10.1016/j.ifacol.2022.08.007
36. Kumbhar, M., Ng, A. H. C., & Bandaru, S. (2023). A digital twin based framework for detection, diagnosis, and improvement of throughput bottlenecks. *Journal of Manufacturing Systems, 66*, 92–106. https://doi.org/10.1016/j.jmsy.2022.11.016
37. Latsou, C., Farsi, M., & Erkoyuncu, J. A. (2023). Digital twin-enabled automated anomaly detection and bottleneck identification in complex manufacturing systems using a multi-agent approach. *Journal of Manufacturing Systems, 67*, 242–264. https://doi.org/10.1016/j.jmsy.2023.02.008
38. Li, X., Luo, J., Li, Y., Wang, W., Hong, W., Liu, M., Li, X., & Lv, Z. (2022). Application of effective water-energy management based on digital twins technology in sustainable cities construction. *Sustainable Cities and Society, 87*, 104241. https://doi.org/10.1016/j.scs.2022.104241
39. Liu, X., Jiang, D., Tao, B., Xiang, F., Jiang, G., Sun, Y., Kong, J., & Li, G. (2023a). A systematic review of digital twin about physical entities, virtual models, twin data, and applications. *Advanced Engineering Informatics, 55*, 101876. https://doi.org/10.1016/j.aei.2023.101876
40. Liu, Y., Zhang, J.-M., Min, Y.-T., Yu, Y., Lin, C., & Hu, Z.-Z. (2023b). A digital twin-based framework for simulation and monitoring analysis of floating wind turbine structures. *Ocean Engineering, 283*, 115009. https://doi.org/10.1016/j.oceaneng.2023.115009
41. Lv, Z., & Xie, S. (2022). Artificial intelligence in the digital twins: State of the art, challenges, and future research topics. *Digital Twin, 1*, 12. https://doi.org/10.12688/digitaltwin.17524.2
42. Macchi, M., Roda, I., Negri, E., & Fumagalli, L. (2018). Exploring the role of digital twin for asset lifecycle management. *IFAC-PapersOnLine, 51*(11), 790–795. https://doi.org/10.1016/j.ifacol.2018.08.415
43. Mo, F., Rehman, H. U., Monetti, F. M., Chaplin, J. C., Sanderson, D., Popov, A., Maffei, A., & Ratchev, S. (2023). A framework for manufacturing system reconfiguration and optimisation utilising digital twins and modular artificial intelligence. *Robotics and Computer-Integrated Manufacturing, 82*, 102524. https://doi.org/10.1016/j.rcim.2022.102524
44. Nasirahmadi, A., & Hensel, O. (2022). Toward the next generation of digitalization in agriculture based on digital twin paradigm. *Sensors, 22*(2), 498. https://doi.org/10.3390/s22020498
45. Okegbile, S. D., Cai, J., Yi, C., & Niyato, D. (2022). Human digital twin for personalized healthcare: Vision, architecture and future directions. *IEEE Network*, 1–7. https://doi.org/10.1109/mnet.118.2200071
46. Osadcha, I., Jurelionis, A., & Fokaides, P. (2023). Geometric parameter updating in digital twin of built assets: A systematic literature review. *Journal of Building Engineering, 73*, 106704. https://doi.org/10.1016/j.jobe.2023.106704
47. Rahman, M. S., Ghosh, T., Aurna, N. F., Kaiser, M. S., Anannya, M., & Hosen, A. S. M. S. (2023). Machine learning and internet of things in industry 4.0: A review. *Measurement: Sensors*, 100822. https://doi.org/10.1016/j.measen.2023.100822
48. Semeraro, C., Aljaghoub, H., Abdelkareem, M. A., Alami, A. H., & Olabi, A. G. (2023). Digital twin in battery energy storage systems: Trends and gaps detection through association rule mining. *Energy, 273*, 127086. https://doi.org/10.1016/j.energy.2023.127086
49. Singh, M., Fuenmayor, E., Hinchy, E., Qiao, Y., Murray, N., & Devine, D. (2021). Digital twin: Origin to future. *Applied System Innovation, 4*(2), 36. https://doi.org/10.3390/asi4020036
50. Stradowski, S., & Madeyski, L. (2023). Exploring the challenges in software testing of the 5G system at Nokia: A survey. *Information and Software Technology, 153*, 107067. https://doi.org/10.1016/j.infsof.2022.107067

51. Tao, F., & Qi, Q. (2019). New it driven service-oriented smart manufacturing: Framework and characteristics. *IEEE Transactions on Systems, Man, and Cybernetics: Systems, 49*(1), 81–91. https://doi.org/10.1109/tsmc.2017.2723764
52. Tao, F., Zhang, M., & Nee, A. Y. C. (2019). Five-dimension digital twin modeling and its key technologies. *Digital Twin Driven Smart Manufacturing*, 63–81. https://doi.org/10.1016/b978-0-12-817630-6.00003-5
53. van der Valk, H., Haße, H., Möller, F., & Otto, B. (2021). Archetypes of digital twins. *Business & Information Systems Engineering, 64*(3), 375–391. https://doi.org/10.1007/s12599-021-00727-7
54. Vicente-Serrano, S. M., Domínguez-Castro, F., Reig, F., Beguería, S., Tomas-Burguera, M., Latorre, B., Peña-Angulo, D., Noguera, I., Rabanaque, I., Luna, Y., Morata, A., & El Kenawy, A. (2022). A near real-time drought monitoring system for Spain using automatic weather station network. *Atmospheric Research, 271*, 106095. https://doi.org/10.1016/j.atmosres.2022.106095
55. Wang, Y., Zhao, C., Dong, D., & Wang, K. (2023). Real-time monitoring of insects based on laser remote sensing. *Ecological Indicators, 151*, 110302. https://doi.org/10.1016/j.ecolind.2023.110302
56. Yang, B., Liu, Y., & Chen, W. (2023a). A twin data-driven approach for user-experience based design innovation. *International Journal of Information Management, 68*, 102595. https://doi.org/10.1016/j.ijinfomgt.2022.102595
57. Yang, X., Liu, X., Zhang, H., Fu, L., & Yu, Y. (2023b). Meta-model-based shop-floor digital twin architecture, modeling and application. *Robotics and Computer-Integrated Manufacturing, 84*, 102595. https://doi.org/10.1016/j.rcim.2023.102595
58. You, Y., Chen, C., Hu, F., Liu, Y., & Ji, Z. (2022). Advances of digital twins for predictive maintenance. *Procedia Computer Science, 200*, 1471–1480. https://doi.org/10.1016/j.procs.2022.01.348
59. Zhang, Z., Wen, F., Sun, Z., Guo, X., He, T., & Lee, C. (2022). Artificial intelligence-enabled sensing technologies in the 5G/internet of things era: From virtual reality/augmented reality to the digital twin. *Advanced Intelligent Systems, 4*(7), 2100228. https://doi.org/10.1002/aisy.202100228
60. Zhong, D., Xia, Z., Zhu, Y., & Duan, J. (2023). Overview of predictive maintenance based on digital twin technology. *Heliyon, 9*(4). https://doi.org/10.1016/j.heliyon.2023.e14534

DT-AXYOLOV5: An Efficient Digital Twin–Assisted Deep-Learning-Based Blockchain Framework for Patient Discomfort Detection in Smart Healthcare System

J. Antony Vijay, C. D. Premkumar, and P. Revathi

Introduction

Enormous developments in technologies have heavily affected many sectors, such as healthcare, teaching, farming, manufacturing, and the IT industry. The combination of machine learning (ML), blockchain, and the Internet of Things (IoT) has significantly improved the healthcare sector by collecting accurate data in real time, analyzing and transmitting more-reliable outcomes, expanding the scope of health information, and combating global epidemics. Especially in healthcare, the Internet of Things is vital for synchronizing smart devices to monitor patients' bodies. A lack of exercise and inconsistent behaviors have been found to be quite widespread among various age groups, leading to major medical problems such as obesity, cholesterol disorders, fractures, heart-related illnesses, and blood pressure. According to a report from the World Health Organization (WHO), the number of people aged 55 and older could exceed 2.5 billion by 2055 [1]. As a result, nearly 50% of all healthcare resources would be used only for elderly people's welfare. Furthermore, the bulk of existing healthcare resources use traditional monitoring approaches, which represents a substantial barrier to facilitating the transfer of health-related resources [2, 3]. In addition to that, another report from WHO revealed various medical errors in medical histories: difficulty in identifying patients' medical records, delayed treatment, and incorrect medical records, which are the prominent causes of 425,000 deaths each year [4]. Furthermore, older people's abnormal body

J. Antony Vijay (✉)
Department of Computer Science and Engineering, Karpagam College of Engineering, Coimbatore, India

C. D. Premkumar · P. Revathi
Department of Information Technology, Hindusthan College of Engineering and Technology, Coimbatore, India
e-mail: premkumar.it@hicet.ac.in

movements can help technology to detect discomfort in certain locations of the body to better classify diseases [5].

Moreover, the Internet of Things platform creates a medium to link up and access all the smart devices on a certain network, which helps doctors to collect medical data from patients. However, the development of a smart healthcare system also presents certain health hazards for patients [6]:

1. Doctors only periodically observe their patients.
2. The connectivity between sensors and the smart healthcare system is not effectively monitored.
3. Smart devices are easily hackable, so criminals could take control patients' bodies, which could lead to unpredictable damage.
4. This system fails to maintain the medical records of each patient throughout their respective lifecycles.

The digital twin (DT) technology was introduced by Michael Grieves in 2002. Initially, it was used only in manufacturing, but a few years later, it was already being applied to various other sectors [7]. The National Aeronautics and Space Administration reintroduced digital twin technology in 2020 to replicate the Earth's system. To detect health issues, digital twin technology creates a virtual model of any physical entity in real time. This technology connects the physical entity and its virtual model to technologies such as IoT, ML, and blockchain [8]. This improves the utilization of DT in various sectors and various real-time applications. DT can also effectively design and monitor a health regimen, especially for older patients. This technology can assist healthcare professionals in taking care of older people's health issues by reviewing their past medical records. In this scenario, blockchain technology is used to protect patients' records [9].

This paper presents an architecture for analyzing older people's physical activity by using intelligent, personalized healthcare concepts and technologies such as digital twin, machine learning, and blockchain. The proposed framework structure is depicted in Fig. 1.

This chapter has the following primary research goals:

1. Acquiring an IoT-based dataset of patients for continuous physical activity monitoring
2. Preparing patients' discomfort data to connect them in the digital twin layer
3. Analyzing patients' irregular movement activities with the machine-learning-based DT-AXYOLOV5 algorithm in the cloud layer
4. Continuously monitoring past medical records is highly protected and maintained by blockchain technology

The chapter is organized as follows: A literature survey is represented in Sect. 2; the proposed system is described in Sect. 3; a performance analysis is elaborated in Sect. 4; and, lastly, a conclusion is reached and discussed in Sect. 5.

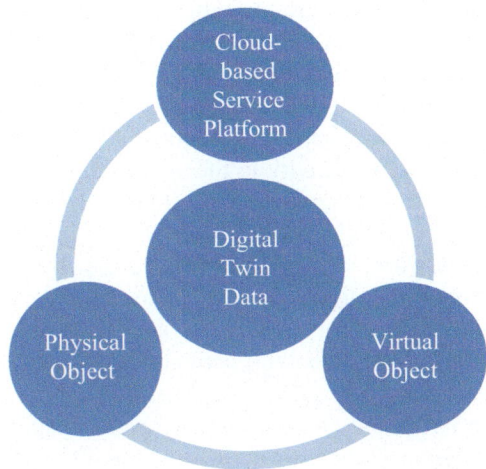

Fig. 1 The proposed framework structure

Literature Survey

Various studies have been carried out on continuous monitoring the abnormal activities of older people's health concerns, and various technologies have been adopted to improvise and analyze the healthcare system. Such continuous monitoring can be approached with various domains to reach appropriate solutions. These approaches include the digital twin–based approach, the deep-learning (DL) cloud-based approach, and the blockchain-based approach.

Digital Twin–Based Approach

In addition to the relevant approaches in the field of healthcare, the study by Rubeis et al. [10] was concerned with ensuring reliable analysis by using automatic and independent estimation criteria to measure patients' health. The researchers used an inspiring setting to demonstrate the usefulness of the proposed technique for controlling unusual medical occurrences produced by type 2 diabetes. Liu et al. [11] presented a DT healthcare architecture for older patients that analyzed patients' past health records stored in the cloud layer. This approach did not perform well in effectively detecting healthcare issues by using patient movement analysis. Eftimie et al. [12] proposed a neural network–based healthcare system to deeply analyze patients' irregular body movements by using data from the AlphaPose Library; this system factored in 18 key points from the human body to identify health issues. In FlexSim HC is a three-dimensional (3D) patient-centric virtual healthcare environment that continuously monitors patients. Myrzashova et al. [13] proposed federated learning (FL), which alone was not accurate in predicting health issues, but when FL was combined with blockchain technology, it overcame all the top-level security threats related with healthcare. Pal [14] represented a system made up of IoT with blockchain technology to confer high protectivity to the system against various vulnerability attacks. A lightweight cryptography algorithm (SHA-512) was incorporated to provide data security. Sasikumar et al. [15] identified complications when nodes are increasing in networks with restricted resources. To handle such a situation, blockchain networks with digital twin can be used to create a sophisticated framework for industrial IoT. Garg et al. [16] suggested using digital twin with various health instruments to monitor patients' activities through smart watches and health bands and thus to reach more-accurate diagnoses. The monitored data is highly protected in the cloud layer to prevent unauthorized access. The data are accessed only via a biometric authentication service and processed by a convolution neural network (CNN) model to determine the authenticity of the data. Elayan et al. [17] proposed using digital twin with a machine-learning-based electrocardiogram pulse dataset to classify heart diseases. Neural network–based algorithms are better at predicting heart diseases on the basis of using an ECG dataset than older algorithms are. Azzaoui et al. [18] represented a digital twin–based blockchain framework for collecting a dataset on COVID-19. This framework created a virtual system to deal with more-complex surgery plans. Chaudhari et al. [19] proposed a digital twin–based blockchain platform to collect and monitor detailed reports on patients' irregular activity, such as evaluating what types of drugs suit individuals and monitoring their medical records to identify the best treatment for each individual. Zhang et al. [20] suggested a deep-learning-based vulnerability detection system that predicts more than a thousand vulnerabilities from harmful code.

Deep-Learning Cloud-Based Approach

Shreshth Tuli et al. [21] proposed a technique for automatically detecting heart-related health concerns by using cloud-computing devices equipped with ensemble learning. The cloud-computing platform didn't perform well when the data storage was high, and it required long processing times. On the other hand, a deep-learning framework performed well when compared with the cloud-computing devices. Sahoo et al. [22] suggested an automatic healthcare recommender, which used a restricted Boltzmann machine (RBM) and a convolution neural network (CNN). This system suggested that patients take certain drugs for certain issues, acting like a clinician. Stephen et al. [23] proposed using a CNN model to train on a chest X-ray images dataset to predict lung diseases. This system didn't perform well when the size of the image dataset was large, so a small dataset and an image-augmentation technique were used instead. Muhammad et al. [24] proposed a CNN-based brain tumor detection model trains through an image dataset, where various medical parameters and threshold values were used to extract the affected region from an image. Brain tumor attributes were clearly identified in the extracted region. Zhou et al. [25] proposed a deep-learning-based system for monitoring irregular human activity. This approach continuously monitored the poses of patients, and by using the AlphaPose Library and 18 key points of the human body, movements were identified and organ distance calculated. This approach easily classified the diseases of patients. Ali et al. [26] proposed two stages of predicting heart issues. In the first stage, features are extracted from sensor data. Noises and repeated information can be removed to exclude them from the dataset. In the second stage, the ensemble-based deep-learning technique effectively classifies heart diseases.

Blockchain-Based Approach

Bhattacharya et al. [27] presented a blockchain-based deep-learning approach as a service framework to store the medical records of patients. This framework consists of two stages: The first stage features an authentication process to prevent attacks, and the second stage takes a deep-learning-based approach to predict diseases. Dwivedi et al. [28] presented an IoT-based blockchain framework by using a verified healthcare system to generate confidential notifications that alert patients to critical health concerns. Zhang et al. [29] proposed a blockchain-based healthcare system to protect healthcare records. This system analyzes users' identity by using the various attributes of certain patients, such as authentication IDs, master keys, and revocation IDs, so that patients' medical records are kept confidential without any leakage. Aujla and Jindal [30] proposed a blockchain-based edge network architecture to link edge devices in a blockchain framework to secure data transmission, to keep patients' health records private.

Proposed Solution

The step-by-step process of Proposed architecture is represented if Fig. 2. The layer-by-layer approach of the whole system architecture is represented in Fig. 3. The steps are summarized in Sections 3.1, 3.2, and 3.3.

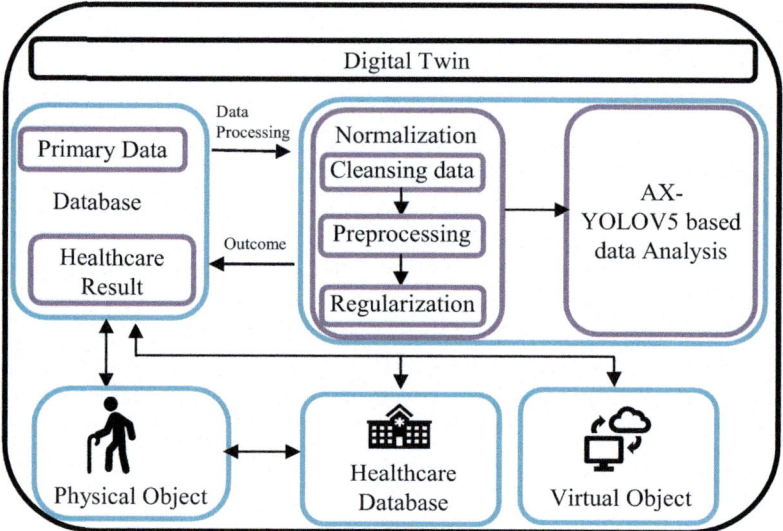

Fig. 2 Step-by-step process of proposed architecture

Data Acquisition

Initially, the data can be acquired from various raw-data sources then converted into the required formats for various contexts. This process collects the signals of data at regular intervals to continuously monitor patients' health. Each activity contains four metrics: T_b, T_a, S_e, and S_s. The duration of every data signal is calculated as $\Delta T = T_b - T_a$. Once the data have been collected from various sources, the data-fusion and data-preprocessing steps take place to remove noise from the dataset.

The Processing of the AX-YOLOV5 Deep-Learning Model

The preprocessing features are extracted after locating the region of interest (ROI) from the dataset, which can be achieved via pattern matching. This can be determined by calculating Eq. 1:

$$Y'_{1k} = Y'_{1n} + \left(Y_{1(n+1)} - Y_{1n}\right) \frac{P_2 \times k - P_1 \times n}{P_1} \tag{1}$$

where Y_1 denotes the event pattern and P_1 and P_2 denote the same activity but in different time periods.

The overall process of a pattern analysis is described in Algorithm 1.

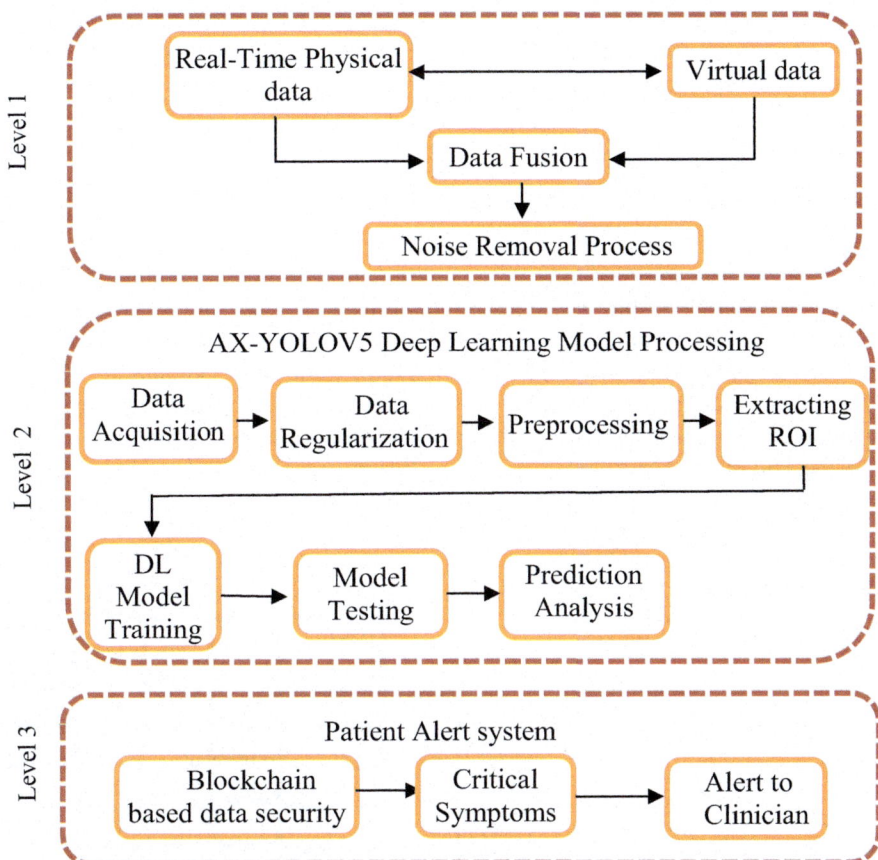

Fig. 3 Outline of various levels in the digital twin architecture

Algorithm 1, for the Process of Patient Discomfort Detection

```
Input: Yj and Yk are a pair of activities, and activity
acquisition periods Pi and Pj correspond to Yi and Yj,
respectively.
When new activity discomfort is termed Yj' for Yj in the database
for each Y1k in Y1 do
Corresponding period instance p2k = P2 × k
Seek for Y1n and Y1o in Y1
where p1n = max (p1j | p1j < p1k)
p1o = min (p1j | p1j > p1k)
end for
for each Y'1k in Y'1 do
Y'1k = Median (Y1n, Y1o)
end for
```

By applying this algorithm, the patterns of the patients' activity were collected in different periods, which are given as inputs for deep-learning-based AX-YOLOV5 model for training and testing.

Blockchain-Based Alert System

The collected data were securely stored by using the practical Byzantine fault tolerance consensus algorithm. Once abnormal activity is found in certain periods of time, the data are accumulated for a set number of transactions. These data are stored as a single block, where every block has a unique ID to facilitate monitoring. The fundamental process of blockchain is represented in Fig. 4. Blockchain is used to preclude unauthorized access to the data. The blockchain consensus algorithm for pBFT is laid out in Algorithm 2.

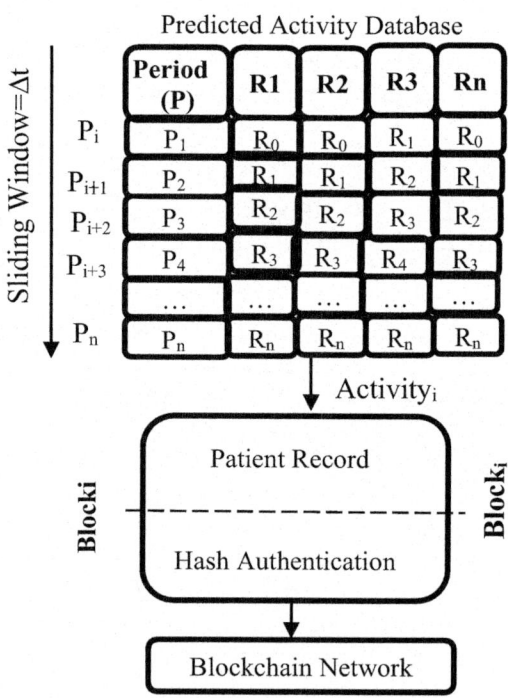

Fig. 4 Step-by-step process of block generation

Algorithm 2, for Practical Byzantine Fault Tolerance (pBFT)

```
Input: TS->TimeStamp, MD-> Message Digest
Outcome: pBFTᵢ
Assure: consent inference
Agreement validation
for each block = I to N do
Compute impression pBFTᵢ
i++
end for
if i ≥ α then
Establishment of NTP server
Activate TS and MD process
Consent = true
Data updating in blockchain network = True
else
Hold for TS activity ΔT
end if
Return consent
```

The symptoms of patients are continuously monitored to alert them and their respective clinicians and to start recovery processes. The sending of alerts to patients follows Algorithm 3.

Algorithm 3, for Alerting Patients to Critical Symptoms

```
Input: Activityⱼ, T is Threshold
Outcome: Critical remainder
if Activityⱼ ≤ T then
Remainder = Critical
else
for j = 1 to n do
Calculate Φⱼ = Activityⱼ — Activityⱼ ₊₁
end for
end if
if Φi ≤ T then
Remainder = Critical
else
Remainder = Not Critical
end if
```

Evaluation of Detecting Patients' Abnormal Activity

The implementation of the abnormal activities of patients is carried out by using Python libraries, and the pBFT algorithm is incorporated into the Python framework. Performance evaluations of discomfort detection can be approached as follows:

1. A performance evaluation of patient discomfort detection
2. Deep-learning-based algorithm analysis
3. Analysis of degrees of delay

Performance Evaluation of Patient Discomfort Detection

The performance evaluation of patient discomfort detection abides by the following steps:

1. Choosing the right metrics
2. Selecting the window size of activation maps
3. Estimating the accuracy of patients' abnormal activity

Choosing the Right Metrics

Various metrics are used for continuously monitoring patient discomfort detection that uses the deep-learning approach, such as hidden layers, selecting large activation maps, down sampling, and selecting a gated recurrent unit (GRU).

Hidden Layers

Deep-learning models have three or four hidden layers. The more hidden layers, the better the learning process of the model, which improves the model's ability to predict discomfort detection. A deep-learning-based model's hidden layers are represented in Fig. 5.

Selecting the Window Size of Activation Maps

The window sizes of the activation maps in the AX-YOLOV5 model vary depending on the type of feature detection. The DL model used various window sized to determine the optimal size for further processing. Activation map sizes of 80, 90, 100, 110, 120, 130, 140, and 150 produced accuracy values of 85.67%, 88.23%, 89.01%, 84.96%, 86.3%, 87.67%, 90.01%, and 91.23%, respectively. These performance values for feature extractions are represented in Fig. 6. The accuracy values, represented in Fig. 7, also depend on the filter size, which varies from 1×7 to 1×9.

Estimation of Accuracy for Predicting Patients' Abnormal Activity

This pooling layer also plays a small part in filtering out information so that only the important information from the last step is retained. The purpose of this layer is mainly to remove irrelevant data from the dataset, which reduces the processing times of all other layers. Estimations of the accuracy values of various pooling layers are represented in Fig. 8. In the last step, the GRU collects all the important cells

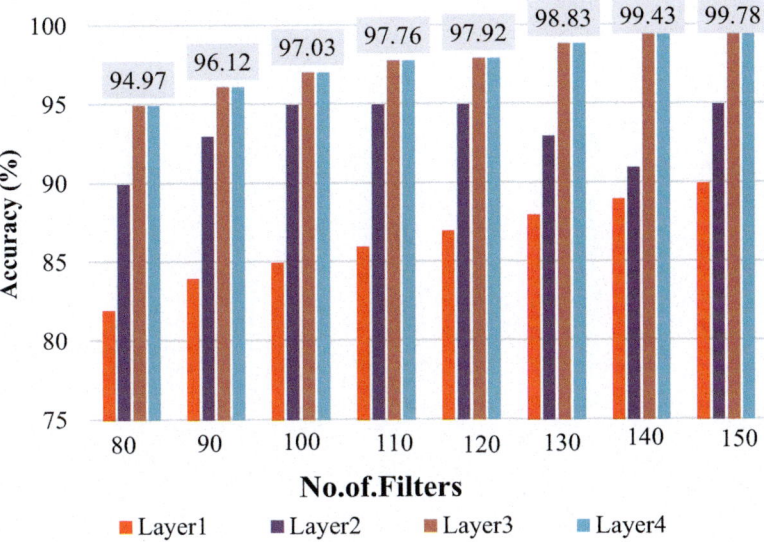

Fig. 5 Size of hidden layers

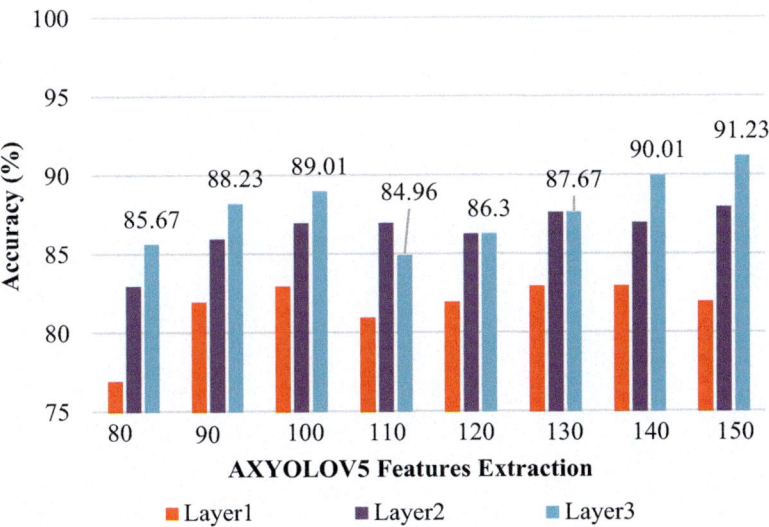

Fig. 6 Performance values of AXYOLOV5 feature extraction

into a single block for predicting accuracy—which varies from 0 to *n* cells. In this, we have included GRUs with 0 to 120 cells. From this analysis, GRUs with 70 cells achieved an accuracy value of 96.78%. The selection of the optimal number of GRU cells is represented in Fig. 9. Performance evaluations based on filter size are presented in Table 1.

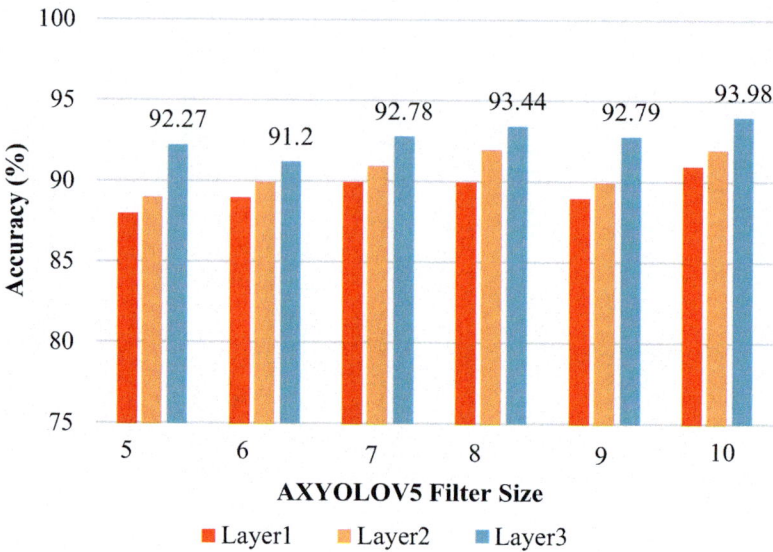

Fig. 7 Accuracy of various filter sizes

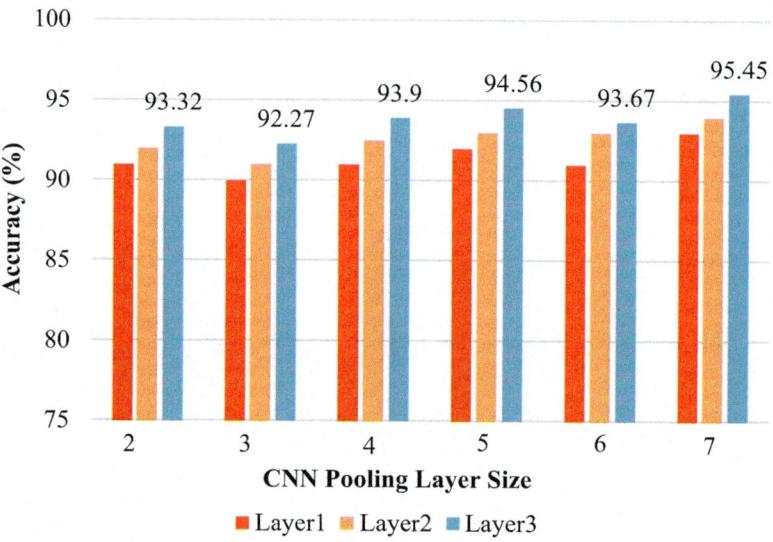

Fig. 8 Estimations of accuracy values of various pooling layers

Performance metrics such as precision, recall, and F1-score were compared with various algorithms and appear in Tables 2, 3, and 4. The performance comparison assessed with four Machine Learning based Models such as region-based convolutional neural network (R-CNN), faster R-CNN (F-RCNN), Bayesian network (BN), and random forest (RF).

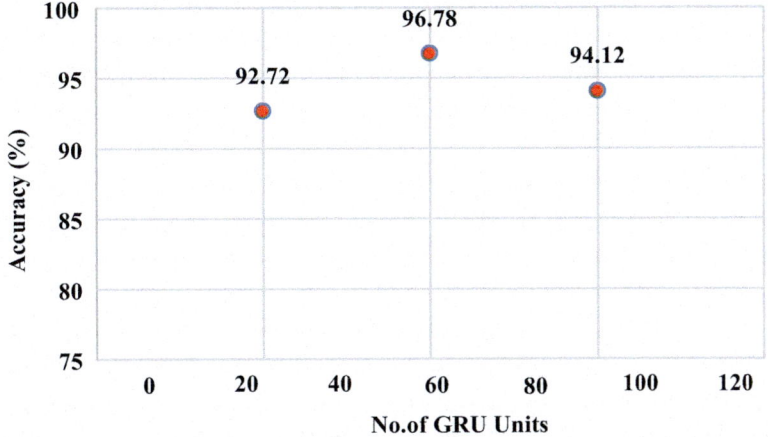

Fig. 9 Choosing the right Optimum GRU Cells

Table 1 Performance evaluations based on filter size

S. No	Activity Filter Size	Performance (%)
1	0.2	87.23
2	0.4	95.67
3	0.6	91.12
4	0.8	92.23
5	1.0	89.14

Table 2 Comparison of precision calculations based on events

S. No	Events	R-CNN (%)	F-RCNN (%)	BN (%)	RF (%)	Proposed System (%)
1	Event 1	89.23	88.71	87.78	89.34	99.97
2	Event 2	82.91	87.11	91.29	92.78	92.87
3	Event 3	78.22	79.34	82.45	85.67	97.23
4	Event 4	90.42	91.15	90.17	90.98	96.54
5	Event 5	88.12	89.06	88.45	87.36	99.34
AVERAGE		85.78	87.07	88.02	89.22	97.19

Table 3 Comparison of recall calculations based on events

S. No	Events	R-CNN (%)	F-RCNN (%)	BN (%)	RF (%)	Proposed System (%)
1	Event 1	87.21	87.91	77.88	79.43	96.87
2	Event 2	78.32	82.91	81.92	87.87	89.21
3	Event 3	81.88	79.34	72.54	75.76	88.43
4	Event 4	87.22	82.25	80.71	82.89	89.71
5	Event 5	81.12	79.24	78.54	79.98	92.34
AVERAGE		83.15	82.33	78.31	81.18	91.31

Table 4 Comparison of F1-score calculations based on events

S. No	Events	R-CNN (%)	F-RCNN (%)	BN (%)	RF (%)	Proposed System (%)
1	Event 1	87.91	85.45	81.23	82.23	97.78
2	Event 2	82.23	87.21	79.29	81.78	90.12
3	Event 3	82.23	79.43	77.45	81.67	89.34
4	Event 4	81.43	84.52	82.17	80.98	92.12
5	Event 5	89.72	81.42	81.45	79.89	94.98
AVERAGE		84.7	83.6	80.31	81.31	92.86

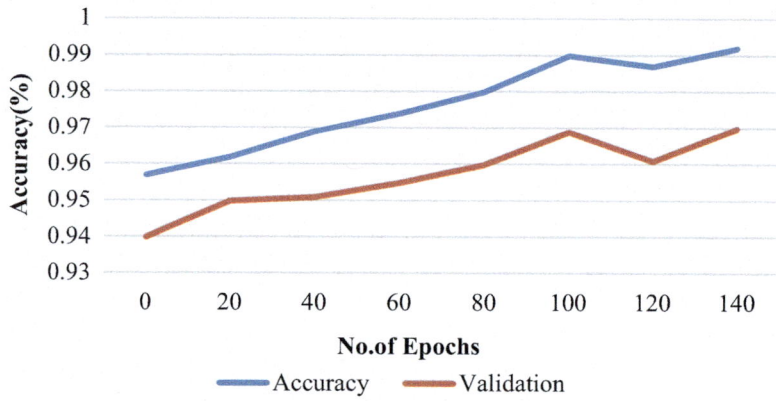

Fig. 10 Evaluation of accuracy based on no. of time periods

Here, we considered five event activities for performance comparisons, namely relaxing, sitting, jogging, working out, and falling.

Analysis of Deep-Learning-Based Algorithm

DL model training and testing takes place for 140 time periods with a learned rate of 0.02, which is represented in Fig. 10.

Figure 11 presents an evaluation of loss that is based on various time periods, where loss percentage and the number of time periods are inversely proportional—i.e., loss percentage decreases when the number of time periods increases.

The GRU was trained on 140 time periods, in which the model reached an accuracy value of 99% for 140 time periods, and loss validation took place for 140 time periods, with a loss percentage of 0.02 after 140 time periods. The accuracy of the GRU training and that of its loss validation appear in Figs. 12 and 13, respectively.

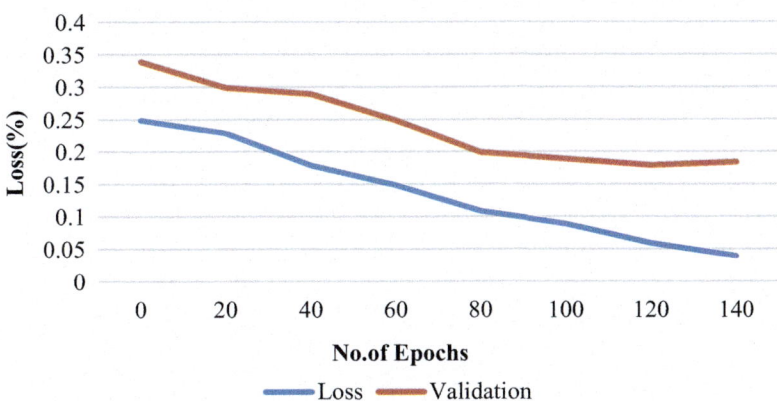

Fig. 11 Evaluation of loss based on number of time periods

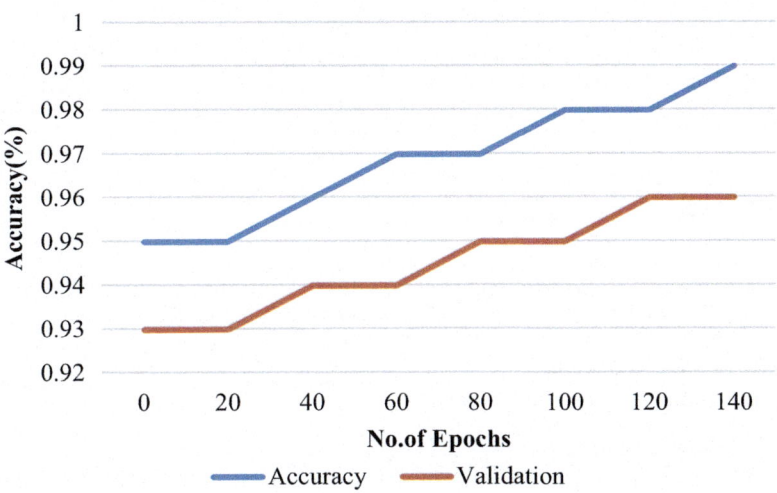

Fig. 12 Accuracy of GRU training

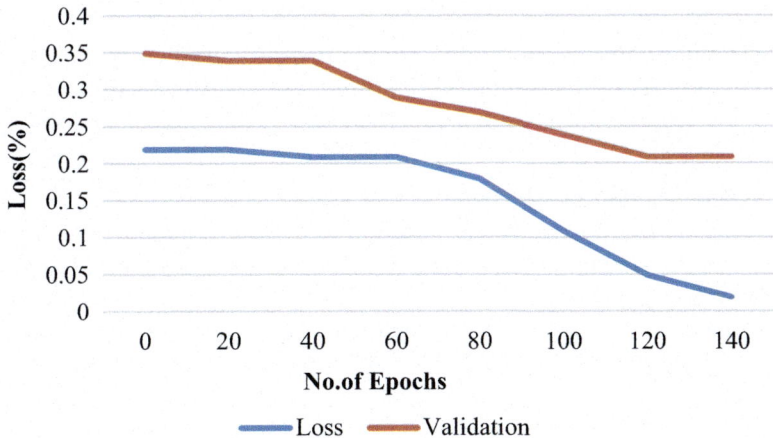

Fig. 13 Accuracy of GRU loss validation

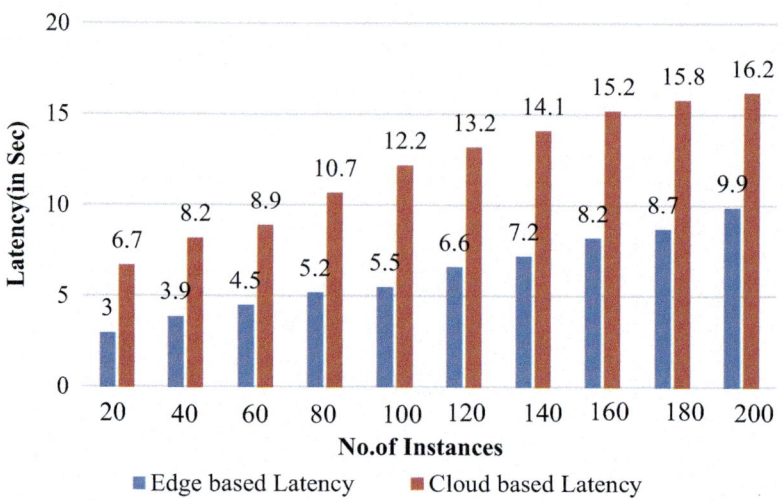

Fig. 14 Degrees of delay in edge and cloud networks

Analysis of Degrees of Delay

Degrees of delay (DOD) are expressed as the differences between the requested times and the response times. They are classified as follows:

1. DOD in a DL framework
2. DOD in a blockchain framework

DOD in Deep-Learning Framework

In the healthcare field, calculating the degrees of delay is difficult. The DOD are expressed by Eq. 2, as follows:

$$DOD = \text{discomfort activity detection} - \text{time of notification delivery} \quad (2)$$

Comparing the performance values of edge and cloud computing, based on various inputs, produces the minimum delay of 3 seconds for 20 instances and the maximum of 9.9 seconds for 200 instances. The DOD of edge and cloud networks are presented in Fig. 14.

DOD in Blockchain Framework

The blockchain-based ledger latency records are classified into minimal, maximal, and average. Depending on the number of users, the delay of a blockchain framework varies, with a maximum delay of 17.22 seconds for 30 users, a minimum delay of 1.1 seconds for 10 users, and an average delay of 9.76 seconds for 20 user counts. The classification of latency records for a blockchain framework is presented in Table 5.

Table 5 Blockchain-based ledger latency record

Number of Users	Positive (%)	Failure (%)	Max. Latency	Min. Latency	Avg Latency
10	100	0	2.81	1.1	2.2
20	100	0	10.73	4.84	9.76
30	100	0	17.22	7.89	14.23

Conclusion

This digital twin–assisted deep-learning-based blockchain framework continuously monitors patient discomfort in real time. The data are collected from various sources, which are formatted, and then the framework is trained and tested by deep-learning-based algorithms. Digital twin technology has various advantages such as data privacy, analyzing the data, accuracy in predicting the patient discomfort movements and health status. This approach overcame all the drawbacks of digital twin technology in, improving the accuracy of predicting patients' health status and reduced the loss of user data thanks to its using a blockchain framework. One of the important drawbacks of digital twin is that only one round of training is not enough to handle the healthcare data; to improve this framework, continuous training is required. Finally, this approach yields better results i.e) Precision (97.19%), Recall (91.31%), F1-Score (92.86) than all other deep-learning methods.

References

1. Kaul, R., Ossai, C., Forkan, A. R. M., Jayaraman, P. P., Zelcer, J., Vaughan, S., & Wickramasinghe, N. (2023). The role of AI for developing digital twins in healthcare: The case of cancer care. *Wiley Interdisciplinary Reviews: Data Mining and Knowledge Discovery, 13*(1), e1480.
2. Haleem, A., Javaid, M., Singh, R. P., & Suman, R. (2023). Exploring the revolution in healthcare systems through the applications of digital twin technology. *Biomedical Technology, 4,* 28–38.
3. Aluvalu, R., Mudrakola, S., Kaladevi, A. C., Sandhya, M. V. S., & Bhat, C. R. (2023). The novel emergency hospital services for patients using digital twins. *Microprocessors and Microsystems, 98,* 104794.
4. Wang, N., Han, W., & Ou, W. (2023). A novel security scheme for mobile healthcare in digital twin. In *International conference on machine learning for cyber security* (pp. 425–441). Springer.
5. Manocha, A., Afaq, Y., & Bhatia, M. (2023). Digital Twin-assisted Blockchain-inspired irregular event analysis for eldercare. *Knowledge-Based Systems, 260,* 110138.
6. Jia, W., Wang, W., & Zhang, Z. (2023). From simple digital twin to complex digital twin part II: Multi-scenario applications of digital twin shop floor. *Advanced Engineering Informatics, 56,* 101915.
7. Chengoden, R., Victor, N., Huynh-The, T., Yenduri, G., Jhaveri, R. H., Alazab, M., et al. (2023). Metaverse for healthcare: A survey on potential applications, challenges and future directions. *IEEE Access, 11,* 12765.
8. Attaran, M., & Celik, B. G. (2023). Digital Twin: Benefits, use cases, challenges, and opportunities. *Decision Analytics Journal, 6,* 100165.

9. Utku, D. H., Catak, F. O., Kuzlu, M., Sarp, S., Jovanovic, V., Cali, U., & Zohrabi, N. (2023). Digital twin applications for smart and connected cities. In *Digital twin driven intelligent systems and emerging metaverse* (pp. 141–154). Springer Nature Singapore.
10. Rubeis, G. (2023). Hyperreal patients. Digital twins as simulacra and their impact on clinical heuristics. In *Medizin–Technik–Ethik: Spannungsfelder zwischen Theorie und Praxis* (pp. 193–207). Springer Berlin Heidelberg.
11. Liu, X., Jiang, D., Tao, B., Xiang, F., Jiang, G., Sun, Y., et al. (2023). A systematic review of digital twin about physical entities, virtual models, twin data, and applications. *Advanced Engineering Informatics, 55*, 101876.
12. Eftimie, R., Mavrodin, A., & Bordas, S. P. (2023). From digital control to digital twins in medicine: A brief review and future perspectives. *Advances in Applied Mechanics, 56*, 323–368.
13. Myrzashova, R., Alsamhi, S. H., Shvetsov, A. V., Hawbani, A., & Wei, X. (2023). Blockchain meets federated learning in healthcare: A systematic review with challenges and opportunities. *IEEE Internet of Things Journal, 10*, 14418.
14. Pal, K. (2023). Security implications of IoT applications with cryptography and blockchain technology in healthcare digital twin design. In *Digital twins and healthcare: Trends, techniques, and challenges* (pp. 229–252). IGI Global.
15. Sasikumar, A., Vairavasundaram, S., Kotecha, K., Indragandhi, V., Ravi, L., Selvachandran, G., & Abraham, A. (2023). Blockchain-based trust mechanism for digital twin empowered industrial internet of things. *Future Generation Computer Systems, 141*, 16–27.
16. Garg, H., Sharma, B., Shekhar, S., & Agarwal, R. (2022). Spoofing detection system for e-health digital twin using EfficientNet Convolution Neural Network. *Multimedia Tools and Applications, 81*(19), 26873–26888.
17. Elayan, H., Aloqaily, M., & Guizani, M. (2021). Digital twin for intelligent context-aware IoT healthcare systems. *IEEE Internet of Things Journal, 8*(23), 16749–16757.
18. Azzaoui, A. E., Kim, T. W., Loia, V., & Park, J. H. (2021). Blockchain-based secure digital twin framework for smart healthy city. In *Advanced multimedia and ubiquitous engineering* (Vol. 716, p. 107). Springer.
19. Chaudhari, P., Gangane, C., & Lahe, A. (2021). Digital twin in industry 4.0 a real-time virtual replica of objects improves digital health monitoring system. In *International conference on information systems and management science* (pp. 506–517). Springer International Publishing.
20. Zhang, J., Li, L., Lin, G., Fang, D., Tai, Y., & Huang, J. (2020). Cyber resilience in healthcare digital twin on lung cancer. *IEEE Access, 8*, 201900–201913.
21. Tuli, S., Basumatary, N., Gill, S. S., Kahani, M., Arya, R. C., Wander, G. S., & Buyya, R. (2020). HealthFog: An ensemble deep-learning-based smart healthcare system for automatic diagnosis of heart diseases in integrated IoT and fog computing environments. *Future Generation Computer Systems, 104*, 187–200.
22. Sahoo, A. K., Pradhan, C., Barik, R. K., & Dubey, H. (2019). DeepReco: Deep-learning-based health recommender system using collaborative filtering. *Computation, 7*(2), 25.
23. Stephen, O., Sain, M., Maduh, U. J., & Jeong, D. U. (2019). An efficient deep learning approach to pneumonia classification in healthcare. *Journal of Healthcare Engineering, 2019*, 1.
24. Muhammad, K., Khan, S., Del Ser, J., & De Albuquerque, V. H. C. (2020). Deep learning for multigrade brain tumor classification in smart healthcare systems: A prospective survey. *IEEE Transactions on Neural Networks and Learning Systems, 32*(2), 507–522.
25. Zhou, X., Liang, W., Kevin, I., Wang, K., Wang, H., Yang, L. T., & Jin, Q. (2020). Deep-learning-enhanced human activity recognition for Internet of healthcare things. *IEEE Internet of Things Journal, 7*(7), 6429–6438.
26. Ali, F., El-Sappagh, S., Islam, S. R., Kwak, D., Ali, A., Imran, M., & Kwak, K. S. (2020). A smart healthcare monitoring system for heart disease prediction based on ensemble deep learning and feature fusion. *Information Fusion, 63*, 208–222.
27. Bhattacharya, P., Tanwar, S., Bodkhe, U., Tyagi, S., & Kumar, N. (2019). Bindaas: Blockchain-based deep-learning as-a-service in healthcare 4.0 applications. *IEEE Transactions on Network Science and Engineering, 8*(2), 1242–1255.

28. Dwivedi, A. D., Malina, L., Dzurenda, P., & Srivastava, G. (2019). Optimized blockchain model for internet of things based healthcare applications. In *2019 42nd international conference on telecommunications and signal processing (TSP)* (pp. 135–139). IEEE.
29. Su, Q., Zhang, R., Xue, R., & Li, P. (2020). Revocable attribute-based signature for blockchain-based healthcare system. *IEEE Access, 8*, 127884–127896.
30. Aujla, G. S., & Jindal, A. (2020). A decoupled blockchain approach for edge-envisioned IoT-based healthcare monitoring. *IEEE Journal on Selected Areas in Communications, 39*(2), 491–499.

Smart Factory Digital Twin for Performance Measurement, Optimization, and Prediction

Suhas D. Joshi

Introduction

Industry 4.0 shows the vision for a smart factory. A smart factory uses a data-driven approach to improve manufacturing efficiency. Application of technologies such as sensors, Supervisory Control and Data Acquisition (SCADA), Data Lake, Manufacturing Execution System (MES), Digital Twin, Internet of Things (IoT), and Augmented Reality make a factory smart. A Digital Twin integrates the design model of a physical object, e.g., a machine, with the real-time operation of the physical object. Sensors or the machine itself generate data. Such data represents machine status and the working of the machine.

A Digital Twin for a machine or a manufacturing line is an enabler for performance management, which includes measurement, optimization, and prediction. This chapter focuses on the requirements and design for a Smart Factory Digital Twin (SFDT) that can do manufacturing performance management.

This chapter starts with an overview of various phases in the production life cycle. Then different types of manufacturing processes are explained. Before jumping into a Smart Factory Digital Twin (SFDT), an overview of the Enterprise Reference Architecture model is presented. This shows various layers starting from the enterprise layer at the top, down to the machines on the shop floor in the bottom and the Digital Twin in the middle. This discussion will help the reader understand the role the Digital Twin plays in the factory. Next, an explanation of Overall Equipment Effectiveness (OEE) is given. OEE is a measure of manufacturing processes performance.

Since IoT and Data Lake are the basic technologies that the SFDT relies on, a solution architecture showing how data from the shop floor can be stored in the Data

---

S. D. Joshi (✉)
Olyphaunt Solutions Private Limited, Pune, Maharashtra, India
e-mail: Suhas.Joshi@olyphaunt.com

Lake is presented. After describing the foundational IoT and Data Lake architecture, we review the requirements manufacturers have regarding performance measurement. In response to the above-mentioned requirements, the SFDT solution architecture for measuring performance is explained. Each architectural layer and components within each layer are described. The operation of each component is made clear by examining the data flow through the components.

Having described the solution architecture for performance measurement, this chapter explains how the SFDT supports optimizing and predicting manufacturing process performance. Results found using SFDT are described followed by a statement about future directions and trends.

The concepts are illustrated using examples from discrete manufacturing. However, what is presented here is broadly applicable to other types of manufacturing as well.

Very often, digital twins are used for simulating a product in the design phase or for monitoring a product when it is in the service phase. In contrast, this chapter is about using a digital twin for managing performance of a manufacturing process.

Production Life Cycle

In this section, we review four major phases of Product Life Cycle Management (PLM). Figure 1 shows four phases of PLM [1].

As shown in Fig. 1, the Conceive phase starts by analyzing business requirements, i.e., "*WHAT*" and creating requirement specifications that the product should meet. High-level design explains "*HOW*" the product will meet the business requirements. Prototyping is done to show how the design will work.

The Design phase picks up after prototyping. During the Design phase, attention is paid to manufacturing the product. The design may be simulated using software techniques. Any tooling required is specified.

The Realize phase of the product life cycle begins with a plan that shows the quantity that will be manufactured daily, weekly, or monthly. Various

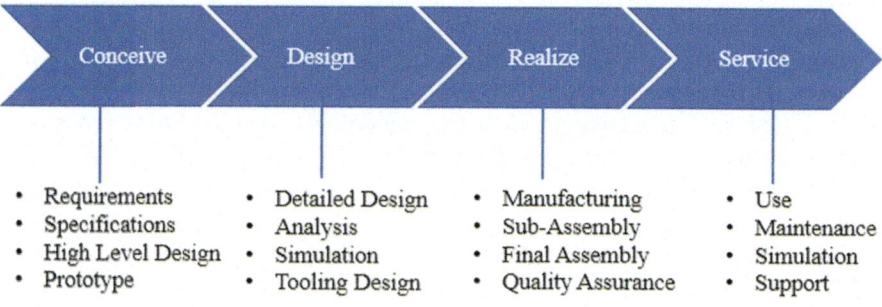

Fig. 1 Product life cycle

manufacturing processes are envisioned that produce various sub-assemblies of the product leading to the final assembly of the product, quality assurance, and final shipment.

During the Service phase, customers use the product and the product maintenance and support team provide the required effort to ensure the product is running smoothly.

Clearly, Digital Twin technology can be used effectively for supporting the Conceive and Design phases. For example, a Digital Twin can be used for simulation of a physical object or process. Similarly, a Digital Twin can be used to simulate and monitor an object that has been deployed in the field. However, the focus of this chapter is on the Digital Twin for performance management during the Realize phase.

Overview of Manufacturing Processes

Before we dive into what a Digital Twin does for a smart factory, it is better to understand the commonly used manufacturing processes.

A factory is designed to implement one or more manufacturing processes that involve a series of steps involving complex activities with machines and people. The following process types are commonly recognized.

Repetitive: In this case, the factory uses manufacturing lines, with each line containing machines and people supervising or operating the machines perform specific functions repeatedly. An example is an automobile factory which makes a certain make and model of a car without stopping.

Discrete: Factory uses manufacturing lines of machines and Machine Operators and makes multiple products to cater to demands from multiple customers. Multiple quantities of one product are made and then changes to machine configuration are made before manufacturing of another product begins. This involves the time required for "changeover" before production begins. An example of discrete manufacturing is a factory that makes wind turbines in many shapes, sizes, and quantities depending upon customer requirements.

Continuous: A product is continuously made without requiring much manual intervention. There is no assembly line involved in the operation. An example is a paint factory that continuously makes paint.

Continuous (Batch): There are no assembly lines, but the product is manufactured in batches. After manufacturing one batch, the machines may have to be cleaned or calibrated. For example, a pharmaceutical company makes a pharmaceutical drug in batches.

Additive Manufacturing: Digital design for a part and material used for manufacturing are fed into a 3-D printer. The 3-D printer adds two-dimensional layers to create the desired part. Traditional manufacturing techniques that involve "subtractive manufacturing," such as cutting, are not used. The parts produced by the 3-D printer may be assembled using a traditional assembly line. An example is that turbine blade is 3-D printed and then assembled with the rest of the final product.

Job Shop: Each product is customized to meet specific customer requirements. There may be some automation involved. An example is a factory that produces customized solutions depending upon specific requirements from customers. An assembly line may be involved depending upon the solution that is created.

It is easy to see that there are common factors and roles across all manufacturing process types.

1. Machines: Whether conventional manufacturing or Additive Manufacturing techniques are used, machines automate manufacturing operations.
2. Manufacturing Lines: Typically, a product is made by through a series of steps which are executed by more than one machine. A complex product may have multiple sub-assemblies, where each sub-assembly is produced by one line and one line that may produce the final assembly.
3. Machine Operators: Human Operators configure, set up, and monitor the machine during the manufacturing process.
4. Maintenance Engineers: When a machine or any other artifact such as a conveyor belt stops working, Maintenance Engineers are called upon to resolve the issue as quickly as possible.
5. Line Managers: Manufacturing Line leadership must make sure that the line is running at a performance level such that target delivery commitments can be met.
6. Plant Managers: Plant leadership is responsible for delivering target product quantity in the committed time. Plant leadership needs to make sure that the plant is performing efficiently, and over time, performance can be improved.

Smart Factory Digital Twin and Enterprise Reference Architecture

A Digital Twin integrates the design model of a physical object such as a machine or a manufacturing line with the real time operation of the object. Let us see how a Smart Factory Digital Twin fits in the enterprise architecture.

The Purdue Enterprise Reference Architecture [2] shows how various factory systems and applications work together. Figure 2 shows a conceptual model with 5 levels.

At level 0, is the physical manufacturing process including machines. At level 1 are the process level sensors, instrumentation, and actuators that work with or monitor machines. Programmable Logic Controllers (PLC), Supervisory Control and Data Acquisition (SCADA), and Human Machine Interface (HMI) are at Level 2 and drive components at Levels 1 and 0. Components in Levels 0, 1, and 2 reside on the factory shop floor.

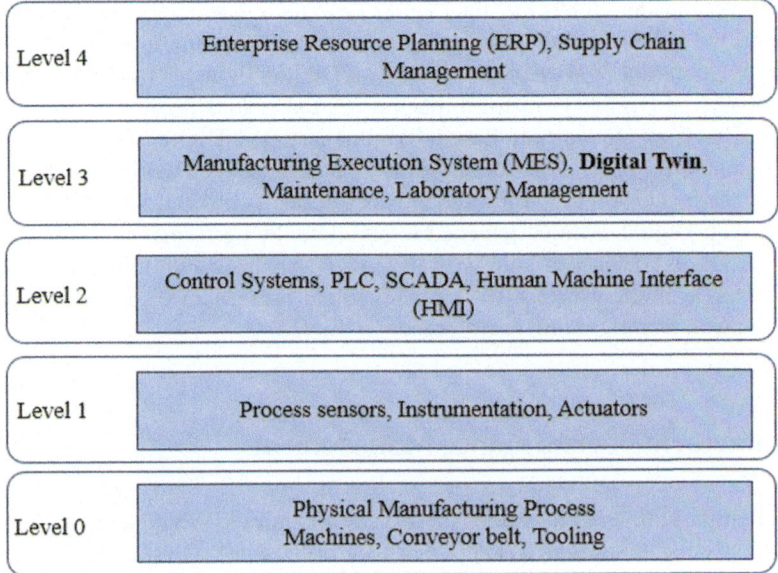

Fig. 2 Smart Factory Digital Twin and its place in the Enterprise Reference Architecture

At Level 3, Manufacturing Execution System (MES) dispatches work to the shop floor based on Production Planning in the Enterprise Resource Planning (ERP) or Supply Chain Management (SCM). MES usually monitors and tracks production based on messages received from ERP or SCM business applications which are in Level 4 of the Enterprise Reference Architecture. A manufacturing Digital Twin can co-exist with MES and support the MES in measuring, optimizing, simulating, and predicting manufacturing performance. The use cases in which a Digital Twin can manage performance include legacy MES which cannot be changed or specific machines and manufacturing lines whose performance cannot be managed by the MES.

Why Digital Twin for Manufacturing Performance Management

Traditionally, MES is responsible for starting production on the shop floor and then tracking manufacturing performance. So, why should Digital Twin manage, i.e., measure, optimize, and predict manufacturing performance? There are several possible answers. First, from a data perspective, the Digital Twin is synchronized with the shop floor machinery. The Digital Twin continuously receives data from the physical object. So why not use that data to measure the operational efficiency and identify bottlenecks?

Another consideration is that the Digital Twin for a machine can send commands to change the behavior of the physical counterpart in real time if the manufacturing process is going out of control. In this way, a Digital Twin can minimize manufacturing losses. A Digital Twin for a machine can also start Predictive Maintenance activities such as "Condition Based Maintenance." A Digital Twin can order a replacement spare part before the machine fails because the Digital Twin knows the status of the machine and it knows the history about its utilization and issues.

Finally, a Digital Twin is designed to simulate the performance of the physical object. A Digital Twin can help with "What-if" Analysis and a Digital Twin provides a single framework for tracking historical performance, improving performance as well as forecasting performance in the future.

Manufacturing Performance Management

Very often the following questions are asked in factories. What is the performance at which my machines are running? Are they utilized fully? Where are the bottlenecks on my manufacturing lines? How can I improve the efficiency at which my plant is running? What can technology do to proactively take actions so that downtime is minimized?

Answers to all these questions can be found through the Digital Twin technology.

The Smart Factory Digital Twin (SFDT) is an enabler for performance measurement, optimization, and prediction. As shown in Fig. 3, data from the shop floor, i.e., data generated by machines, data from sensors on the lines, Machine Operator entered data, or data from PLC and SCADA will enter a data store. Conceptually, SFDT implements three business processes described below.

Performance Measurement: Data from the shop floor is used to calculate the efficiency with which an object is running. Such an object could be a machine, a manufacturing line, a group of lines, or the whole plant. A common performance measurement standard for manufacturing is Overall Equipment Effectiveness

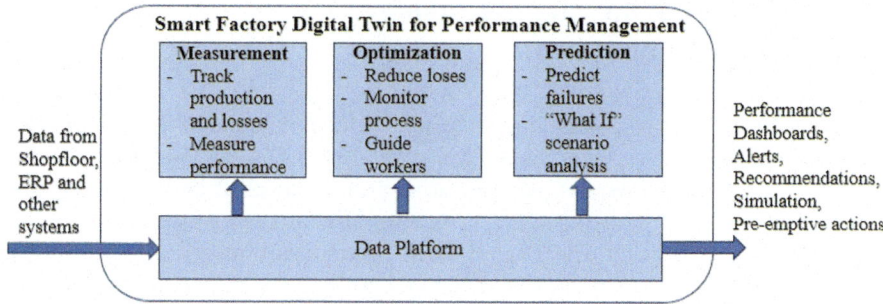

Fig. 3 Smart Factory Performance Management with Digital Twin

(OEE). An explanation of OEE calculation and the solution architecture for Performance Measurement is explained in detail in this chapter.

Performance Optimization: Performance Optimization ensures that the manufacturing process is operating correctly with minimal losses. Use of Statistical Process Control (SPC) is described to reduce defects. Automated work instructions and recommendations are generated for the manufacturing team to reduce time to repair and improve manufacturing performance. Detailed requirements as well as the solution architecture for Performance Optimization is explained in this chapter.

Performance Prediction: Performance Prediction focuses on analyzing recent events, predicting what might happen in the future and taking actions proactively to avoid failures. In addition to such predictive analytics, simulation tools are provided so that factory leadership can do "What-If" analysis and analyze what the factory throughput will be, if production requirements change. A detailed set of requirements along with the solution architecture for Performance Prediction is given in this chapter.

IoT and Data Platform Requirements

A Smart Factory is data driven. Data is generated by various sensors, machines, and applications running in the manufacturing environment. Performance Management services and applications will utilize the platform that will provide IoT and Data services. The Data Platform should be designed to meet the following main requirements.

- *Scalability*: The volume of data that needs to be stored will grow over time. It should be possible to scale the storage without impacting the applications and services.
- *Connectivity*: It should be possible to connect to the data stores and store the data generated by a variety of machines and sensors. Also, it should be possible to connect to the data store either locally from within the factory or from outside of the factory, by following proper cyber-security methods.
- *Versatility*: Shop floor devices generate different types of data including continuous data streams, events, or bulk data. Data stores should allow storage of time series, relational (SQL) data, non-SQL data and batch files.
- *Analytics*: Support for both descriptive and predictive analytics will be required. This includes answering "what happened" through aggregation, grouping, and reporting. It also includes identifying trends and "what will happen" type of predictive analytics.
- *Open architecture*: The Smart Factory Digital Twin will be one of the clients of the data, but the Data Platform architecture should be open to other types of clients such as Business Intelligence tools.

Architecture Considerations

SFDT users will focus primarily on the SFDT functional requirements. But solution architects need to keep in mind the following important considerations.

- *Ability to Work for Different Physical Objects*: The SFDT solution will be able to work for multiple objects and not just one object.
- *Scalability*: SFDT should not be limited to a fixed number of instances. It may be only constrained by resources such as memory and storage.
- *Data Sharing*: SFDT should be able to share performance data with other systems such as MES. Reporting applications should be able to query the performance data created by SFDT.
- *Openness*: SFDT should integrate with other applications in the ecosystem. It should receive and send data to other applications such as Material Handling System, Spare Part Management system, and Operator Skill Tracking through well-defined Application Programming Interfaces (APIs).
- *Event Driven*: SFDT solution architecture should be designed to process events in real time. For example, as data is generated, it should get processed.
- *Cyber-Security*: The solution architecture should implement security of data at rest and data in motion and Role Based Access Control (RBAC).
- *User Experience*: The Digital Twin solution should be designed to support local as well as remote users. Commonly used browsers and mobile devices should be supported.

IoT and Data Platform Solution Architecture

Figure 4 shows the Data Platform solution architecture that will be designed to meet the above-mentioned IoT and Data Platform requirements. The layers shown in this figure do not contain "Performance Management" related functionality. Those layers are described later in this chapter.

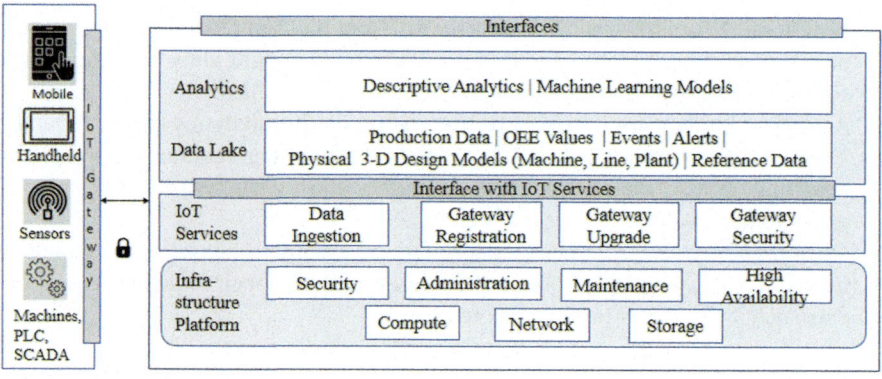

Fig. 4 IoT and Data Platform Solution Architecture

Shop Floor Components

Typically, the factory shop floor has machines, Machine Operators, manufacturing lines, Maintenance Engineers and systems such as PLC and SCADA that generate data. Machine Operators and Maintenance Engineers may use mobile applications, scanners, or handheld devices for entering data. The IoT Gateway is responsible for taking the data generated from shopfloor devices and sending it to a common data store. The IoT Gateway will connect to equipment on the shop floor through field bus protocols such as OPC/UA [3], ModBus [4], ProfiNet [5], ProfiBus [6], WiFi [7], BlueTooth [8], and Ethernet IP [9] and will send the data to the Data Stores through an industry standard protocol such as secure AMQP [10], MQTT [11], or HTTP [12]. The IoT Gateway will also be able to receive commands from the SFDT applications and services and will pass them to the target shop floor components.

Infrastructure Platform and IoT Services Layer

The infrastructure layer provides compute, network, and storage capabilities. Cybersecurity, High Availability, Platform Administration, and Maintenance are required to keep the basic platform operational.

IoT Services Layer

The IoT Services layer supports various functions such as registration and configuring of IoT gateways, upgrading the firmware within the IoT gateways, and enabling or disabling an IoT gateway. The Data ingestion function includes support for an industry standard data communication protocol such as secure HTTP, MQTT, or AMQP. Valid data is passed to the Data Lake for storage in the right data store.

Data Lake

Above the IoT layer are the Data Lake and Analytics layers. Data Lake contains data stores including relational database, time-series database, flat files, and non-SQL document style database. Data from shopfloor as well as performance data will be stored in the Data Lake. Components in the Data Lake layer are described as follows.

1. *Production Data*: This contains details about the actual operations that were made for manufacturing a part, e.g., grinding, welding, and turning.

2. *OEE Values*: After OEE values are calculated, they are stored so that tracking can be done on a daily, weekly, and monthly basis.
3. *Events*: Events such as machine changeover started and changeover complete will be generated as part of the production process. These events are stored in the Data Lake.
4. *Alerts*: Notifications regarding various conditions such as material not available on the manufacturing line, unscheduled operator stops, unscheduled machine stops will be generated from the shopfloor equipment, which are stored in the Data Lake.
5. *Physical 3-D Design Models*: A 3-D Drawing for the machine or a manufacturing line or a plant will serve as the supporting model, as the base for animation and simulation tasks. It will give additional insight into the operation of the physical object.
6. *Reference*: Data that does not change often is called Reference Data. Examples of Reference Data include the following:

    - *Plant Layout*: This includes the number of manufacturing lines, identification of machines in each line, and a layout of the physical shop floor area.
    - *Operator Skills*: For each Operator that works on the shop floor, a list of skills the Operator has and the proficiency level in each skill is maintained. For example, the Operator may have beginner skills for operating a grinding machine but advanced skills for operating a welding machine.
    - *Context model*: A model showing the relationship of the object to other objects. More about this is explained later in this chapter.
    - *Ideal Cycle Time*: This is used for producing a part.
    - *Approximate Price of the Part That Is Manufactured*: This helps in estimating the financial impact of the downtime.

Analytics Services

1. *Descriptive Analytics*: Descriptive Analytics services include filtering, aggregating, grouping, sorting, querying, and updating data. Tools for report writing and execution are also provided.
2. *Machine Learning Models*: Commonly used machine learning models including forecasting through linear and multi-linear regression analysis and anomaly detection. Models analyze trends in the data to predict outcomes.

Application Programming Interface (API) is provided so that the SFDT applications and services can access the data in the Data Lake or the analytics services without any knowledge of the physical model of the data they are accessing.

The IoT and Data Platform described above can be hosted with a cloud services provider in an instance dedicated to the SFDT. An alternative is to host the IoT and Data Platform on premises. Many criteria are used in deciding between a cloud-based infrastructure and on-premise infrastructure. A detailed discussion of such criteria is beyond the scope of this chapter.

Overall Equipment Effectiveness (OEE)

Overall Equipment Effectiveness (OEE) is a universally accepted measure of the ability to run equipment without a failure, at the defined speed with zero defects [13, 14].
Mathematically,

$$OEE = \text{Availability Efficiency } (AE) * \text{Performance Efficiency } (PE) * \text{Quality Efficiency } (QE)$$

where,

$$AE = (\text{Total Time} * \text{Unplanned Downtime}) / \text{Total Time}$$

$$PE = (\text{Ideal Cycle Time} / \text{Actual Cycle Time}) * (\text{Actual Cycle Time} * \text{Output} / \text{Operating Time})$$

$$QE = (\text{Output} * \text{Quality Defects}) / \text{Output}$$

In the above equations,

Total time is the Operational time of equipment, excluding planned stops for holidays, maintenance, and lack of orders.
Unplanned Downtime is the time for which the machine is scheduled for production but cannot work.
Operating time is Total time minus the Unplanned Downtime.
Ideal cycle time is the average theoretical processing rate at which a part can be produced.
Actual cycle time is the weighted average cycle time for a machine for a given part.
Output is the gross number of parts produced.
Quality defects are the number of parts rejected for not passing Quality inspections.

The above equations related to the OEE help us shape the requirements for the SFDT for Performance Management. OEE is expressed as a percentage. If availability efficiency (AE) is more than 90%, performance efficiency (PE) is than 95% and quality efficiency (QE) is more than 99%, with the three measures multiplied leads to a world-class OEE of 84% as mentioned by Nakajima in [14]. In a study of 23 companies, food and beverage companies are found to have a higher median OEE (74%) than other types of companies, while automated discrete production companies have the lowest OEE (59%) as described in [15].

Performance Measurement

This section begins with what various factory persona require from a performance measurement perspective. Then data requirements are described, followed by a reference architecture. In describing the reference architecture, the function of each component is explained. Description of the data flow through shows how various components work together for implementing performance measurement.

Use Cases

Regardless of the manufacturing process, various factory worker roles have the following high-level requirements from a performance management perspective (Table 1).

Table 1 Performance Measurement use cases

	Persona	Digital Twin use case
1	Machine Operator	1. Need a tool to measure, monitor, and visualize in real time the operation of a machine. 2. Remote monitoring of the machine whether it is in service or running in degraded mode or out of service, should be possible. 3. View any active alerts and the financial impact of that alert at the machine level. 4. Need performance metrics along with a comparison between planned and actual performance.
2	Line Manager	5. All the above use cases for the Machine Operator. 6. Need a tool to monitor and visualize in real time the operation of a Manufacturing Line. 7. Need a tool to monitor the performance (OEE) of each machine within the line and the OEE for the line on a daily, monthly, and weekly basis. 8. Should be able to drill down to the components within the line, see the input queue length at each machine, Work in Progress (WIP), and quantity of parts produced at the end of the line. 9. Need to view any alerts and the financial impact of the alerts for all the machines in the line.
3	Maintenance Engineer	10. Should be able to drill down to the critical components in the machine and see values of specific properties. For example, in a grinder machine, spindle speed, temperature, and thickness should be visible as the machine is operating. 11. Should be able to visualize values of certain physical machine properties through "Virtual Sensors." For example, the grinder machine may not report the thickness of the grinding wheel, but SFDT should calculate the value based on the physical model of the machine and the usage of the grinding wheel so far.
4	Plant Manager	12. Visualize the performance of a specific manufacturing line, a group of lines, and the entire plant. 13. Track performance of a machine, specific line, a group of lines, and the entire plant on a daily, weekly, and monthly basis. 14. Compare the performance of a specific machine, line, group of lines against another machine, line, or group of lines within another plant in the enterprise. 15. Show a prioritized list of pending alerts across the entire plant. 16. Show the financial impact of any pending alerts across the plant.

Data Required for Performance Measurement

Based on the abovementioned use cases, various data streams will be required by SFDT. Such data will have to come from sensors, machines, or people on the shop floor. Reference Data such as the design model for a machine will have to be provided by the machine manufacturer. Also, applications such as ERP or Production Planning will have to provide data such as ideal cycle time for a manufacturing a part. Table 2 summarizes the data required against each major use case.

As described earlier, OEE is a product of Availability Efficiency (AE), Performance Efficiency (PE), and Quality Efficiency (QE). Formulas for calculating the three efficiencies, namely, AE, PE, and QE have been mentioned earlier in this chapter. With reference to those formulas, Table 3 shows the data that will be required to calculate the efficiency values, and consequently the OEE.

Table 2 Data required for Performance Measurement

Use case	Data required	Data source	Data generation frequency
Visualize values of critical properties at a component level	Machine Physical 3-D Design Model	Machine Manufacturer	One time
	Values of critical properties	Sensors (If sensors are not available, values will be calculated based on machine model and values of other parameters.)	Either periodically generated by the sensor or periodically calculated
	Machine status	Machine firmware or Operator	Status changes will be notified by the machine firmware. Alternatively, Operator will have to enter it using a Mobile App or another suitable tool
Visualize values of critical properties at a Manufacturing Line level	Manufacturing Line Physical 3-D Design Model	Manufacturing Line design	One time
	Values of critical properties, e.g., finished product at end of the line	Sensor at the input and a sensor at the output of each machine	Sensors generate notification messages when the status changes
	Operator Identification	Mobile App or a similar tool used by the Operator. Biometric login/logout	Operator ID provided when the Operator logs in our out
Track performance at a daily, weekly, and monthly basis	OEE values	OEE values as calculated	Calculated at the end of each shift and stored in a database for analysis
Alerts	Alert data	Generated by machine or PLC or SCADA	As generated

Table 3 Data required for OEE calculation

OEE formula variable	Data required	Data source	Data generation frequency
Total Time	Shift Duration	Production Planning System	SFDT will retrieve before shift begins
	Planned Downtime	Production Planning System	Retrieved before shift begins
Unplanned Downtime	Machine setup time	Machine or Operator	Events: Start, Finish
	Machine changeover time	Machine or Operator	Events: Start, Finish
	Machine breakdown time	Machine or Operator	Events: Start, Finish
	Material unavailable time	Machine or Operator	Events: Start, Finish
	Tooling unavailable	Operator	Events: Start, Finish
	Operator unavailable	Operator	Events: Start, Finish
Ideal Cycle Time	Expected time required to produce the part	Production Planning System or ERP	SFDT will retrieve before manufacturing for a part number begins
Actual Cycle Time	Time required to produce the part	Machine	SFDT will retrieve when manufacturing for a part number begins
Output	Number of parts that were produced	Machine or Operator	Event generated when part is complete
Quality Defects	Number of defective parts	Machine or Operator	Event generated defect is detected

An explanation of Table 3 is as follows. Ideally, data will be generated by the machine. For that to happen, the machine needs to have sensors to detect conditions, and through the firmware, it generates events. If the machine and associated sensors cannot automatically generate data, then the other option is that through the Human Machine Interface (HMI), the machine prompts the Operator to enter certain data which the machine transmits. If that is not possible, then data will be entered by the Machine Operator through a variety of sources. For example, the Operator can use a Mobile App, an app running on a tablet computer or a handheld scanner.

Solution Architecture for Performance Measurement

The foundational layers such as the Infrastructure platform, IoT Services, Data Lake, and Analytics have been described earlier. In what follows, the Process Models layer is described, followed by interfaces with external applications.

Smart Factory Digital Twin for Performance Measurement, Optimization, and Prediction

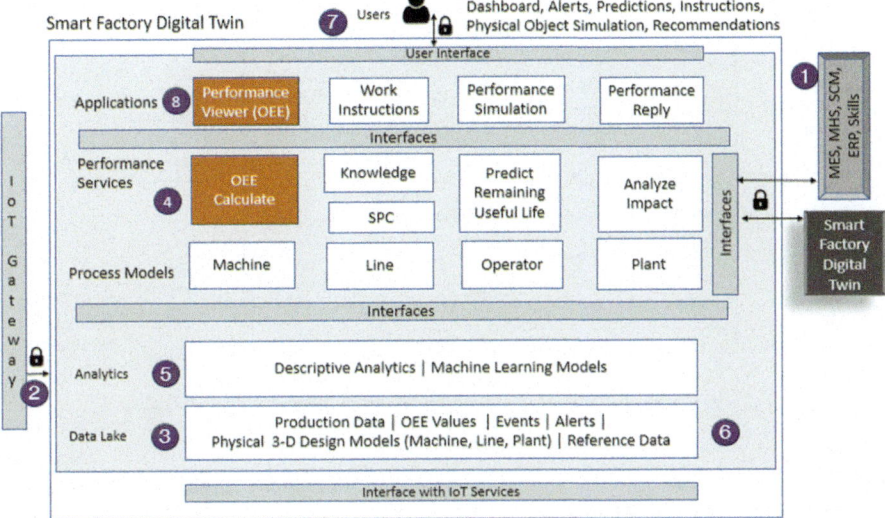

Fig. 5 SFDT Architecture with Performance Measurement high-lighted

Process Models

The process model represents what the physical object does.

1. *Machine Model*: The machine model implemented in software will "simulate" the physical machine. The machine model will also include a "Virtual Sensor" capability. As an example, for the grinding machine, the thickness of the grinding wheel can be calculated based on the previous value of the thickness and knowing for the time for which the machine has been running. See Fig. 3 which shows the model for a grinding machine. Even though there is no thickness sensor, the "Virtual Thickness Sensor" will algorithmically calculate the thickness value and will make it available for consumption by Operators or Maintenance Engineers.
2. *Line Model*: The line object implemented in software will simulate the physical manufacturing line operations. This object contains the overall status of the manufacturing line, up time and downtime, number of parts produced by the line, etc.
3. *Operator Model*: The Operator object contains Operator identification, the duration for which the Operator is working on the machine, etc. This model can algorithmically prevent the Operator from starting the machine if the Operator does not have the required skills.
4. *Plant Model*: The plant object simulates the entire plant. It contains data such as quantity of parts produced grouped by part number (Planned and Actual), Work In Progress (WIP), pending alerts, and if certain alerts are causing loss of production, the financial impact of that alert (Fig. 6).

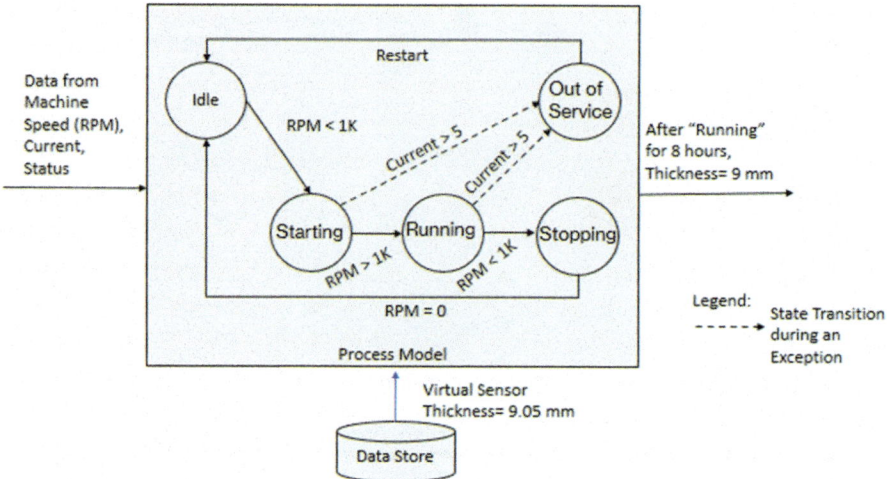

Fig. 6 Model for a grinding machine

External Interfaces

The Smart Factory Digital Twin will interoperate with external applications such as Manufacturing Execution System (MES), Material Handling System (MHS), Supply Chain Management (SCM), Enterprise Resource Planning (ERP), Maintenance Management, and Operator Skills Tracking system. The Digital Twin will also be able to work with other Digital Twins through well-defined interfaces.

Services for Performance Measurement

The services described in this chapter are continuous processes that run either continuously or periodically. On the other hand, applications described in this chapter run when invoked by a user.
OEE Calculate: Generally, OEE is calculated at the end of the shift. This service will run at the end of the shift and will calculate the OEE based on the OEE data received in the data store. All the calculated values are stored in a database so that they can be used for analysis by the Performance Optimization algorithms.

Applications for Performance Measurement

SFDT applications use underlying services and databases. Applications interface with the Digital Twin users through a user interface. Applications may also interface with other systems through APIs. Examples of possible applications are described as follows.

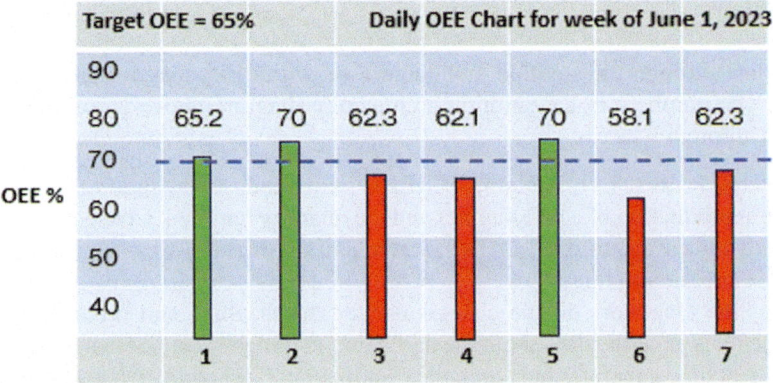

Fig. 7 Example of an OEE chart

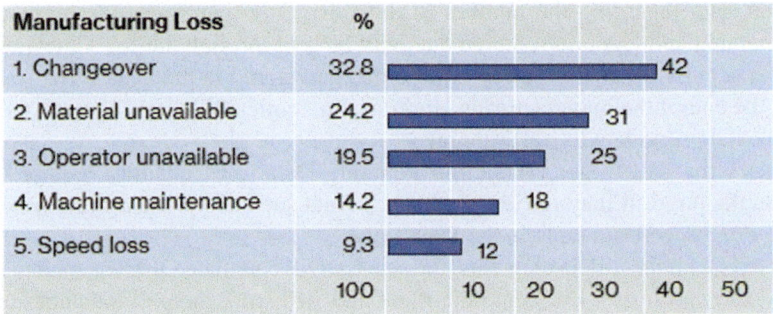

Fig. 8 Example of a Pareto Chart

Performance Viewer (OEE): By using this application, the user will be able to visualize the performance (OEE) and the trend for a plant, manufacturing line, manufacturing line groups, or a specific machine. OEE data for the previous day, current week, month, and year as well performance comparison with a previous period is also possible through this application. Figure 7 is an example of a daily OEE chart for a manufacturing line.

Pareto Analysis charts for Availability, Productivity, and Quality losses are presented. By reviewing these charts, the users can identify which are the most frequently occurring losses and can prioritize their efforts to reduce the most frequent losses. Figure 8 shows an example of manufacturing losses per day and the top five loss categories.

Data Flow

Data flow through the solution architecture shown in Fig. 5 is described below in steps. The numbers in the solution architecture diagram above correspond to the steps 1 through 8 below.

1. Enterprise Resource Planning (ERP) system sends a message to the MES to begin production of a part number and the quantity required. Consequently, MES sends a message to the SCADA system and production work begins on one or more production lines.
2. machine Operators do the changeover for the machine and production starts. Machines generate data about the part being manufactured. For instance, a grinding machine generates how long the grinding operation took, what was the thickness of the part when grinding started, and what was thickness of the part when grinding started. Operators will enter data about any unscheduled stops, such as stops due to machine failures and material unavailability. Quality assurance will enter data regarding the number of defective parts per shift. All the machine-generated and Operator-entered data will be sent through the IoT Gateway.
3. Data sent by the IoT Gateway is validated and stored in the Data Lake.
4. At the end of the manufacturing workday, i.e., typically every 24 hours, the OEE Calculate service will start running.
5. It uses the API to get Production Planning Data, e.g., quantity required to be manufactured. It may retrieve Reference Data such as the ideal cycle time from the Reference Data table in the Data Lake.
6. Based on all these data elements, the service will calculate OEE for each machine and line, groups of lines, and the plant and will store them in a database in the Data Lake.
7. A user wants to visualize the performance through the Performance Viewer from their mobile phone or browser.
8. The Performance Viewer (OEE) will retrieve the OEE results from the database. Pareto analysis of the losses will also be displayed. Comparisons with past reporting periods will also be made available. The Performance Viewer application can also run a query about the OEE before a manufacturing shift starts and push the query results to users by email.

Performance Optimization

This section describes the Performance Optimization use cases, followed by the data model, reference architecture and data flow for a Smart Factory Digital Twin that enables Performance Optimization.

Use Cases

Use cases for Performance Optimization are shown below in Table 4. Use cases corresponding to specific Manufacturing losses are shown in Table 5.

Table 4 Use Cases for Performance Optimization

	Persona	Digital Twin Use case
1	Machine Operator, Line Manager	1. Need a mechanism to confirm that the manufacturing process is operating within control limits. For example, during the screwing operation, the torque applied is within the minimum and maximum value. 2. If values of certain manufacturing properties are outside of control limits, receive automatic recommendations about how to recover from an error, malfunction, or a condition that is causing a machine to underperform. 3. View Work Instructions regarding how to complete a specific operation with a machine.
2	Plant Manager	4. Need ways to reduce availability, productivity, and quality losses. See table below. 5. Get recommendations regarding the priority order in which production losses must be addressed. 6. Ability to compare one plant's performance against another.
3	Maintenance Engineer	7. Need work instructions regarding how to complete a repair task superimposed over the physical view of the object. 8. Need the ability to see values of critical parameters on a historical basis to troubleshoot and isolate the problem.

Table 5 Manufacturing losses and Digital Twin use cases

Loss category	Loss	Digital Twin use case
Availability	Machine setup time	Guide the Operator through video Work Instructions that are superimposed on the physical model of the machine.
	Machine changeover time	Same as above.
	Time lost due to machine adjustment	Same as above.
	Machine breakdown time	Guide the Maintenance Engineer through video Work Instructions that are superimposed on the physical model of the machine.
Productivity	Operator running slower than expected because of lack of experience or training	Provide guidance to the Operator through an interactive video.
	Unscheduled stops and interruptions	Analyze the causes and automatically make recommendations.
	Machine idling	Identify the queue depth at each machine in the manufacturing line. Identify bottlenecks.
Quality	Machine operations outside of control limits	Monitor machine operations in real time. If outside of control limits, raise alert and make recommendations.

Data Required for Performance Optimization

Considering the use cases for Performance Optimization, the following data sources are envisioned (Table 6).

Table 6 Data required for Performance Optimization

Use case summary	Data required	Data source	Data generation frequency
Work Instructions for either the Machine Operator or Maintenance Engineer	Machine design model	Machine Manufacturer	Once
Optimize losses	Historical data regarding Availability, Productivity, and Quality Losses	Data regarding machine failures, difference between planned and actual cycle time, count of defective parts produced per shift	Generated at the end of every shift
Process Control	Process parameter values	Machine	Generated in real time

Solution Architecture for Performance Optimization

Services for Performance Optimization

The following reference architecture is used to support the Performance Optimization use cases. This section explains primarily the following services: Statistical Process Control, Knowledge Management (recommendations based on exceptions detected through Statistical Process Control), and the Performance Optimization application (Prioritized handling of losses and recommendations).

1. *Knowledge*: The Knowledge service continuously reviews exceptions, alerts, and process control violations. The Knowledge service uses a Rules Engine to determine what recommendation to take. The rules essentially represent knowledge of experts who have the experience of what to do when a particular situation arises. For example, if the torque sample in a screwing process indicates that the process is out of control, the recommendation is to stop the machine, recalibrate it, and then restart it. The Knowledge service uses a rule base, which is populated by Subject Matter Experts.
2. *Statistical Process Control (SPC)*: The performance of a manufacturing process can be observed through sampling and sample values are analyzed to see trends and issues. Detection of a particular process variable being out of control is possible using control charts and Nelson's Rules [16]. For example, suppose that the

expected value of torque in a screwing machine is 10 Newton meters (Nm). Consider the value of torque delivered by a screwing machine. Let us suppose that the standard deviation is .05 Nm. Upper and Lower Control Limits are established to be plus 3 standard deviations above the mean value of torque and minus 3 standard deviations below the mean value of torque, respectively. So, if the value of the torque exceeds 10.15 Nm or goes below 9.85 Nm, the sample is out of control. Figure 10 shows the value how torque value may vary with time. A value higher than Upper and Lower Control Limits is considered a violation.

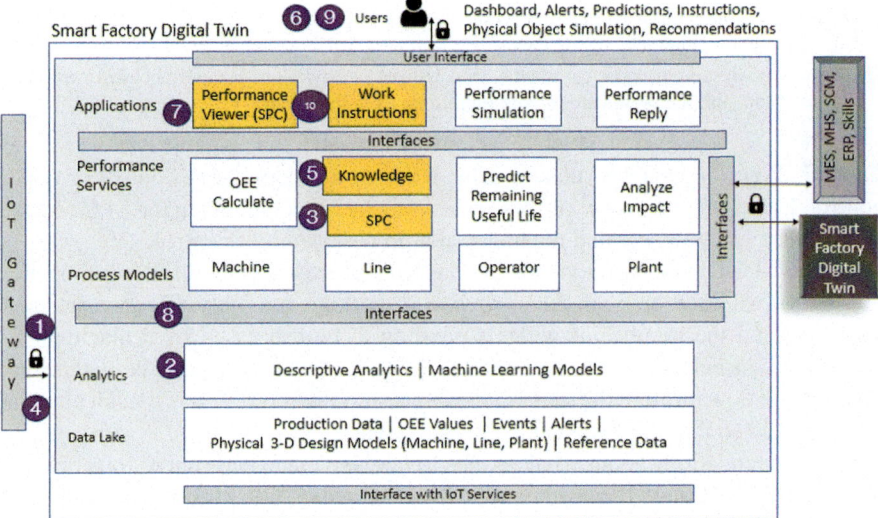

Fig. 9 SFDT Architecture with Performance Optimization highlighted

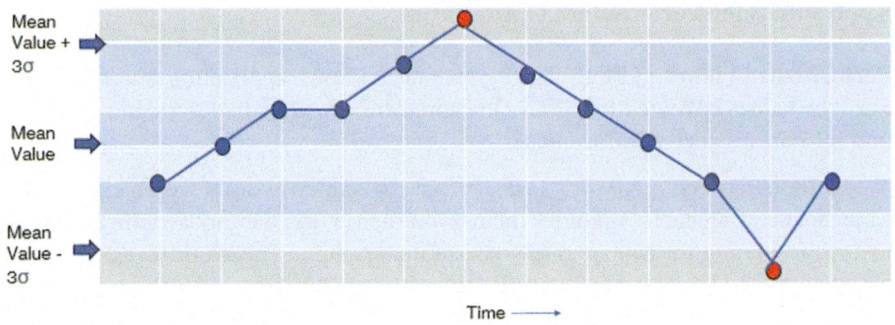

Fig. 10 Example showing SPC analysis

Table 7 Nelson's Rules

	Condition	Issue with process
1	Sampled value outside of Upper or Lower Control Limit	Sample out of control
2	9 consecutive samples are above or below the mean	Prolonged bias
3	6 or more samples are either increasing or decreasing	Increasing or Decreasing Trend
4	14 or more samples alternate between increasing and decreasing	Oscillations in the process
5	2 or 3 consecutive samples are more than 2 standard deviations from the mean in the same direction	Medium tendency for mediumly out of control
6	4 or 5 consecutive samples are more than 1 standard deviation from the mean in the same direction	Strong tendency to be slightly out of control
7	15 or more consecutive samples are within 1 standard deviation of the mean on either side of the mean	Greater variation may be expected
8	8 consecutive samples exist but none within 1 standard deviation of the mean, and the points are in both directions from the mean	Random behavior expected

Table 7 shows Nelson's 8 rules to infer whether a process is in control or out of control. Either all or a subset of these rules may be implemented in the SPC service, and an alert can be generated to show if the process is out of control.

Just as SPC can be used to monitor whether a process is in control, it can be used to monitor whether the manufacturing line is working smoothly. For instance, sensors can send the number of widgets pending to be processed by a machine. By analyzing consecutive values of number of pending widgets, it can be determined whether the flow through the manufacturing is smooth or not. If it is not, an alert can be generated so that the issue can be investigated.

Recommendations about what actions to take are made and sent to the appropriate individuals through the notification service. For example, if SPC detects that the torque applied by the screwing machine is higher than the upper limit, it sends a recommendation about what to do so that such violations do not occur.

Applications for Performance Optimization

Performance Viewer (SPC): The Performance Viewer application is used to visualize the values collected for SPC. This application is already described in the section for Performance Measurement.

Work Instructions: A Work Instructions application is used by Operators and Maintenance Engineers. It guides them to do their work, thereby avoiding the time lost while waiting for an expert to arrive at the machine. The Work Instructions are based on the 3-D drawing as reference information. Augmented Reality (AR) technology [17] is used to develop such applications because work instructions are superimposed over the image of the physical object. Work Instruction applications will typically be used on a mobile device or a hands-free device such as smart glasses.

Data Flow

Data flow through the solution architecture (Fig. 9) is described below in steps. The numbers in the solution architecture diagram above correspond to the step numbers 1 through 10 below.

1. As the machine starts manufacturing a part, the machine generates data and the IoT Gateway sends that data. As an example, a screwing machine periodically sends the torque value.
2. Machine-generated data is validated and stored in the Data Lake.
3. The SPC Service will retrieve the data from the Data Lake.
4. Stability of the manufacturing process is determined by applying the rules mentioned earlier. If the process is out of control, an alert is generated and stored in the Data Lake. Secondly, a command is sent to the machine through the IoT Gateway.
5. The Knowledge service will process any alerts generated by the SPC Service and based on a Rule Base and will generate a recommendation to be sent to appropriate users.
6. Suppose a Quality Control engineer wants to visualize the value of a parameter, e.g., value of the torque applied by the screwing machine, then she or he will access the Performance Viewer (SPC).
7. Performance Viewer (SPC) will get inputs from the user such as machine and process identifier.
8. Performance Viewer (SPC) will retrieve the torque value stream from the Data Lake and display it.
9. If the Knowledge service recommends a maintenance action on the screwing machine but the Maintenance Engineer is not sure how, the engineer will start the Work Instruction application.
10. The Work Instructions are created using the 3-D design model of the screwing machine. It provides step-by-step instructions to the Maintenance Engineer.

Performance Prediction

This section describes the performance prediction use cases, followed by the data model, reference architecture, and data flow for a Smart Factory Digital Twin that enables performance prediction.

Use Cases

Use Cases for Performance Prediction are given below (Table 8).

Table 8 Use Cases for Performance Prediction

	Persona	Digital Twin use case
1	Line Manager	1. Predict when material will run out and place a request to the Material Handling System. 2. Predict that the Operator will need training in a specific area and order it. 3. Predict what the performance will be given the skill levels of the Operators who are assigned to the line. 4. Given a specific production requirement, suggest the Operators who should be assigned to the manufacturing line.
2	Maintenance Engineer	5. Know "Remaining Useful Life (RUL)" for a machine or a spare part. Place an order for a spare part if the RUL is very low. 6. Automatically start "Condition Based Maintenance" based on the usage of a part. 7. "Predictive Maintenance." Receive an alert about replacing a part before the part fails. 8. Simulation of what the object (machine or line) current operation will be.
3	Plant Leadership	9. Need a tool to simulate what happened with the machine and the line in a certain shift in the past. 10. Need a tool to simulate what will happen in the manufacturing line for a certain production requirement.

Data Required for Performance Prediction

Considering the use cases for Performance Optimization, the following data sources are envisioned (Table 9).

Table 9 Data required for Performance Prediction

Use case summary	Data required	Data source	Data generation frequency
Determine Remaining Useful Life	Machine or component utilization	Machine Start and Stop events, Alerts, Critical component properties	Events and alerts as they occur
Take Proactive action based on production events	Events such as material consumption on the manufacturing line, Number of items waiting to be processed	Sensors on the manufacturing line, Machine, Operator	Events as they occur
Performance Simulation	Machine Physical 3-D Design Model, Manufacturing Line Physical Design Model, Log file containing past events	User provided	Provided by user as needed

Solution Architecture for Performance Prediction

Services for Performance Prediction

1. *Predict Remaining Useful Life*: Predict service can be used to monitor values of specific sets of variables and trigger an action. For example, Remaining Useful Life (RUL) for a part can be predicted based on Production data and events in the Data Lake. Based on the number of hours the machine has been used for, as well as values of critical component properties, a regression model can be used to predict the hours of remaining life. For example, the grinding wheel in a grinder is rated for 3000 hours of use. However, the grinding wheel thickness of 8 mm is recommended. Available data shows that the wheel has been used for 2200 hours. So, it may seem like 800 hours of life is available. A virtual sensor shows that the thickness of the wheel is 9.5 mm. Based on the regression analysis, the wheel will wear down to 8 mm which is the minimum thickness expected in 500 hours. So, in this case, the remaining useful life is 500 hours.
2. *Analyze Impact*: Impact of alerts, downtime, and events such as lack of material, part failure, machine changeover time, takes too long which indicates an improvement is needed in Operator skills are processed by the Analyze Impact service. This service determines the impact of the event and acts based on the impact. For example, if the event is that a part is about to fail in a certain number of hours, the Analyze Impact service uses the API for the spare part management system to order a spare part before a part fails. Similarly, the Analyze Impact service can use the API for the Material Handling Service to order material to avert the machine shutting down due to lack of material.

Based on the alerts about a machine, Analyze Impact service may start Predictive Maintenance. For example, it can start "Condition Based Maintenance" for a part for which the RUL is low.

Applications for Performance Prediction

1. *Performance Viewer (RUL)*: This application is used by the Maintenance Engineer to see the Remaining Useful Life (RUL) for a machine and any of its key components. The user can start at the level of the machine and then drill down to its components to see the RUL for each key component.
2. *Performance Simulation*: The Performance Simulation application allows the user to model the throughput and do a "What-If" analysis of a machine or a manufacturing line for manufacturing a specific part. This application will simulate all the operations such as initial set up, changeover, and quality inspections and determine how many parts can be produced.
3. *Performance Replay*: Given a log file of events, the Performance Replay application can produce a visual display of what happened within the machine and the line. The user will be able to pause the replay and analyze. This will help the user

reconstruct a specific scenario and analyze the steps which led to a specific outcome such as a machine failure. This will help in determining what actions to avoid and will help prevent performance losses.

Data Flow

Data flow through the solution architecture (Fig. 11) is described below in steps. The numbers in the solution architecture diagram above correspond to the step numbers 1 through 12 below.

1. The Predict Remaining Useful Life (RUL) service starts running periodically.
2. Based on machine utilization data, machine process model and alerts data from the Data Lake, the Predict RUL service determines the RUL. It creates alerts if action is needed. Such alerts are stored in the Data Lake. The Predict RUL service updates the RUL for a machine or for a part of a machine, in the data store.
3. The Analyze Impact service springs into action and takes action as mentioned below.
4. If Predict RUL service has generated an alert for a part with a low RUL, the Analyze Impact uses the API to order with Spare Parts Management and orders a spare.
5. If the Plant Manager wants to see the RUL and actions taken, then she or he invokes the Performance Viewer (OEE) application.
6. The Performance Viewer (RUL) displays RUL values from the data store mentioned in 2 above.

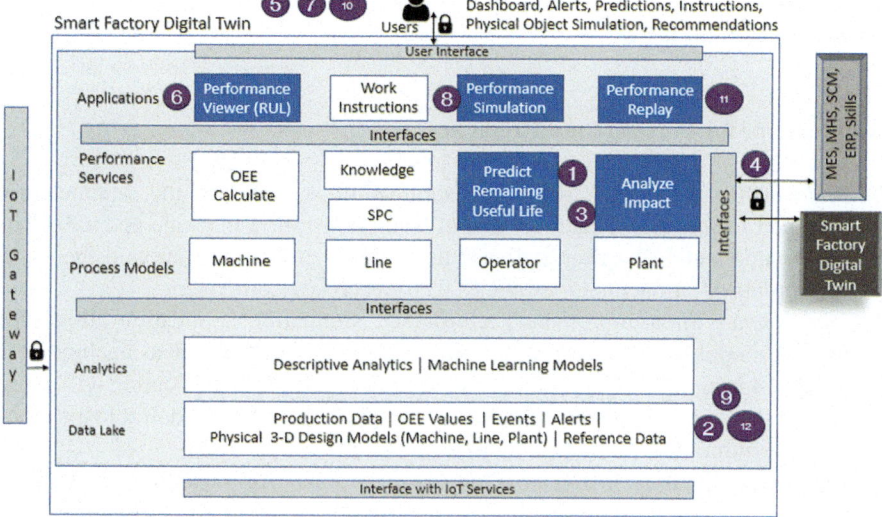

Fig. 11 SFDT Solution Architecture with Performance Prediction highlighted

Smart Factory Digital Twin for Performance Measurement, Optimization, and Prediction

7. The Plant Manager wants to simulate the production performance for a specific business requirement.
8. The Performance Simulation application is invoked.
9. Performance Simulation accesses the Machine and Manufacturing Line models as well as Reference Data to simulate the performance.
10. A Maintenance Engineer wants to replay the events that led to a machine failure.
11. The Performance Replay application is invoked.
12. Performance Replay will access historical event data from the Data Lake as well as the machine model. The simulation will start running. The Maintenance Engineer will be able to pause the simulation and inspect values of critical machine parameters at that point in time.

Results

SFDT has been implemented in a case study in several plants for a company that uses discrete manufacturing. So far, Performance Measurement and Optimization have been implemented and development for Performance Prediction is in progress. The investment made toward creating the SFDT has already paid dividends in many areas.

- All the plants follow a common standard for measuring performance. Comparison between plants is now easier.
- Visualization of the OEE has empowered the manufacturing team to understand the current performance level and take steps to improve it.
- Statistical Process Control (SPC) has been implemented to confirm that the manufacturing process is producing quality parts.
- The SFDT implementation has been in place for 1 year and an average of 6–8% increase in OEE has been observed already. After the Performance Prediction component is deployed, significant performance OEE gains are expected, primarily through the "Condition Based Maintenance" and ordering of spare parts and materials in advance, prior to the occurrence of downtime.

Conclusion and Future Directions

This chapter describes how smart factories can use the Digital Twin technology to measure, optimize, and predict manufacturing performance. A smart factory is data driven. A Smart Factory Digital Twin (SFDT) is ideal for managing manufacturing performance because it is always in sync with events, alerts, and operations related to physical objects such as machines and manufacturing lines. In this chapter, we have described various use cases, data sources, solution architecture, and the data flow across various architectural components of the SFDT. We have shown a pragmatic approach that harnesses several Industry 4.0 technologies such as Sensors, SCADA, Internet of Things (IoT), Data Lake, Analytics, and Augmented Reality. This chapter will help manufacturing and IT leaders to make the right design and implementation choices.

Rapid advances in cloud, networking, and software technologies have accelerated digitization in manufacturing. Secondly, use of software-controlled machines for manufacturing is increasing. For example, use of "Additive Manufacturing," i.e., using software-controlled 3-D Printers to convert a 3D design of a physical object into the manufacturing of that object is on the rise. A few years ago, machine manufacturers provided machines that did not generate data. Now most modern machines generate data showing details of the operations performed. As a next step, machine manufacturers will provide a Digital Twin that is shipped with the machine. Such a Digital Twin will have standard interfaces and will run as a virtual machine in industry standard environments. It will be possible for the Digital Twin software to control the physical machine. The manufacturing line may have different types of machines and the Smart Factory Digital Twin will orchestrate across multiple digital twins.

As reported in [18], in a comparison of different manufacturing modes, the performance of a Digital Twin that controlled CNC machines was very close to an automated manufacturing line. But with the Digital Twin, machine downtime was greatly reduced, and the effort required for quality inspections was significantly reduced.

As the use of AI including machine learning becomes common, the manufacturing software landscape will change. For example, in [19], the authors have explained how AI can dynamically re-plan the initial manufacturing plan made in the MES and then put it into effect through the Digital Twin. The authors have demonstrated a Digital Twin will take a revised manufacturing plan and put it into effect by re-programming the robots in the manufacturing line. This goes to show the increasing scope of AI and Digital Twins in manufacturing.

In conclusion, the use of Digital Twins in manufacturing is expected to grow which will lead to improved manufacturing performance, in terms of high quality, high speed, and low waste.

References

1. Elangovan, U. *Product life cycle management (PLM)* (p. 102). CRC Press, 202.
2. Williams Theodore, J. (1992). *The Purdue enterprise reference architecture: A technical guide for CIM planning and implementation*. Instrument Society of America.
3. OPC/UA, https://opcfoundation.org/about/opc-technologies/opc-ua/
4. ModBus protocol, https://modbus.org/
5. ProfiNet protocol, https://us.profinet.com/technology/profinet/

6. ProfiBus protocol, https://www.profibus.com/
7. WiFi protocol, https://www.wi-fi.org
8. BlueTooth protocol, https://www.bluetooth.com
9. Ethernet IP protocol, https://www.odva.org/technology-standards/key-technologies/ethernet-ip/
10. AMQP protocol, https://www.amqp.org/resources/specifications
11. MQTT protocol, https://mqtt.org/mqtt-specification/
12. HTTP protocol, https://www.w3.org/History/19921103hypertext/hypertext/WWW/Protocols/HTTP.html
13. Trattner, A. L., Hvam, L., & Haug, A. (2020). Why slow down? Factors affecting speed loss in process manufacturing. *International Journal of Advanced Manufacturing Technology, 106*, 2021–2034. https://doi.org/10.1007/s00170-019-04559-4
14. Nakajima, S. (1988). *Introduction to total productive maintenance (TPM)*. Productivity Press.
15. Hedman, R., Subramaniyan, M., & Almström, P. (2016). Analysis of critical factors for automatic measurement of OEE. *Procedia CIRP, 57*, 128–133. https://doi.org/10.1016/j.procir.2016.11.023
16. Nelson, L. S. (1984). The Shewhart control chart—Tests for special causes. *Journal of Quality Technology, 16*(4), 238–239. https://doi.org/10.1080/00224065.1984.11978921
17. Overview of Augmented Reality, Ronald Azuma, ACM SIGGRAPH 2004 Course Notes, https://dl.acm.org/doi/abs/10.1145/1103900.1103926
18. Bao, J., Guo, D., Li, J., & Zhang, J. (2018). The modelling and operations for the digital twin in the context of manufacturing. *Enterprise Information Systems*. https://doi.org/10.1080/17517575.2018.1526324
19. Vyskoˇcil, J., Douda, P., Novák, P., & Wally, B. (2023). A digital twin-based distributed manufacturing, execution system for Industry 4.0 with AI-powered on-the-fly replanning capabilities. *Sustainability, 15*, 6251. https://doi.org/10.3390/su15076251

Blockchain and Digital Twin

Durga Vinay Balla, Sravya Sri Kadiyala, and Nanda Kiran Kante

Introduction

Blockchain: Revolutionizing Trust and Security

The advent of Blockchain technology has ushered in a transformative era, reshaping diverse industries and traditional systems [1]. At its core, Blockchain serves as a digital ledger, meticulously tracking data transactions among the participants or computers constituting the Blockchain network. These transactions encompass a broad spectrum of information, including financial transactions, contracts, assets, identities, and various other digital entities. Within the network, each node, representing a computer, maintains its own copy of the information, necessitating collective validation for any updates to occur.

Notably, the surge of interest in Blockchain technology since 2014 has been partly fueled by financial institutions' curiosity and engagement with Cryptocurrencies. The result is a series of pilot projects that focus on cross-border payments and settlements, securities trading, capital lending, identity management, and other compelling use cases. As of June 2023, there exists a staggering 8832 active Cryptocurrencies worldwide, illustrating the expansive growth of this field [1]. Forecasts suggest that global expenditures on Blockchain solutions are anticipated to reach $19 billion by 2024 [2].

When a transaction is initiated between any two participants, it is broadcasted to a distributed network of nodes, encompassing computers, responsible for its validation through a consensus mechanism. Once a transaction is verified, it is attached to

---

D. V. Balla (✉) · S. S. Kadiyala · N. K. Kante
SRM University—AP, Neerukonda, Mangalagiri Mandal, Guntur District,
Andhra Pradesh, India
e-mail: durgavinay_balla@srmap.edu.in; sravyasri_kadiyala@srmap.edu.in; nandakiran_kante@srmap.edu.in

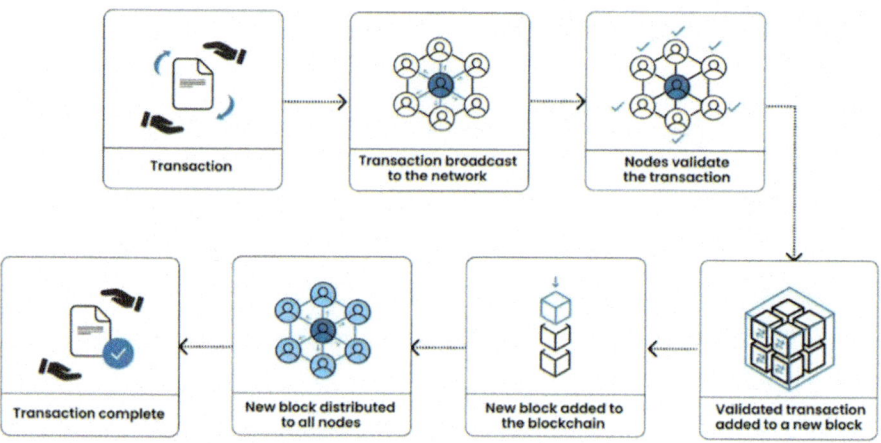

Fig. 1 The process of Blockchain

the end of the existing "chain" as a new "block," which is then distributed to all other nodes within the network (see Fig. 1).

The blocks are meticulously created and linked to previous ones using cryptographic techniques. The most well-known consensus mechanism used in Blockchain is "proof of work" (PoW), which depends on the computing or processing power of nodes, often referred to as "miners," who endeavor to solve intricate mathematical puzzles with utmost speed and efficiency.

The realm of Blockchain encompasses a diverse array of Blockchains, each possessing distinct functionalities and architectural characteristics. Blockchains can be classified as either "open" or "public" when they allow anyone to read and access their contents. Conversely, "closed" or "private" Blockchains restrict access solely to authorized entities. Furthermore, these Blockchains can be further classified as "permissionless" if anyone can send and validate transactions, or "permissioned" if entities must acquire authorization to execute or validate transactions, or both. This classification framework enables the selection of the most suitable Blockchain type for a particular solution's requirements. In contrast to traditional databases governed by a central authority, Blockchain technology operates on a distributed network, wherein each node or participant retains a full replica of the entire ledger. This distributed architecture guarantees that no individual entity possesses absolute dominance over the system, significantly mitigating the risk of targeted attacks. Therefore, the lack of a singular point of failure enhances the resilience of the Blockchain system significantly.

The ledger, commonly referred to as the Blockchain, is accessible to all participants or a predefined set of participants, granting them the ability to observe all transactions ever executed across the network. This transparency in transaction history enhances auditability and fosters trust within the network. A block consists of a header and a body. The header includes the hash of the previous block, a timestamp, Nonce, and the Merkle root (a unique fixed-size alphanumeric representation of data generated through a cryptographic algorithm).

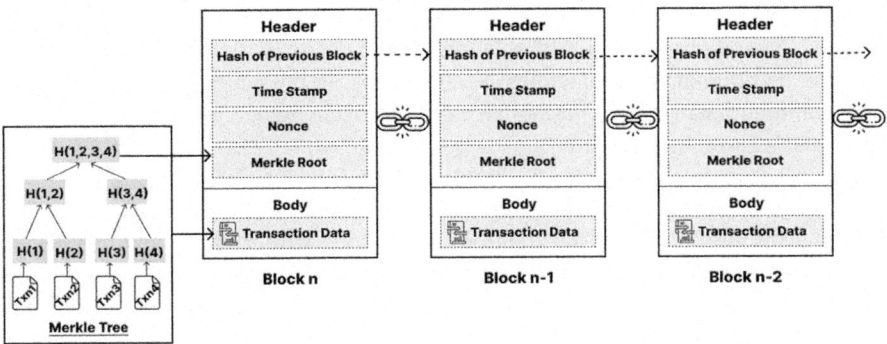

Fig. 2 The structure of Blockchain

The Merkle root serves as the root hash of a Merkle tree, residing in the block body alongside the transaction data. The root hash holds the hashed value of all the valid transactions within the same block as shown in Fig. 2.

One remarkable characteristic of Blockchains is their robust resistance to tampering. Altering or erasing transaction records is a formidable challenge, barring exceptional circumstances such as a "51% attack" wherein an entity gains control over more than 50% of the network's mining power. Any attempt to modify a block necessitates modifying all subsequent blocks, an arduous task demanding immense computational power. Moreover, such modifications are transparent to all participants, ensuring the Blockchain's high level of security. The utilization of public-private keys or cryptographic signatures serves as an additional layer of assurance, guaranteeing the integrity and authentication of transactions and rendering the records virtually impervious to tampering.

In the realm of Blockchain, all transactions bear a timestamp, imbuing the technology with valuable chronological information enhancing its utility as a reliable source for establishing transaction chronology and historical data integrity. Another significant contribution of Blockchain technology lies in its decentralized services, notably exemplified by smart contracts [3]. Smart contracts represent self-executing computer programs capable of autonomously executing terms of agreements stored on the Blockchain without human intervention. This automation feature streamlines and accelerates transaction processes while minimizing the potential for errors or manipulation amplifying the efficiency and reliability of Blockchain-based operations.

Digital Twin

The concept of Digital Twin encompasses the representation of a digital asset's anatomy in a virtual domain, mirroring the physical phenomena occurring in the physical space. It is an advanced system that synchronizes the digital and physical realms, acquiring cognitive knowledge about the physical environment in the process [4]. The central purpose of a Digital Twin involves the interaction between physical and digital representations. With a focus on the entire life cycle, it leverages a unified data source to establish a bidirectional connection between physical space and information space.

The construction of a Digital Twin relies on the generation of physical and functional models during the design phase, which continually enhance its integrity and accuracy through data and information interactions with physical entities during subsequent manufacturing and utilization phases. Ultimately, the Digital Twin achieves a comprehensive and concise portrayal of the associated physical entity (Fig. 3).

In the digital era, Digital Twin technology has emerged as a transformative force. A key advantage of Digital Twins lies in their ability to receive continuous updates from their physical counterparts through sensors or IoT devices. This data can be harnessed for a multitude of purposes, including real-time assessments, diagnostics, and future predictions. Experiments can now be conducted directly on the Digital Twin, eliminating the need for physical replication and mitigating resource consumption. Traditional trial-and-error approaches and chaotic experiments can now be readily conducted using Digital Twins, circumventing the limitations associated with physical replicas. As a result, Digital Twins enable clever, cost-effective, and accelerated innovation, transforming the landscape of technological advancement. By empowering three critical drivers—real-time assessment, diagnostics, and innovative experimentation—Digital Twins are reshaping the innovation paradigm. Figure 4 [5] provides insights into the high-level perspective of Digital Twin.

- ***Continuous evaluation***—Sensors affixed to the product continuously update its Digital Twin over its lifespan. Similarly, vehicles incorporate numerous sensors to capture operational data from essential components. This enables real-time data transmission to the vehicle's Digital Twin, facilitating prompt processing and analysis for actionable insights.

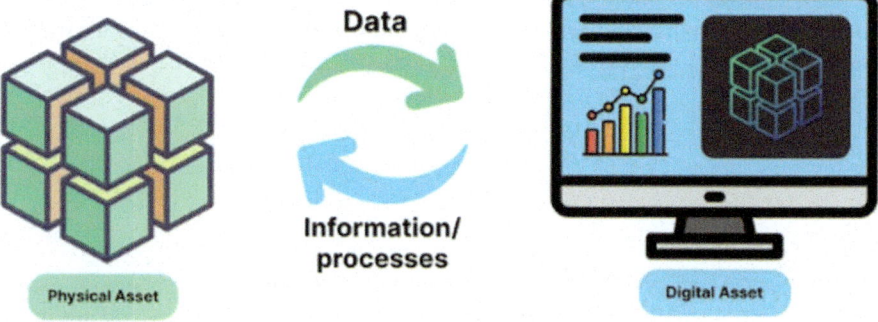

Fig. 3 Mirroring the physical and virtual spaces

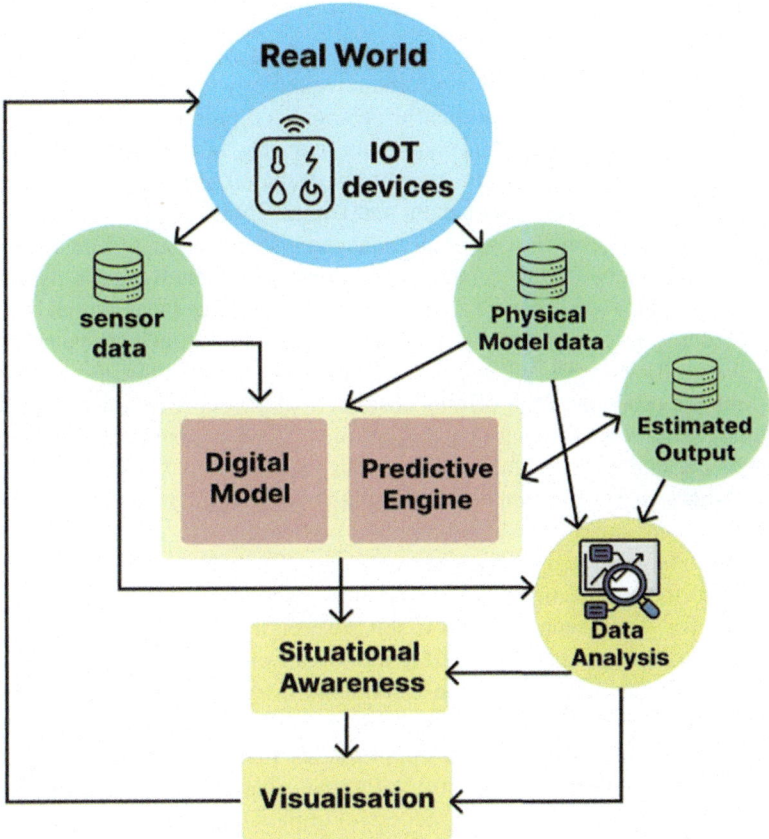

Fig. 4 High-level component view of a digital twin

- *Faster and cheaper prototyping*—Digital Twins enable cost-effective and rapid prototyping. Oklahoma University's Digital Twin increased the efficiency of an aerosol drug, targeting tumor cells from 20% to 90%. Such case studies demonstrate the strategic value of Digital Twin implementation.
- *Uninhibited innovation*—Digital Twin technology has ushered in a new era of innovation, with companies embracing its potential for experiential and experimental breakthroughs. By harnessing Digital Twins, the ability to predict and manage traffic congestion in specific locations has become a tangible reality. As Digital Twin technology continues to gain traction, its transformative potential spans industries, ensuring seamless integration across instrumented, implemented, and operated systems [6].

Literature Review

The advent of Digital Twins and Blockchain technology has revolutionized several sectors, offering innovative solutions for personalized production, smart city management, IoT integration, and the construction industry. In this literature review, we aim to provide an overview of the current research and highlight the implications and prospects of integrating Blockchain and Digital Twins.

The authors in [7] proposed a data management method based on Blockchain and peer-to-peer networks to enhance data sharing efficiency among participants involved in Digital Twin creation. The study in [8] examines the potential of Digital Twins in the field of personalized production. It presents a framework that combines Digital Twins, Blockchain, and additive manufacturing to meet the growing demand for customized products. By involving customers in the entire product lifecycle and leveraging Digital Twins, this framework enables seamless customization. However, challenges pertaining to this framework are also discussed.

The authors in [9] presented benefits with the integration of Blockchain and Digital Twins in smart city environments, addressing the management of vast amounts of data generated by these technologies. The study emphasizes the use of Digital Twins to simulate urban scenarios and provide efficient services. Leveraging Blockchain for data security offers a self-sovereign identity model that ensures higher authentication levels compared to traditional smart city models. The study in [10] emphasizes the need for the construction industry to embrace digitization and automation to keep pace with Industry Revolution 4.0. The Methodology of the paper discusses the challenges and opportunities presented by technologies like augmented reality (AR), virtual reality (VR), and mixed reality (MR), emphasizing the importance of social factors for successful implementation.

The study in [11] addresses challenges in the building industry focusing on leveraging Blockchain and Digital Twins to improve data management throughout the building lifecycle. The adoption of Digital Twins enables efficient communication and real-time observation of actual assets. Key issues such as low productivity, delayed payments, and lack of trust among stakeholders are identified. The authors of [12] proposed Blockchain-based creation process of DTs that guarantee the secure and trusted traceability, accessibility, and immutability of transactions, logs, and data provenance using smart contracts to govern and track transactions initiated by participants involved in the creation of DTs along with decentralized storage of interplanetary file systems to store and share DTs data. The authors in [13] address challenges in the BECOM industry 4.0, the literature of the paper identifies gaps in adopting Blockchain technology for Digital Twins. It proposes a technological framework called the Decentralized Digital Twin Cycle (DDTC) to bridge these gaps.

This chapter delves into the integration of Blockchain and Digital Twin technology, exploring various aspects in distinct sections. Section "Integration of Blockchain and Digital Twin" provides insights into the integration process itself, highlighting the intricate relationship between Blockchain and Digital Twin. Section "Challenges" delves into the challenges associated with this integration, addressing the complexities that must be navigated. To showcase the practical applications of this fusion, Section "Real-Time Applications/Results" delves into real-time implementations across various sectors. Looking ahead, Section "Future Research" outlines the future research prospects, opening doors for further exploration. Finally, Section "Conclusion" concludes this chapter, summarizing the key findings and implications. Capping off this scholarly discourse, a comprehensive list of references is provided to guide readers toward additional resources.

Integration of Blockchain and Digital Twin

The process of creating and deploying a Digital Twin entails complexity, involving multiple entities, data transmissions, and processes. This complexity results in the generation of vast and diverse product data, thereby giving rise to challenges related to data access control, data authenticity, data storage, and data traceability. Efficient and secure management of these data-related issues is crucial to ensure the integrity and reliability of Digital Twin deployments. Traditional technologies that rely on centralized storage systems fall short in addressing these concerns, as they expose vulnerabilities to unauthorized access and single points of failure. Moreover, they lack the necessary traceability and trusted data sources.

To address these challenges and secure Digital Twins, one can harness the transformative capabilities of Blockchain technology. Blockchain technology offers inherent properties such as decentralization, security, traceability, transparency, and immutability, which can effectively address the security constraints faced by Digital Twins [1, 6, 14]. By employing Blockchain, the Digital Twin ecosystem can be elevated to new heights, enabling comprehensive data management within the intricate network. The integration of Blockchain technology empowers Digital Twins to operate efficiently and securely, instilling trust and confidence in the management of data assets (Fig. 6).

Data Management

The advent of Digital Twin technology has revolutionized the monitoring and optimization of product activities throughout their entire lifecycle. The product lifecycle encompasses various stages such as design, manufacturing, and maintenance, resulting in the generation of vast and diverse data known as product lifecycle data. This data plays a crucial role in maintaining the Digital Twin's accuracy and vitality. Continuous updates are required to ensure that the Digital Twin remains synchronized with its physical counterpart, enabling performance monitoring, optimization, maintenance scheduling, and more. However, inaccurate data can lead to flawed analysis and decision-making, resulting in undesired outcomes. Therefore, effective data management is paramount for the successful implementation of a product's

Digital Twin. Researchers acknowledge Blockchain technology's potential for enhancing data management of digital assets, providing advantages in terms of data storage, security, and sharing [7, 15].

The fundamental concept of Blockchain technology revolves around the notion of a block. A Blockchain consists of a series of authenticated blocks organized chronologically according to the adopted consensus algorithm. Every block incorporates a cryptographic hash of the previous block, forming a chain-like structure. Additionally, blocks include transaction data, a root hash, and a timestamp. Modifying any block is extremely difficult due to its connection with its subsequent blocks, ensuring integrity and tamper-proof records. Introduction of timestamps enables traceability of block creation time. Transactions are central to data recording and sharing. Participants possess public and private keys for encryption and decryption, respectively. This process guarantees data authenticity and security, and the keys can help track the origin. Only authorized participants can initiate transactions ensuring origin tracking. Changes in the product lifecycle trigger transactions, which are immutable once verified. Even a minute change in the data creates a new transaction and will be broadcast to all the participants of the network.

Digital Twins Identity and Legitimacy: Digital twins serve as digital certificates tailored for the digital era, offering a secure and perpetual means of storing and issuing certificates. Leveraging Blockchain technology, these digital certificates become immune to theft, tampering, and misuse, enabling businesses to safeguard information about their products effectively [3]. Moreover, by utilizing the Blockchain network, transaction data associated with these products can also be securely stored, allowing potential buyers to access comprehensive information about the product's origin and its journey, including details such as the genuine manufacturer and previous owners. This transparency in the purchase process serves as a powerful tool in combating counterfeit products within the market. It establishes a reliable proof of authenticity and identity for products, instilling a much-needed sense of trust and credibility in business operations [16].

Data Sharing and Data access control

The product lifecycle involves various participants, such as product designers and manufacturers, each consisting of one or more individuals [7]. Effective communication among these diverse participant types, including intra-participant communication, is crucial but can be challenging due to the complex nature of the network. For improved communication efficiency, one can utilize a decentralized peer-to-peer network of computers, enabling resource sharing and communication without relying on a centralized server. Participants join the network after validation to ensure the exclusion of nodes that may pose a threat to the network's integrity. Participants can freely exit the network when needed. The lifecycle data of a product gets recorded as transactions and is stored in dedicated blocks, which are subsequently linked together to create a Blockchain. By distributing copies of the Blockchain to eligible participants, data authenticity can be maintained, and data fraud can be prevented.

The Blockchain relies on a peer-to-peer network as its foundation. To optimize the storage and sharing of large files, the integration of IPFS (Interplanetary File System) with Blockchain can be employed [7]. Storing data on IPFS is more cost-effective compared to storing it directly on-chain. IPFS ensures data integrity through the generation of a unique IPFS hash for each uploaded file, which is stored alongside the Digital Twin's information and data on IPFS.

These distinct hashes are also embedded in smart contracts, autonomous programs stored on the Blockchain and triggered when predefined conditions are fulfilled [3, 12]. A consortium Blockchain system, where public and private Blockchains coexist, can be employed to manage different transaction data circulation within specific interest groups or make data public to all participants. This decentralized approach allows for differentiated data sharing and data access control schemes to be implemented effectively (Fig. 5).

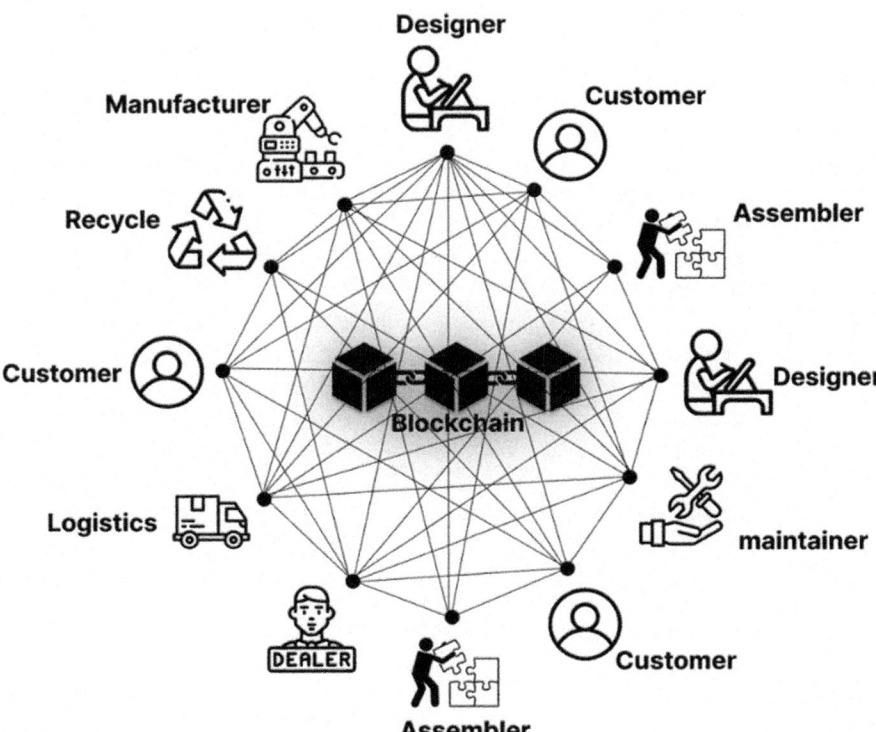

Fig. 5 Peer-to-peer network for data management

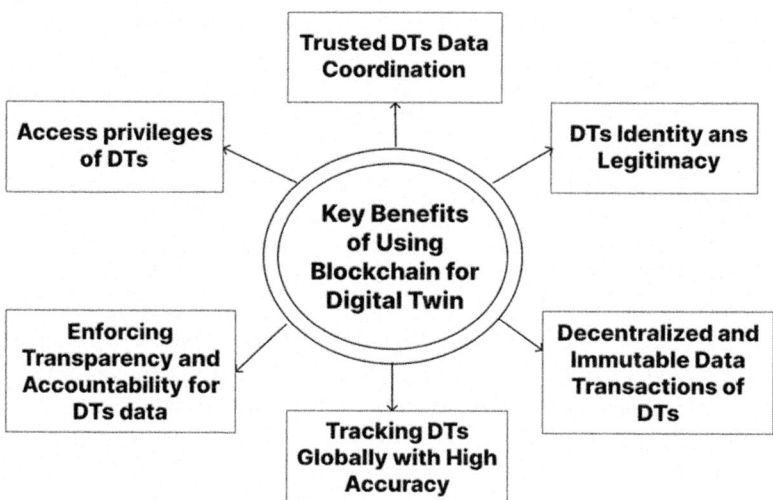

Fig. 6 Key benefits of integration of Blockchain and Digital Twin

Smart Contracts and Automation

With the expansion of the Digital Twin system, it becomes imperative to establish an autonomous data management mechanism to ensure efficient and optimized operations within the Blockchain-enabled Digital Twin. To achieve this, data transactions initiated by network participants and access control can be autonomously executed with the support of smart contracts. Smart contracts are protocols within Blockchain networks that enable the autonomous proposal, negotiation, and execution of contracts.

The inclusion of smart contracts brings substantial convenience to the departments involved in the Digital Twin creation process, particularly in decision-making regarding future material flow. For instance, in the validation of product design feasibility, different departments can sign the smart contract to indicate their approval. After obtaining signatures from all pertinent departments, such as Designer 1 and Designer 2, the smart contract enables the automatic transmission of the product design to the manufacturer for production [7]. Furthermore, as mentioned earlier, smart contracts play a role in ensuring the integrity of data stored using IPFS.

Challenges

The integration of Blockchain and Digital Twins presents various challenges that need to be addressed for successful implementation. To gain a comprehensive understanding of these challenges, the PESTELS approach is employed. This approach allows for the analysis of macro-environmental factors that impact the technological fusion of Blockchain and Digital Twins. Among the PESTELS factors, social factors are deemed highly influential and critical, while political factors have relatively minimal influence.

Analyzing Challenges Through the PESTELS Approach

The PESTELS approach is applied to examine the challenges associated with the incorporation of Blockchain and Digital Twin technologies. Each factor is explored to shed light on the specific challenges within the technological fusion.

Political Factors:
Political factors are found to have a minor influence on the challenges of integrating Blockchain and Digital Twins. These factors may include government regulations, policies, and standards that impact the implementation and adoption of these technologies.

Economic Factors:
The economic aspects related to the fusion of Blockchain and Digital Twins are analyzed to identify challenges such as the cost of implementation, financial feasibility, and return on investment. Economic factors play a significant role in determining the viability and sustainability of the integration.

Social Factors:
Social factors emerge as the most influential and critical challenges in the implementation of Blockchain and Digital Twin fusion. These factors encompass social acceptance, user adoption, cultural norms, workforce readiness, and the impact on labor dynamics. Addressing these social challenges is crucial for ensuring successful implementation and widespread acceptance of the integrated technologies.

Technological Factors:
Technological challenges encompass the compatibility, scalability, interoperability, and complexity of integrating Blockchain and Digital Twins. These factors may include the need for robust infrastructure, data management capabilities, and technical expertise to effectively merge the technologies.

Environmental Factors:
Environmental factors refer to the impact of the integration on sustainability, energy consumption, and environmental footprint. Ensuring that the fusion of Blockchain and Digital Twins aligns with environmental goals and promotes responsible resource utilization is essential.

Legal Factors:
Legal challenges include regulatory frameworks, intellectual property rights, data protection, and privacy concerns. Adhering to legal requirements and ensuring compliance in the integration process are vital for building trust and mitigating risks.

Security Factors:
Security challenges pertain to data integrity, cybersecurity, and protection against unauthorized access or malicious activities. Implementing robust security measures and encryption protocols is crucial to safeguard the integrated system and prevent data breaches.

Technological Challenges

Scalability is the primary issue in terms of Blockchain. The size of Blockchain is a topic to be discussed. It is increasing enormously and today it is around 120 GB [11]. There is a considerable delay in the transactions due to the size of the chain and the blocks. Suppose we check on the existing applications of Blockchain like the Bitcoin network. A few observed challenges are the reduction of blocks that can eliminate Blockchain. This is possible in case of a global blackout of the internet. Integrating Blockchain and Digital Twin requires alignment of protocols, data formats, and APIs for interoperability. Efficient synchronization mechanisms are required for smooth data transmission. The integration incurs computational overhead due to complex operations. Compatibility between smart contracts and Digital Twins should be verified.

Legal Challenges

Incorporating Digital Twins requires adherence to data protection regulations like GDPR (General Data Protection Regulation) to ensure privacy and proper data management. Legal validity and rights management for intellectual property and data access among multiple parties are crucial. Accountability for financial losses resulting from transmission or disruptions should be assigned. International usage may pose challenges due to jurisdictional differences and strict data transmission laws.

Security Challenges

Individually, Digital Twins have limitations in terms of security. To overcome this, we are here to integrate Blockchain technology. But the overall integration process also faces some security challenges. Ensuring the data fed to Digital Twins is secure. That is securing data transmission from physical entities to Digital Twin is crucial. Monitoring and ensuring data integrity during the process is prominent to prevent manipulation. Access level management and network participation can be achieved through different architectures, allowing control over who can access and participate in the networks. Proper design and implementation of smart contracts is crucial to avoid contract failures and malicious attacks. Monitoring the security of the consensus algorithm is important to prevent breaches.

Additionally, there are challenges that need to be addressed (see Fig. 7).

Though the emergence of private Blockchains provides security, it deviates from the intention of a fully decentralized concept and therefore raises concerns about its usage for big companies or banks. Also, smart contracts aid in secure transactions and automation, but, there is a potential loss of human rights protection caused as the result of removing third parties.

Authors of [17] added that there are two kinds of challenges, one kind requires the passage of time so that this technology gets adapted and imbibed into people, and adequate skilled developers, and operators are forged. This kind also includes

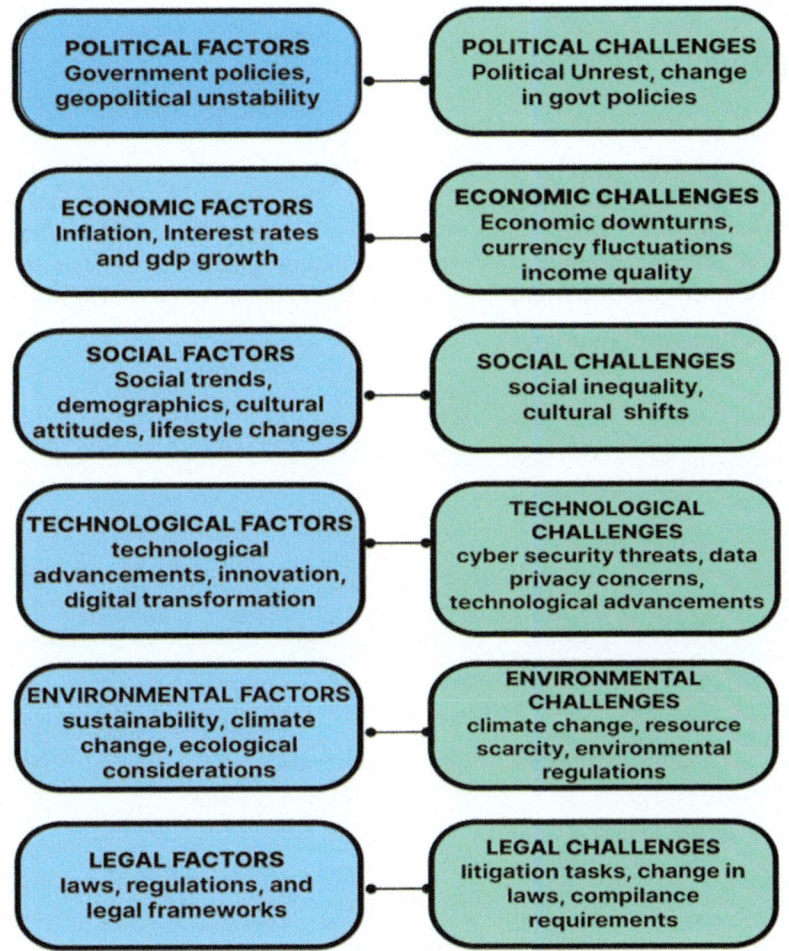

Fig. 7 Challenges in the integration of blockchain and digital twins

acknowledgment and trust in the public. The second kind of challenge would be regarding the inherited technical features and technologies including data privacy, scalability, and energy consumption. One of the critical issues is the threat of quantum attacks [11].

There is a lack of international standard definitions and conceptualizations of the technologies including Digital Twin technology and Blockchain technology. If they were made in that way, then it promotes shared understanding. Most importantly, it is essential for researchers, policymakers, designers, and also industry professionals to prioritize empathy, privacy, and ethics which are currently lacking in smart-built environments.

Real-Time Applications/Results

Supply Chain Management

By recording the lifecycle events of Digital Twins of the physical products or assets on the Blockchain, supply chain stakeholders gain real-time visibility into the movement, location, and condition of goods. This integration allows for precise tracking and tracing of products throughout the supply chain and notifies back whenever attention is required or whenever there is an issue as the data from the sensors connected to the Digital Twin assess data continuously as in Fig. 8 [18, 19].

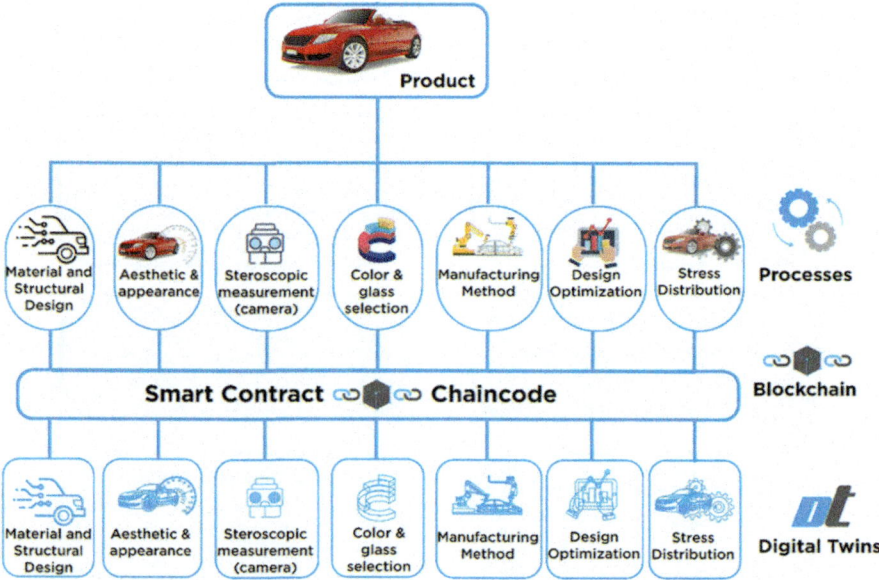

Fig. 8 Blockchain-based Digital Twins in manufacturing

Additionally, it facilitates the verification of product authenticity by securely storing product information and transaction history on the Blockchain, which is immutable. This transparency enhances trust among stakeholders, streamlines supply chain operations, reduces counterfeit products, and improves overall supply chain efficiency. Blockchain-based access control avoids unauthorized data manipulation and only lets authorized supply chain parties have access privileges to view, modify, and simulate the DT data and keeps it secure. On the other hand, Digital twin and Blockchain are combined together to ensure that counterfeit products are identified by providing product authenticity. For instance, there are a great number of replicas of luxury products with fake documents to prove their authenticity [6]. Nevertheless, DTs can act as digital certificates stored on the Blockchain to make them unable to be tampered with or stolen [4].

Manufacturing

Combining manufacturing with Blockchain has the potential to improve manufacturing sector, facilitating real-time monitoring of manufacturing processes, production control, and design changes and enhancing trust and security. Blockchain can be used to create a secure and transparent platform for collaboration between different stakeholders in the manufacturing chain [20]. This can help to improve communication and coordination, which can lead to faster product development and delivery. Digital twins enable engineers to enhance asset performance and quality of manufacturing machines. Also, applying advanced analytics and visualization tools to the data of Digital Twins enhances efficiency in manufacturing asset management. When integrated, these technologies offer benefits such as enhanced supply chain traceability, improved quality assurance, protection of intellectual property, streamlined maintenance and servicing, comprehensive product lifecycle management, and automation of contractual agreements. Digital Twins often contain valuable intellectual property (IP) information, such as design specifications and manufacturing processes. By storing this IP on the Blockchain, manufacturers can establish proof of ownership and protect their innovations from unauthorized access, tampering, or theft. Also, the increasing disturbances in manufacturing phase call for intelligent dynamic production planning. Digital Twin makes it possible for a globally optimized production plan according to real-time status change. The impact of both these in different phases of manufacturing can be observed as follows:

- ***Design***—The virtual replica of the physical product is created to capture and simulate the behavior of the physical counterpart in this phase. Also, the Digital Twin's parameters and specifications, including its structure, connectivity, and information flow, are defined.
- ***Build phase***—In this phase, development of hardware and software components that enable this integration is done. A Digital Twin is built using simulation and modeling techniques, and a Blockchain is set up for sharing and transaction recording.

Test Phase—In this phase, integrated system is tested, and the accuracy of the Digital Twin, the security and reliability of the Blockchain, and the overall functionality, performance, and interoperability of the integrated system are tested.

Deliver and Manage Phase—The integrated system is deployed into the manufacturing environment in this phase. Accurate monitoring and analysis take place.

Smart Cities

Smart Cities are built with IoT, CCTV, drones, Intelligent transportation, autonomous vehicles, and diverse technologies of Industry 4.0. They aim to optimize the functions of cities and boost economic growth with better quality of life using smart technologies [9]. Now, there are enormous amounts of data generated by the usage of all these technologies, and hence management of this data becomes crucial and also urges to provide security to this data. To improve service quality while solving social problems, interoperability and coordination among different components are prominent.

In the making of a smart city, Blockchain can be used as a framework for the connection between services and technologies. In the energy sector, Blockchain enables direct transactions between energy producers and consumers, eliminating intermediaries, and thus reducing cost. Energy transactions can be tracked. They can also improve self-sovereignty among individuals. Analysis of urban situations with Digital Twin simulation, of the data regarding cities, like terrain, transportation, and social problems can be done. Precautions can be taken as unhappening conditions can be observed through Digital Twins.

Healthcare and Medical Records

Healthcare is the most effective industry that has benefited from using Digital Twins principles. DTs technologies are availed in the health industry to estimate and evaluate the operation of examination equipment and then recommend medication, recognize lifestyle changes, enhance hospital activities, provide remote surgery, and assist governments with patient healthcare. Inpatient care stresses that the development of individual DT and information derived from it may be utilized in rehabilitation, augmentation, dietary behaviors, and illness prediction [18]. Blockchain-based Digital Twins enhance the security, privacy, and accessibility of medical records, enabling real-time sharing of patient data [21, 22]. Patient information, including medical history, test results, and treatment plans, can be securely stored on the Blockchain. This integration provides a thorough and up-to-date view of a patient's health status, enabling healthcare providers to make better decisions from the available health status and provide personalized treatments. Additionally, Blockchain ensures data integrity and privacy, allowing patients to have control over their own health data and grant access to authorized parties. Real-time sharing of medical records reduces redundancies, improves care coordination, and enhances patient outcomes.

Digital Identity and Authentication

Blockchain and Digital Twins enhance real-time identity management and authentication systems. Digital twins represent individuals or entities, while Blockchain ensures identity records' integrity, security, and immutability. This integration enables secure and decentralized identity verification, reducing the risk of identity fraud and enhancing user privacy. Blockchain's transparency allows for efficient and reliable authentication processes without relying on centralized authorities. Real-time updates to digital identities enable seamless and secure access to services and platforms, while users maintain control over their personal information.

Energy Grid Optimization

Digital twins representing energy infrastructure, integrated with Blockchain technology, support real-time energy management. By using Blockchain's decentralized consensus and smart contracts, energy transactions can be automated and optimized based on fluctuating supply and demand conditions. This integration enables peer-to-peer energy trading, where excess energy produced by one party can be directly sold to another party in need. Blockchain ensures transparency and trust in these transactions, eliminating the need for intermediaries and reducing transaction costs. Additionally, it facilitates efficient grid balancing by optimizing energy distribution and storage, allowing for the integration of renewable energy sources into the grid.

By combining Blockchain and Digital Twin technology, these applications offer extensive benefits, including real-time data exchange, enhanced visibility, transparency, security, automation, improved decision-making, operational efficiency, and trust among stakeholders.

Future Research

Blockchain's issue of scalability exists. There are a few potential solutions like an increase in the block size in a Blockchain, and partitioning of databases into several smaller ones also termed sharing. But they require further research in order to shrink this problem. There is a significant scope of research in the standardization of Digital Twins. A lack of standardized frameworks and protocols raises collaboration, interoperability, and data exchange challenges. Standardization can unlock the full potential of Digital Twins. Exploration of strategies for fostering collaboration, establishing Industry standards and guidelines, to facilitate knowledge sharing and innovation in the context of IR 4.0, especially in construction Industry, should be pursued.

Assessing the current level of technology adoption and maturity is a research scope. This can include the progress and impact of different technologies on integrating Blockchain and Digital Twins like AR, VR, and Metaverse. Understanding this impact can guide further effective implementation. Addressing implementation challenges is also vital, despite the accessibility to Industry 4.0. There are areas where there are challenges in the implementation process. Developing strategies for successful implementation and resistance to change is required.

There are many cryptographic algorithms that are working really well and ensuring the safety factory of Blockchain. But it is essential to note that the widely used cryptographic algorithms are vulnerable to quantum attacks. They are basically done through quantum computers, though in practice, they are not yet available, this is a significant threat. The development of quantum-resistant cryptographic algorithms is salient. There are currently synergies not only in integrating Blockchain and Digital Twin technologies but also in any kind of incorporation among technologies. This can be overcome through further research on making Unified System Architecture for these technologies.

Giving theoretical implications and giving them stronger bases and backgrounds is definitely the prominent foundation and the primary step to be taken, which is being done magnificently by the researchers. But, the research on implementing these technologies needs to be enhanced. As we expect people to adapt to these technologies, it is crucial to understand the necessity of user-friendly interfaces and the need for providing better technical inputs for a finer experience that could be improved and should be worked on.

There is a requirement for research regarding legal challenges that arise in integration. Individually too these technologies have challenges that are yet to be resolved. But, research on this area would provide a better understanding of what can be done. Considering the challenges and making further research on them, could lead to the discovery of fruitful solutions and provide implications.

Conclusion

In conclusion, this chapter highlights the critical need for the security system, and protection mechanism for the vulnerable aspect of a Digital Twin and this void is immaculately filled by the Blockchain technology. This chapter began by providing deep insights of Blockchain and digital technologies and their fundamental characteristics. Further, the crucial aspects of the integration such as data management, data sharing, access control, smart contracts, and automation are discussed. Additionally, literature review showcases existing research on different aspects of both these technologies, exploring individual strengths as well as potential synergies that arise from their integration. In the process of incorporating them, the challenges that occur were mentioned employing the PESTELS approach. The real-time applications across diverse domains further illustrate the transformative power they bring together. Finally, it concluded by shedding light on the future research opportunities in this domain that are yet to be addressed. By focusing on the paramount aspect of security, this chapter contributes to the advancement of Digital Twin technology highlighting the benefits from its integration with Blockchain.

References

1. Swan, M. (2015). *Blockchain: Blueprint for a new economy*. O'Reilly Media, Inc.
2. Statista. (2021). Statista Research Department. [Online]. Available: https://www.statista.com/statistics/800426/worldwide-blockchain-solutions-spending/
3. Khan, S., Loukil, F., & Ghedira-Guegan, C. (2021). Blockchain smart contracts: Applications, challenges, and future trends. *Peer-to-Peer Networking and Applications, 14*, 2901–2925.
4. Liu, J., Yeoh, W., & Gao, L. (2022). *Blockchain-based digital twin for supply chain management*. Cornell University.
5. Stanford-Clark, A., Frank-Schultz, E., & Harris, M. (2019, November). IBM Developer. [Online]. Available: https://developer.ibm.com/articles/what-are-digital-twins/
6. Pethuru, R. (2020). Empowering digital twins with blockchain. In *Advances in computers*. Elsevier.
7. Huang, S., Wang, G., Yan, Y., & Fang, X. (2020). Blockchain-based data management for digital twin of product. *Journal of Manufacturing Systems, 54*, 361–371.
8. Guo, D., Ling, S., Li, H., Ao, D., Zhang, T., Rong, Y., & Huang, G. Q. (2020). A framework for personalized production based on digital twin, blockchain and additive manufacturing in the context of Industry 4.0. https://doi.org/10.13140/RG.2.2.13587.71202
9. Song, Y., & Hong, S. (2021). Build a secure smart city by using blockchain and digital twin. *International Journal of Advanced Science and Convergence, 3*(3), 9–13.
10. Alaloul, W. S., Liew, M. S., Zawawi, N. A. W. A., & Kennedy, I. B. (2020). Industrial Revolution 4.0 in the construction industry: Challenges and opportunities for stakeholders. *Ain Shams Engineering Journal, 11*(1), 225–230.
11. Sadri, H., Yitmen, I., Tagliabue, L. C., Westphal, F., Tezel, A., Taheri, A., & Sibenik, G. (2023). Integration of blockchain and digital twins in the smart built environment adopting disruptive technologies—A systematic review. *Sustainability, 15*(4), 3713.
12. Hasan, H. R., Salah, K., Jayaraman, R., Omar, M., Yaqoob, I., Pesic, S., & Boscovic, D. (2020). A blockchain-based approach for the creation of digital twins. *IEEE Access, 8*, 34113–34126.
13. Teisserenc, B., & Sepasgozar, S. (2021). Adoption of blockchain technology through digital twins in the construction industry 4.0: A PESTELS approach. *Buildings, 11*(12), 670.
14. Nakamoto, S. (2008). *Bitcoin: A peer-to-peer electronic cash system* [Online]. Available: https://bitcoin.org/bitcoin.pdf
15. Chen, J., Lv, Z., & Song, H. (2019). Design of personnel big data management system based on blockchain. *Future Generation Computer Systems, 101*, 1122–1129.
16. Jadhav, R., Shaikh, A., Jawale, M. A., Pawar, A. B., & William, P. (2022). System for identifying fake product using blockchain technology. In *2022 7th International conference on communication and electronics systems*. IEEE.
17. Gračanin, D., Lasisi, R. O., Azab, M., & Eltoweissy, M. (2019). Next generation smart built environments: The fusion of empathy, privacy and ethics. In *2019 First IEEE international conference on trust, privacy and security in intelligent systems and applications (TPS-ISA)*. IEEE.

18. Hemdan, E. E. D., El-Shafai, W., & Sayed, A. (2023). Integrating digital twins with IoT-based blockchain: Concept, architecture, challenges, and future scope. *Wireless Personal Communications, 131*, 1–24.
19. Yaqoob, I., Salah, K., Uddin, M., Jayaraman, R., Omar, M., & Imran, M. (2020). Blockchain for digital twins: Recent advances and future research challenges. *IEEE Network, 34*(5), 290–298.
20. Mandolla, C., Petruzzelli, A. M., Percoco, G., & Urbinati, A. (2019). Building a digital twin for additive manufacturing through the exploitation of blockchain: A case analysis of the aircraft industry. *Computers in Industry, 109*, 134–152.
21. Egala, B. S., Pradhan, A. K., Dey, P., Badarla, V., & Mohanty, S. P. (2023). Fortified-Chain 2.0: Intelligent blockchain for decentralized smart healthcare system. *IEEE Internet of Things Journal, 10*, 12308.
22. Egala, B., Pradhan, A., Badarla, V., & Mohanty, S. (2019). Fortified-Chain: A blockchain-based framework for security and privacy-assured internet of medical things with effective access control. *IEEE Internet of Things Journal, 8*(14), 11717–11731.

Personalize Learning Experience in Education Using Digital Twins with Human-Centered Design and Pedagogy

A. Reethika and P. Kanaga Priya

Introduction

A digital twin is an actual representation that mimics the properties and behavior of a physical thing and is produced by combining data from sensors and Internet of Things devices. The concept of a digital twin emerged from the growing availability of data, advancements in computing power, and the need to optimize and improve the performance of physical assets and systems. By creating a virtual replica of a physical entity, organizations can gain valuable insights, monitor performance, and simulate scenarios to enhance efficiency, productivity, and decision-making.

Role of Digital Twin

Design and Development: Digital twins, which are utilized during the design and development phase of simulate and optimize performance, identify potential issues and test different scenarios before physical implementation. This helps in reducing costs, accelerating time-to-market, and improving overall product quality.

A. Reethika (✉)
Department of Electronics and Communication Engineering, Sri Ramakrishna Engineering College, Coimbatore, Tamil Nadu, India

P. Kanaga Priya
Department of Computer Science and Engineering, KPR Institute of Engineering and Technology, Coimbatore, Tamil Nadu, India

© The Author(s), under exclusive license to Springer Nature Switzerland AG 2024
A. Mishra et al. (eds.), *Transforming Industry using Digital Twin Technology*,
https://doi.org/10.1007/978-3-031-58523-4_9

Monitoring and Maintenance: Digital twins enable continuous monitoring of real-time data from sensors embedded in physical assets. By comparing the data collected from the physical asset with its digital twin, organizations can identify anomalies, predict failures, schedule maintenance proactively, and optimize asset performance.

Remote Operation and Control: Digital twins can be accessed and operated remotely, allowing organizations to monitor and control physical assets and systems from a centralized location. This is particularly useful in scenarios where physical access is limited, hazardous, or expensive.

Lifecycle Optimization: Digital twins provide valuable insights throughout the entire lifecycle of a physical asset. From design and development to operation and maintenance, digital twins help optimize performance, reduce downtime, extend asset life, and support informed decision-making.

Applications of Digital Twins

Digital twins are finding applications in various industries, including manufacturing, healthcare, energy, transportation, and smart cities. They have the potential to revolutionize how organizations design, operate, and maintain their assets, leading to improved efficiency, productivity, and sustainability.

Manufacturing: Digital twins are extensively used in manufacturing to optimize production processes, monitor equipment performance, and improve quality control. By creating virtual replicas of manufacturing systems, organizations can simulate different production scenarios, identify bottlenecks, and optimize resource allocation to enhance productivity and efficiency.

Healthcare: In healthcare, digital twins can be created to represent individual patients or entire healthcare systems. They can aid in personalized medicine, allowing doctors to simulate treatment options, predict outcomes, and optimize therapies based on a patient's specific characteristics. Digital twins can also be used to simulate and optimize the design and operation of healthcare facilities.

Energy and Utilities: Digital twins play a crucial role in optimizing energy generation, transmission, and distribution. They can be used to model power plants, electrical grids, and renewable energy systems. By monitoring real-time data from sensors, digital twins help identify inefficiencies, predict maintenance needs, and optimize energy usage for better resource management.

Smart Cities: Digital twins are employed in the development and management of smart cities. They provide a holistic view of urban infrastructure, including transportation networks, buildings, and utilities. Digital twins enable city planners and administrators to simulate different urban development scenarios, optimize traffic flow, monitor environmental conditions, and enhance overall urban sustainability.

Aerospace and Defense: In the aerospace and defense industry, digital twins are used to simulate and optimize the performance of aircraft, missiles, and other complex systems. They aid in design validation, predictive maintenance, and operational optimization. Digital twins can help identify potential issues before they occur, optimize fuel consumption, and enhance mission success rates.

Retail and Supply Chain: Digital twins find application in optimizing retail and supply chain operations. By creating virtual representations of supply chain networks, organizations can simulate and analyze different scenarios, optimize inventory management, and improve logistics efficiency. Digital twins can also be used to model customer behavior and preferences, enabling personalized marketing and improved customer experiences.

Construction and Infrastructure: Digital twins are employed in the construction industry to enhance project planning, design, and management. By creating virtual models of buildings, bridges, and infrastructure projects, organizations can simulate construction processes, identify potential clashes or errors, and optimize resource allocation. Digital twins also assist in ongoing maintenance and facility management by providing real-time data on the condition of assets.

These are just a few examples, and digital twin technology is being applied in many other industries and domains to drive innovation, improve operations, and optimize performance.

Education

Pedagogy Learning

Pedagogy learning refers to the principles, methods, and practices employed in teaching and learning processes. It focuses on the effective design and delivery of educational experiences to facilitate the acquisition of knowledge, skills, and competencies by learners. Pedagogy encompasses various instructional strategies, assessment techniques, and classroom management approaches that aim to optimize the learning experience and promote student engagement and achievement.

The goal of pedagogy is to create an inclusive and supportive learning environment where learners can actively participate, construct meaning, and develop critical thinking and problem-solving abilities. It involves understanding the needs and characteristics of learners, adapting instructional techniques to suit different learning styles, and promoting student-centered learning experiences.

Teaching and Learning Process

Curriculum Design: Pedagogy involves designing and organizing the curriculum to ensure coherence, relevance, and progression in learning. It entails selecting and sequencing learning objectives, content, and activities to facilitate the acquisition of knowledge and skills in a logical and meaningful manner.

Instructional Strategies: Pedagogy encompasses a wide range of instructional strategies that educators use to facilitate learning. These strategies may include lectures, discussions, demonstrations, group work, hands-on activities, simulations, and technology-enhanced learning, among others. The choice of instructional strategies depends on the learning objectives, content, and the needs of the learners.

Assessment and Feedback: Pedagogy involves designing effective assessment methods to measure learners' progress and provide feedback. It includes formative assessments, such as quizzes, class discussions, and projects, to monitor ongoing learning and provide feedback for improvement. Summative assessments, such as exams or presentations, are used to evaluate learners' achievement of learning outcomes.

Active and Experiential Learning: Pedagogy promotes active engagement of learners in the learning process. It emphasizes hands-on activities, problem-solving tasks, and real-world applications to foster deeper understanding and retention of knowledge. Experiential learning approaches, such as internships, field trips, and project-based learning, provide opportunities for learners to apply their knowledge and skills in authentic contexts.

Classroom Management: Pedagogy includes methods for efficiently managing the classroom to establish a constructive and efficient learning atmosphere. To create a supportive learning environment, it entails laying out clear guidelines, norms, and procedures, encouraging constructive interactions between students, efficiently managing time, and attending to both individual and group dynamics.

Continuous Professional Development: Pedagogy recognizes the importance of ongoing professional development for educators. It involves staying updated with research and best practices in teaching and learning, reflecting on teaching experiences, seeking feedback, and engaging in professional learning communities to enhance instructional strategies and promote effective learning outcomes.

Overall, pedagogy learning places learners at the center of the educational experience, emphasizing active engagement, critical thinking, and meaningful interactions. It aims to foster a love for lifelong learning, equipping learners with the necessary knowledge, skills, and competencies to succeed in a rapidly evolving world.

Digital Twin Education

Digital twins have the potential to play a transformative role in the education field by leveraging learning technologies. Here are five learning technologies that can be integrated into digital twins to enhance educational experiences:

Virtual Reality (VR) and Augmented Reality (AR): Digital twins combined with VR and AR technologies can create immersive and interactive learning environments. Students can explore virtual replicas of real-world objects, locations, or scenarios, enhancing their understanding and engagement. For example, a digital twin of a historical site can be explored using VR, allowing students to virtually visit and experience the site firsthand.

Simulation and Gamification: Digital twins can incorporate simulation and gamification elements to make learning more interactive and enjoyable. By simulating real-world processes or systems, students can apply their knowledge, experiment with different variables, and observe the outcomes. Gamification elements, such as challenges, rewards, and leaderboards, can further motivate and engage students in the learning process.

Data Analytics and Visualization: Digital twins can integrate data analytics and visualization tools to help students analyze and interpret complex data sets. By providing real-time data from the digital twin environment, students can explore patterns, trends, and relationships, enhancing their data literacy skills. Visualization techniques can be used to represent data in meaningful and accessible ways, aiding comprehension and critical thinking.

Collaborative Learning Platforms: Digital twins can be integrated into collaborative learning platforms, enabling students to work together on projects and problem-solving tasks. These platforms can facilitate communication, knowledge sharing, and collaborative decision-making within the digital twin environment. Students can engage in virtual teamwork, share ideas, and collectively analyze and solve problems.

Adaptive Learning Systems: Adaptive learning systems, which tailor the learning process to each student's requirements and progress, can be integrated with digital twins. The adaptive system may offer feedback, resources, and information that are specifically suited to each student by evaluating data about their performance, preferences, and learning styles. It may dynamically modify the learning route and speed to provide every learner with the challenges and help they need.

These learning technologies, when integrated into digital twins, can create dynamic and interactive educational experiences. They provide opportunities for active engagement, problem-solving, collaboration, and personalized learning, fostering deeper understanding and knowledge retention. Digital twins in education hold great potential to transform traditional learning approaches and prepare students for the challenges of the digital age.

Literature Review

A digital twin is typically a software model created as a customer-ordered prototype for a product. It might be applied to all phases of this ordered product's industrial production, comparing the real condition to the model and adjusting for discrepancies. During manufacturing, the "real world" will be used to construct a "virtual world," and both "worlds" will communicate with one another. The digital twins are software models of industrial facilities that are replicated, visualized, and synchronized with their industrial originals at engineering educational institutions like universities, training centers, or schools. The article that is being presented provides examples from the master's degree program at the University of Applied Sciences to illustrate the design process for digital twins [1].

It will be discussed why and how using digital twins is feasible. We will draw the paths from an original gadget to its model. It will be demonstrated how software models that have varying degrees of source code are developed at various phases [2]. It will be discussed how to integrate hardware into a software model, often known as rapid control prototyping and hardware-in-the-loop. Finally, the benefits and drawbacks of using digital twins for engineering research, including any economic implications, will be covered.

The idea of a "Digital Twin" is becoming more and more important for achieving Industry 4.0. The stages of the product lifecycle can be optimized with Digital Twin, and the industries are supported in making wise choices and finding economical business solutions [3]. Numerous pieces of literature have been written about this hot topic, and numerous implementation strategies have been created. However, the majority of current implementations fall short of the scope and specifications set forth for a digital twin, mostly because there is a lack of agreement over the characteristics of a digital twin and its accompanying components. By improving the accompanying modeling and simulation practices, this study proposes a toolset for realizing a Digital Twin in domain-specific applications [4].

Emerging technologies and the digital transformation open up new business prospects that deviate from conventional needs. Digital twins, also known as digital reproductions of physical entities, are evolving quickly and opening up new avenues for innovation. In order to better grasp the implications of using digital twins for innovation, this study conducted case studies to gather data. The paper makes a valuable contribution to the literature on digital twins by providing a framework that illustrates how digital twins are defined, their impact on innovation processes, and the range of applications to which they can be put (scope) [5]. The results also have intriguing ramifications for other practitioners who are interested in using digital twins.

Existing Method

Traditional Learning

The term "traditional learning" describes the age-old, customary approaches to teaching and learning that have been around for ages. It usually entails direct communication between educators and learners in a traditional classroom environment [6]. Conventional learning methods have long served as the cornerstone of education and are still in use today.

Here are different types of traditional learning commonly found in educational settings:

Lectures: During a lecture, the instructor gives the pupils organized presentations of material. With the use of visual aids like slideshows or whiteboards, the instructor uses spoken communication to impart knowledge, ideas, and concepts. Students participate in passive learning by listening and taking notes.

Textbook-Based Learning: This method uses textbooks as its main source of data and educational resources. Students perform exercises, read chapters or portions that have been assigned, and study on their own. When needed, teachers offer direction and clarity.

Group Discussions: Group discussions encourage students to actively participate in learning by sharing their ideas, perspectives, and insights. Students engage in conversations, debates, and collaborative problem-solving activities, fostering critical thinking and communication skills.

Laboratory Experiments: In subjects like science and engineering, laboratory experiments provide hands-on learning experiences. Students conduct experiments, collect data, analyze results, and draw conclusions. Teachers guide students through the process and ensure safety protocols are followed.

Project-Based Learning: Project-based learning involves students working on extended projects that require research, planning, and problem-solving. They investigate a topic or issue, propose solutions, and present their findings. This approach promotes independent learning, teamwork, and application of knowledge.

Case Studies: Case studies present real or hypothetical scenarios that require students to analyze and apply their knowledge to solve problems or make decisions. Students examine the facts, consider different perspectives, and propose solutions based on critical thinking and problem-solving skills.

Field Trips: Field trips involve students visiting places outside the classroom to gain firsthand experience and contextual understanding. They observe and learn from real-world examples, such as museums, historical sites, businesses, or nature reserves. Teachers facilitate discussions and reflections on the field trip experiences.

Teacher-Led Instruction: Teacher-led instruction involves teachers leading the learning process, structuring the curriculum, delivering content, and providing guidance. Teachers play a central role in facilitating learning, answering questions, and assessing student progress.

These traditional learning approaches have their own strengths and limitations. While they provide structure, face-to-face interaction, and guidance from teachers [7, 8]; they may vary in terms of student engagement, active participation, and personalized learning. As education evolves, traditional learning approaches are often combined with modern instructional methods and technologies to create blended learning environments that leverage the benefits of both traditional and innovative approaches.

Traditional Teaching

The conventional teaching technique is a teacher-centric approach in which the instructor sets the curriculum, practices teaching methods, and maintains complete control over the learning environment [9]. This approach limits critical thinking and fosters information acquisition in the classroom by encouraging drill and rote memorization.

Methodology

In traditional education, students are taught in a physical classroom under the direct supervision of the teacher. The learning environment is built around textbooks, and the pace and schedule are predetermined. Thanks to this technique, students will experience a well-organized and fruitful learning process.

Innovative Learning

Innovative learning refers to the use of modern technologies, pedagogical approaches, and creative strategies to enhance the teaching and learning experience [10]. It leverages innovative tools and methods to engage learners, foster critical thinking, promote collaboration, and personalize learning experiences. Innovative learning approaches aim to adapt to the needs of the digital age and prepare students for the challenges of a rapidly evolving world.

Advantages of Pedagogy Using Digital Twin for Better Understanding

Optimized learning encounter or augmented educational journey: Students' understanding of complex topics and ideas is improved by the highly interactive and immersive learning environment provided by digital twins.

Expanded availability: Digital twins are an adaptable and easily accessed tool for remote learning, providing flexible access at any time and from any place.

Enhanced security: With the use of digital twins, risky or impracticable experiments or activities may be simulated in a real-world setting.

Budget-friendly: Since digital twins are less expensive to construct than physical counterparts, they provide an economical option for educational institutions.

Digital Twin Challenges

Technological constraints: Digital twin creation necessitates advanced technology, which may not be accessible in all educational settings.

Learning prerequisites: The production of digital twins requires sophisticated technology, which not all learning environments have access to.

Preparatory requirements: Certain resources, including as software, technology, and experience, are required for digital twins and might not be easily available in every school setting.

Pedagogy Learning in Digital Twin

Teachers need to continually adjust their techniques to match the expectations of digitally literate students, who demand rich, immersive digital learning experiences. These pupils are not like their teachers, who are frequently poorly tech-savvy and self- or peer-taught. Despite their frequent use of technology, their knowledge may not be as current as what their pupils or the demands of their teaching require. Technical proficiency is not what digital pedagogy is about; rather, it is about having a positive attitude toward and affinity for digital tools, being ready to utilize them in the classroom, and knowing when and why to use them. This method focuses on understanding how and why digital technologies should be utilized, rather than on being an experienced programmer or web page designer.

Providing relevant syllabuses and comprehending the ways in which students learn are two facets of classroom teaching. It demands active participation from students and helps teachers understand best practices for learning environments. Instructors who are aware of the various ways in which students receive and process knowledge might modify their lessons to better suit the needs of their students. This approach eventually improves the quality of the educational process overall by raising the bar for both instruction and student response.

Teacher Role in Pedagogy Using Digital Twin

The focus of teacher pedagogy is on the instructor's responsibilities for offering relevant course content, reliable information, tracking students' progress in their learning, and modeling good language usage.

Types of Approach in Pedagogy
The five main pedagogical learning styles are as follows:

Constructivist: This method promotes critical thinking, allows for active student participation in comprehension and information acquisition, and creates a learning atmosphere in which students can relate to what they hear.

Collaborative: Students work in groups to solve issues, develop methods, and finish assignments while working separately or in conjunction with teachers.

Integrative: With the goal of helping students comprehend the learning process, distinguish topics based on their relevance, apply lessons in real-world situations, and correlate concepts with everyday life; the integrative method offers them a learning environment that is connected to their curriculum.

Reflective: With the help of the reflective method, students are encouraged to examine their own motives and efficacy in order to assess themselves.

Inquiry-Based Learning: Teachers who use inquiry-based learning not only respond to students' questions but also provide an environment in which ideas are tested, polished, and enhanced in order to help students move from asking questions to comprehending them and asking even more.

Need of Pedagogical Teaching in Digital Twin

Enhancing the Quality of Instruction

When a well-considered pedagogy is used in the classroom, learning outcomes may be improved and students' comprehension of the subject matter can be completely realized.

Collaborative Learning

By fostering collaborative learning, improving students' perspectives, and adjusting cooperative learning settings, the use of pedagogy in education eventually gets students ready for leadership positions in the future.

Interactive Learning

In addition to encouraging complex learning processes like analysis, creativity, and assessment, pedagogy and child development also increase students' receptivity to teacher teaching.

Individualized Learning Paths Are Available to Students

A carefully thought-out pedagogy accommodates a range of learning styles and allows students to use the learning strategies that best suit them. This improves students' comprehension of the material, as well as their skills and learning objectives.

Universal Learning Method

Adopting an appropriate teaching strategy can improve the educational process and encourage inclusion of students with special needs in the general education group.

Enhancing Teacher-Student Interaction

By learning more about the student, the instructor is better equipped to pinpoint the student's areas of weakness and offer constructive criticism.

Digital Twin Role in Education Field

Digital twins have the potential to play a transformative role in the education field by leveraging learning technologies. Here are five learning technologies that can be integrated into digital twins to enhance educational experiences:

Simulation and Gamification: Digital twins can incorporate simulation and gamification elements to make learning more interactive and enjoyable. By simulating real-world processes or systems, students can apply their knowledge, experiment with different variables, and observe the outcomes. Gamification elements, such as challenges, rewards, and leaderboards, can further motivate and engage students in the learning process.

Data Analytics and Visualization: Digital twins can integrate data analytics and visualization tools to help students analyze and interpret complex data sets. By providing real-time data from the digital twin environment, students can explore patterns, trends, and relationships, enhancing their data literacy skills. Visualization techniques can be used to represent data in meaningful and accessible ways, aiding comprehension and critical thinking.

Collaborative Learning Platforms: Digital twins can be integrated into collaborative learning platforms, enabling students to work together on projects and problem-solving tasks. These platforms can facilitate communication, knowledge sharing, and collaborative decision-making within the digital twin environment. Students can engage in virtual teamwork, share ideas, and collectively analyze and solve problems.

Adaptive Learning Systems: Digital twins can be connected to adaptive learning systems that personalize the learning experience based on individual student needs and progress. By analyzing student data, such as performance, preferences, and learning styles, the adaptive system can provide tailored content, resources, and feedback. It can dynamically adjust the learning path and pace, and challenges.

These learning technologies, when integrated into digital twins, can create dynamic and interactive educational experiences. They provide opportunities for active engagement, problem-solving, collaboration, and personalized learning, fostering deeper understanding and knowledge retention.

Proposed Work

When incorporating digital twin technology into education with the help of AI, several tools can be used to enhance the learning experience. Here are some examples:

Simulation Software: Simulation software allows educators to create digital replicas of real-world systems or phenomena. By leveraging AI, these digital twins can provide realistic and dynamic simulations that students can interact with and explore.

Machine Learning Platforms: Machine learning platforms equipped with AI algorithms can be integrated into digital twins to enable predictive analytics and decision-making. Students can learn about machine learning techniques by working with real or simulated datasets, training and evaluating AI models, and observing the impact of different algorithms on the system's behavior.

Natural Language Processing (NLP) Tools: NLP tools can be utilized within digital twins to enhance communication and interaction. Students can engage in dialogue with the digital twin using natural language, ask questions, and receive responses. NLP algorithms can process and understand the student's queries, allowing for a more conversational and intuitive learning experience.

Data Analytics and Visualization Tools: Digital twins integrated with AI can generate a vast amount of data. The tools enable students to analyze and interpret this data effectively. They can explore patterns, trends, and correlations within the system, gain insights, and communicate their findings through visualizations and interactive dashboards.

Intelligent Tutoring Systems: AI algorithms provide personalized and adaptive instruction. These systems can be integrated into digital twins to offer individualized feedback, recommendations, and scaffolding based on the student's learning progress. They can assess the student's strengths and weaknesses, tailor the learning path, and provide targeted support to optimize learning outcomes.

Predictive Analytics and Early Warning Systems: Digital twins integrated with AI can use predictive analytics to identify potential issues or challenges faced by students. Early warning systems can detect patterns in student behavior and performance, allowing educators to intervene and provide targeted support at an early stage. This proactive approach can help prevent learning gaps and enhance student success.

These tools, when integrated into digital twins using AI, create interactive and personalized learning environments. They offer opportunities for hands-on experimentation, data-driven analysis, adaptive instruction, and real-time feedback, promoting a deeper understanding of complex concepts and fostering critical thinking and problem-solving skills.

Result

Real-Time Teaching

Real-time teaching refers to a teaching approach where the instruction and learning activities occur simultaneously and interactively. It involves live interaction between teachers and students, allowing for immediate feedback, discussions, and collaborative engagement. Real-time teaching can take place in physical classrooms, virtual classrooms, or through live online sessions.

Here are some key aspects and benefits of real-time teaching:

Synchronous Interaction: Real-time teaching enables synchronous interaction between teachers and students. This immediate interaction facilitates clarifications, discussions, and the exchange of ideas in real-time, promoting engagement and active participation.

Immediate Feedback: With real-time teaching, teachers can provide immediate feedback to students. They can address misconceptions, correct errors, and offer guidance while the learning is happening. This timely feedback helps students understand and correct their mistakes promptly.

Collaboration and Discussion: Real-time teaching allows for collaborative learning experiences. Students can engage in group discussions, brainstorming sessions, and peer-to-peer interactions.

Q&A and Clarifications: Students can ask questions and seek clarifications from teachers during real-time teaching. Teachers can respond to queries, provide explanations, and offer additional examples or demonstrations to enhance understanding.

Immediate Assessment: Real-time teaching use quizzes, polls, or questioning techniques to assess comprehension and adjust instruction accordingly. This ongoing assessment helps teachers adapt their teaching methods to address individual or collective learning needs.

Flexibility and Adaptability: Real-time teaching can be conducted in various settings, including physical classrooms, virtual classrooms, or live online sessions. This flexibility enables educators to adapt to different learning environments, including blended learning approaches that combine in-person and online interactions.

Engaging Learning Experiences: Real-time teaching promotes dynamic and engaging learning experiences. Through live demonstrations, interactive activities, multimedia presentations, and real-time discussions, students are actively involved in the learning process, leading to increased motivation and retention of knowledge.

Building a Learning Community: Real-time teaching allows for the establishment of a learning community where students can connect with their peers and build relationships. Students can collaborate, share ideas, and support each other's learning journey, creating a sense of belonging and fostering a positive learning environment.

Fig. 1 Teaching and learning process in pedagogical way

Figure 1 shows the real-time teaching, whether conducted in physical or virtual settings, offers numerous benefits in terms of immediate feedback, active engagement, collaboration, and dynamic interactions. It can be complemented with appropriate technologies and instructional strategies to create impactful and effective learning experiences for students.

Tools Used for Pedagogy Learning

Gamification in learning systems using digital twins combines the principles of game design with the interactive and immersive capabilities of digital twin technology mechanics, and dynamics into the digital twin environment to increase student engagement, motivation, and knowledge retention. Here is how gamification can be applied in learning systems using digital twins:

Points, Badges, and Leaderboards: Gamification can introduce students to, achieving milestones, or demonstrating mastery. Badges can be awarded for specific accomplishments, and leaderboards can display students' rankings based on their progress and achievements. This creates a sense of competition, encourages participation, and motivates students to excel.

Challenges and Quests: Digital twins can be designed with challenges and quests that require students to solve problems, complete missions, or accomplish specific learning objectives. These challenges can be embedded within the digital twin environment, offering interactive scenarios or simulations. By engaging in these quests, students actively apply their knowledge, develop critical thinking skills, and achieve a sense of accomplishment upon completion.

Progression and Unlockable Content: Gamification can introduce a progression system where students advance through levels or stages as they demonstrate proficiency in the digital twin environment. Each level can unlock new content, features, or challenges, providing a sense of progression and reward for their efforts. This approach keeps students motivated and encourages them to explore further.

Narrative and Storytelling: Incorporating a narrative or storytelling element into the digital twin environment adds an engaging and immersive experience. The digital twin can be designed with a compelling storyline, characters, and plot that students follow as they progress. This narrative framework helps contextualize learning activities, making them more meaningful and memorable.

Collaboration and Social Interaction: Gamification in digital twin learning systems can promote collaboration and social interaction among students. Multiplayer elements can be incorporated, allowing students to work together, compete, or cooperate within the digital twin environment. This fosters teamwork, communication, and the development of interpersonal skills.

Rewards and Feedback: Gamification provides opportunities for immediate feedback and rewards. Students receive instant feedback on their performance and can earn rewards, such as virtual currency, unlockable content, or virtual items. This feedback loop reinforces positive behavior, encourages continuous learning, and provides a sense of achievement.

By integrating gamification into digital twin learning systems, educators can produce engaging and immersive learning experiences that motivate, foster collaboration, also enhance knowledge acquisition. Gamification taps into the intrinsic motivation of students and provides a dynamic and interactive approach to learning within the digital twin environment as in Fig. 2.

Outcome

Project-based learning (PBL) in the setting of digital twins combines the principles called project-based learning with the immersive and interactive capabilities of digital twin technology. It involves designing and implementing projects that leverage

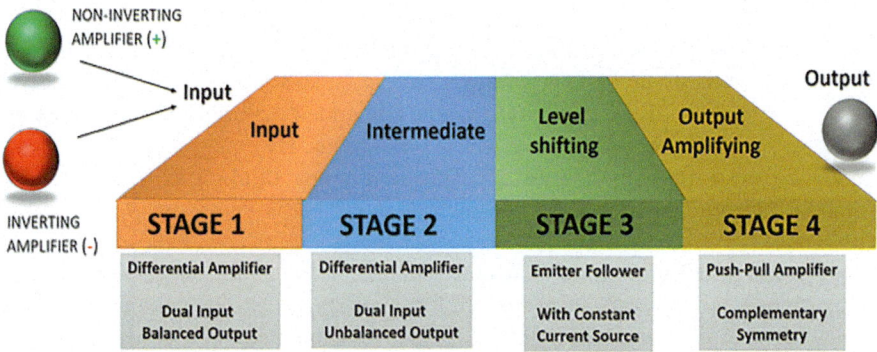

Fig. 2 Gamification-based teaching process using basic tools

digital twin environments to engage students in authentic, hands-on learning experiences. Here is how project-based learning can be applied in the context of digital twins:

Project Design: In project-based learning using digital twins, educators design projects that require students to investigate, analyze, and solve real-world problems or challenges within the digital twin environment. Projects can be interdisciplinary, integrating multiple subject areas and skills. The design phase involves defining project goals, identifying learning objectives, and determining the scope and resources needed as in Fig. 3.

Digital Twin Creation: The digital twin environment is created or utilized as a platform for students to explore and interact with. Educators may develop digital twins that replicate complex systems or phenomena, such as a smart city, a manufacturing process, or an environmental ecosystem.

Authentic Investigations: Students engage in authentic investigations within the digital twin environment, conducting research, collecting data, and analyzing information to understand the problem or challenge at hand. They can explore different variables, simulate scenarios, and observe the impact of their actions on the digital twin system.

Collaboration and Communication: PBL using digital twins encourages collaboration among students. They can work in teams, sharing responsibilities, collaborating on tasks, and communicating their findings and insights shown in Figs. 4 and 5.

Integration of Technologies: PBL using digital twins integrates various technologies to enhance the learning experience. Students can use data analytics tools, visualization software, and modeling platforms to analyze and interpret the data

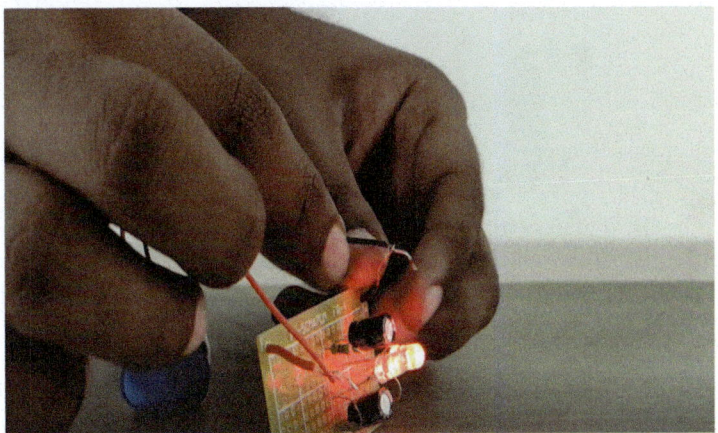

Fig. 3 Project implementation and demo

Fig. 4 Interaction with friends by explaining the project

Fig. 5 Project demonstration

generated by the digital twin. They can leverage AI algorithms to make predictions or optimize decision-making. Students can also utilize communication tools, project management software, or programming platforms to design and present their project outcomes.

Reflection and Evaluation: Throughout the project, students engage in reflective practices to evaluate their progress, learning, and challenges faced. They can document their findings, analyze their decision-making processes, and assess the effectiveness of their solutions. Reflection helps students develop metacognitive skills, self-awareness, and the ability to learn from their experiences.

Fig. 6 Project-based learning achieved from pedagogy learning

Presentations and Showcases: PBL using digital twins culminates in presentations or showcases where students demonstrate their project outcomes, findings, and insights. Students can present their digital twin manipulations, share their analysis, and discuss the implications of their solutions. Presentations can be made to peers, educators, or external stakeholders, providing opportunities for feedback and public speaking skills development shown in Fig. 6.

Digital twins enable project-based learning, fostering student engagement, critical thinking, collaboration, and problem-solving skills through real-world contexts, enhancing practical knowledge beyond the classroom.

Conclusion

The outcomes of pedagogy learning can be diverse and multifaceted, impacting both students and the overall educational process. Here are some key outcomes that can result from pedagogy learning:

Knowledge and Understanding: Pedagogy learning aims to facilitate the acquisition and deep understanding of knowledge and concepts. Students gain factual information, theoretical frameworks, and subject-specific skills relevant to the discipline or topic being taught. They develop a solid foundation of knowledge that forms the basis for further learning and application.

Communication and Collaboration: Pedagogy learning emphasizes effective communication and collaboration skills. Students engage in discussions, group activities, and projects that require them to articulate their ideas, actively listen to others, and work collectively. They learn to express themselves clearly, respect diverse perspectives, and collaborate with peers to achieve common goals.

Creativity and Innovation: Pedagogy learning encourages creativity and fosters an environment that nurtures innovation. Students are encouraged to think creatively, generate new ideas, and explore alternative approaches. They develop the ability to think outside the box, embrace experimentation, and propose innovative solutions to problems.

Social and Emotional Development: Pedagogy learning considers the holistic development of students, including their social and emotional well-being. It provides opportunities for building positive relationships, empathy, and emotional intelligence. Students learn to navigate social dynamics, develop resilience, and manage stress.

Personal Growth and Motivation: Pedagogy learning aims to cultivate students' personal growth and intrinsic motivation. It recognizes and builds on individual strengths and interests, providing opportunities for personalized learning experiences. Students feel a sense of accomplishment, develop a growth mindset, and are motivated to continue learning beyond the classroom.

Active Citizenship and Ethical Awareness: Pedagogy learning instills a sense of civic responsibility and ethical awareness in students. It encourages them to become active citizens, understand societal issues, and engage in discussions about social justice, sustainability, and ethical dilemmas.

These outcomes of pedagogy learning contribute to the overall development of students, equipping them with knowledge, skills, and attitudes necessary for success in academia, career, and life. They foster lifelong learners who are critical thinkers, effective communicators, and socially responsible individuals.

References

1. Peshkova, M., & Yumasheva, V. (2023). Digital twin concept: Healthcare, education, research. *Journal of Pathology Informatics, 14*, 100313.
2. Tao, Z., & Xu, G. (2022). *Digital Twin technology in the field of education: Take the management of the HTC Vive as an example*. Springer Nature Singapore Pte Ltd.
3. Addanki, K., & Corrin, L. (2023). Unveiling the potential of Digital Twin technology for Higher Education. In *People, partnerships and pedagogies* (Proceedings ASCILITE 2023). Authors.
4. Arantes, J. (2023). Digital twins and the terminology of "personalization" or "personalized learning" in educational policy: A discussion paper. *Policy Futures in Education*, 1–20. https://doi.org/10.1177/14782103231176357
5. Holopainen, M., Saunila, M., Rantala, T., & Ukko, J. (2022). Digital twins' implications for innovation. *Technology Analysis & Strategic Management*. https://doi.org/10.1080/09537325.2022.2115881
6. VanDerHorn, E., & Mahadevan, S. (2021). Digital Twin: Generalization, characterization and implementation. *Decision Support Systems, 145*, 113524.
7. Eriksson, K., Alsaleh, A., Behzad Far, S., & Stjern, D. (2022). Applying Digital Twin technology in higher education. In *SPS2022*. IOS Press. https://doi.org/10.3233/ATDE220165

8. Wang, M., Bi, X., & Liu, F. (2023). Research on Digital Twin technology and application in campus. *The Frontiers of Society, Science and Technology, 5*(5), 118–123, ISSN 2616-7433. https://doi.org/10.25236/FSST.2023.050519
9. Kartashova, L. A., & Gurzhii, A. M. (2022). Digital Twin of an Educational Institution: An innovative concept of blended learning. In *First symposium on Advances in Educational Technology (AET 2020)* (Vol. 2, pp. 300–310). SCITEPRESS – Science and Technology Publications. ISBN: 978-989-758-558-6.
10. Liljaniemi, A., & Paavilainen, H. (2020). Using Digital Twin technology in engineering education – Course concept to explore benefits and barriers. *Open Engineering, 10*, 377–385. https://doi.org/10.1515/eng-2020-0040

Human Digital Twin Processes and their Future

R. Hepziba Gnanamalar

Introduction

A human digital twin is a virtual representation or simulation of an individual, in the digital world. It is a concept derived from the broader concept of digital twins, which involves creating a virtual replica of a physical object, system, or process. Human digital twins aim to capture and simulate various aspects of human traits, behaviors, and experiences. This may include both physical characteristics, such as body structure and movement, and psychological and cognitive aspects, such as personality traits and decision-making patterns. Digital twins can be created by using data collected from sensors, wearable devices, social media, or other sources. The purpose of the human digital twin depends on the situation. In some cases, it may be used for research or scientific purposes to gain insights into human behavior, health, and performance. It can also be used for personalized medicine and healthcare applications, allowing physicians and researchers to explore possible treatment options and predict outcomes of specific interventions.

Additionally, human digital twins can be deployed in virtual reality (VR) or augmented reality (AR) environments to enable immersive experiences and training simulations. For example, in sports, digital twins of athletes can be used to analyze technique, identify areas for improvement, and even recreate competition scenarios for training. However, considering ethical and data-protection aspects is important when working with human digital twins. Because HDTs require the collection and analysis of personal data, data security, consent, and privacy issues must be addressed to ensure the responsible and ethical use of this technology.

In recent years, the concept of digital twins has gained a lot of attention in various industries, such as manufacturing, aerospace, and healthcare. In essence, a

R. Hepziba Gnanamalar (✉)
Assistant Professor in Computer Science, PSGR Krishnammal College for Women, Coimbatore, Tamilnadu, India

© The Author(s), under exclusive license to Springer Nature Switzerland AG 2024
A. Mishra et al. (eds.), *Transforming Industry using Digital Twin Technology*,
https://doi.org/10.1007/978-3-031-58523-4_10

digital twin is a virtual replica of a physical object, system, or process that can be used to optimize and predict performance. As the world continues to embrace digital transformation, the idea of creating digital twins for humans has evolved into an interesting and potentially revolutionary concept.

Human digital twins (also known as personal digital twins) are virtual representations of people that can be used to monitor, predict, and optimize health and well-being. These digital replicas can collect and analyze data from various sources, such as wearable devices, medical records, and even social media, to provide valuable insights into a person's physical and mental health. This information can be used to make informed decisions about healthcare, lifestyle choices, and even individual therapy [1].

One of the key benefits of human digital twins is their potential to revolutionize healthcare [2]. By providing a comprehensive, real-time view of an individual's health, digital twins help medical professionals identify potential health risks and intervene before they become serious problems. For example, digital twins can detect early signs of heart disease by analyzing data from wearable devices and medical records, enabling doctors to take preventive measures before the disease worsens. This practical method of delivering healthcare could lead to enhanced patient results and reduced healthcare expenses.

In addition, human digital twins can also play important roles in the development of personalized medicine. By analyzing an individual's genetic makeup, lifestyle, and environmental factors, digital twins can help researchers and medical professionals design targeted treatments and therapies tailored to each person's unique needs. This level of individualization may significantly improve treatment efficacy and reduce the risk of unwanted side effects.

Beyond healthcare, human digital twins can affect many aspects of our daily lives. For example, they can be used to optimize physical and mental health by providing personalized recommendations for exercise, diet, and stress management. By analyzing data from wearable devices and other sources, digital twins can help us make more-informed decisions about our lifestyles and habits, ultimately leading to healthier, happier lives.

Despite the many potential benefits of human digital twins, some challenges and ethical considerations also need to be addressed. Two of the main anxieties are the confidentiality and the safety of the data. Creating a digital twin requires collecting and analyzing large numbers of personal data, raising questions about how that information is stored, used, and protected. As digital twins of humans are developed and deployed, individual privacy must be respected and data kept secure. Additionally, digital twins may exacerbate existing inequalities in healthcare and other areas. For example, access to the technology and resources required to create and maintain digital twins may be restricted to certain tiers, resulting in different benefits that can be provided. Solving these issues is critical to ensuring that human digital twins are used in ways that are fair and beneficial to all.

In the workplace, a human digital twin can be deployed to improve employee productivity and happiness. By monitoring factors such as stress levels, sleep patterns, and cognitive performance levels, digital twins can help employers uncover potential problems and implement targeted interventions to improve employee health and well-being. This can increase employee engagement and productivity, ultimately benefiting both employees and the organization that hired them.

The human digital twin represents an interesting and potentially transformative concept that could revolutionize healthcare, improve well-being, and impact many aspects of our daily lives. However, to realize the full potential of this technology, the challenges and ethical considerations that come with it need to be addressed. In this way, we can harness the power of human digital twins to create a healthier, happier, and more productive world.

History of Human Digital Twin Computing

Looking back over the past 30 years or so, we can see that the focus of digitization shifts to people and "things" roughly every 10 years (Fig. 1). The shape of the digital world of the future will be changed by human digital twin technology (Table 1).

Evolution of the Human Digital Twin: From Concept to Reality

Once confined to science fiction, the concept of a human digital twin is fast becoming a reality. As technology continues to development at an unparalleled pace, the potentials for creating virtual models of ourselves are increasing, with far-reaching consequences for numerous businesses and industries, including healthcare and entertainment, and even for social interaction. This article examines the evolution of

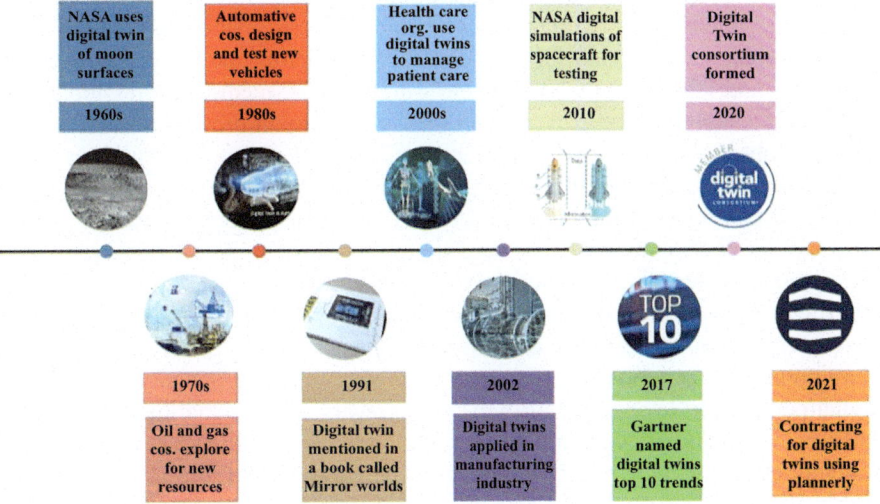

Fig. 1 Digital development for human beings and things

Table 1 Development of human digital twins

Year	Development
1985	Digitization focuses on people (people—first generation) Digitization of "human" communication through the introduction and spread of e-mail
1995	Digitization focuses on "things" (things—first generation) Digitization of "things" such as timetables and maps due to the spread of the Internet
2005	Digitization focuses on people (humans—second generation) Digitization of human connections and networks through the introduction and spread of social media
2015	Digitization focuses on "things" (things—second generation) Digitalization of various "things" such as parking lots through the introduction of IoT and Artificial Intelligence

the human digital twin from its early conceptualization to its current state and possible future developments.

The idea of digital twins, virtual replicas of physical objects and systems, has been around for decades. It was first introduced by NASA in the 1960s, when digital twins were to simulate and monitor spacecraft systems. Over time, this concept has evolved and expanded to include people as well as physical objects. The term *human digital twin* was coined in the early 2000s, and since then, the concept has received a great deal of attention thanks to technological advances such as artificial intelligence (AI), the Internet of Things (IoT), machine learning, and deep learning.

One of the earliest examples of human digital twins was the creation of avatars in virtual worlds like *Second Life*. These digital representations allow users to interact with others in virtual environments, paving the way for the development of more-sophisticated and more-realistic digital twins. As technology has progressed, so too have these virtual replicas, some of which are even able to mimic human emotions and behaviors. In recent years, the concept of human digital twins has gained a lot of attention in the healthcare industry. The ability to create a virtual replica of a patient's body with unique physiological and anatomical features offers great potential for personalized medicine. By simulating the effects of different treatments and interventions on a patient's digital twin, medical professionals can better predict and optimize patient outcomes, ultimately improving patient care.

A notable example of this application is the Living Heart project, an initiative that Dassault Systems started in 2014. The aim of this project is to create a highly accurate digital twin of the human heart that can be used to study cardiovascular diseases and develop new treatments. The Living Heart project has already made great strides, where its digital twins are used in clinical trials to test the safety and efficacy of new medical devices. The entertainment industry has also embraced the concept of human digital twins, and virtual influencers and celebrities are becoming more popular. Created by using AI and computer-generated imagery (CGI), these digital personas have amassed millions of followers across social media platforms and even secured lucrative brand deals. The rise of virtual influencers highlights the potential of digital twins to revolutionize the way that people consume and interact with content.

In the future, the possibilities for human digital twins seem almost limitless. Some experts believe that one day people may be able to create digital twins that can simulate our entire lives, from birth to death. These virtual replicas allow people to explore alternate life paths and see the possible consequences of different choices. Moreover, the integration of human digital twins with new technologies, such as VR and AR, could lead to entirely new forms of social interaction. Imagine being able to interact with a digital twin of a distant loved one or deceased person in a realistic and immersive virtual environment.

In summary, the evolution of the human digital twin has come a long way since its early conceptualization. As technology continues to advance, the potential uses and effects of these virtual replicas will continue to grow, transforming industries and reshaping the ways that people live, work, and interact. The future of human digital twins is undoubtedly exciting and promising, but people are just beginning to scratch the surface of its possibilities.

Advantages of Human Digital Twins

Human digital twins offer multiple benefits and potential benefits in several areas.

- *Personalized healthcare*: Human digital twins help medical professionals develop personalized treatment plans by analyzing personal healthcare data to develop simulations. They can be used to predict disease progression, identify potential risks, and optimize treatment strategies. Digital twins enable precision medicine by taking into account an individual's unique characteristics, genetics, and lifestyle factors.
- *Virtual training and skill development*: Human digital twins can be used in VR or AR environments to create realistic training simulations. They allow individuals to practice and develop their skills in safe and controlled virtual environments. For example, surgeons can use digital twins to train and improve surgical techniques without putting real patients at risk.
- *Performance optimization*: Athletes, artists, and professionals in many fields can benefit from human digital twins for performance optimization. Digital twins can provide insights into engineering, geometry, and biomechanics by analyzing data from sensors and wearable devices. This allows individuals to identify areas for improvement and make informed adjustments to improve their performance.
- *Predictive analytics and decision support*: Human digital twins can simulate scenarios and predict outcomes. Data from multiple sources can be integrated to support decision-making processes. For example, in business, digital twins can analyze customer behavior, market trends, and operational data to provide predictive insights and support strategic decision-making.
- *Behavioral analysis and research*: Researchers can use human digital twins to study and analyze human behavior in a controlled and ethical manner. Digital

twins help researchers gain insights into cognitive processes, social interactions, and psychological phenomena by simulating different conditions, scenarios, and interventions.
- *Telemonitoring and telemedicine*: Human digital twins facilitate telemonitoring and telemedicine applications. Digital twins continuously collect and analyze real-time data such as vital signs, movement patterns, and lifestyle information to provide remote healthcare providers with valuable information for diagnosis, monitoring, and early intervention.
- *Human–machine collaboration*: Digital twins can improve collaboration between humans and machines. Human–machine interaction can be improved by integrating human digital twins with artificial intelligence (AI) systems. In manufacturing, for example, digital twins can leverage AI algorithms to optimize processes, predict maintenance needs, and improve efficiency.

These benefits highlight the potential of human digital twins to improve healthcare, education, decision-making, and cross-disciplinary collaboration [1]. However, addressing ethical and privacy issues and ensuring the responsible and transparent use of this technology are crucial.

Disadvantages of Human Digital Twin

Human digital twins have potential advantages and applications, but they also present certain drawbacks and challenges.

- *Data protection and data security*: Creating human digital twins requires collecting and analyzing large numbers of personal data. This raises privacy and data security concerns. Storing and protecting these data are essential to preventing unauthorized access and misuse. Some people may object to having their personal information used to create virtual replicas of themselves.
- *Ethical considerations*: The ethical implications of human digital twins are complex. Issues such as informed consent, data ownership, and transparency need to be addressed. The technology could be abused, and digital twins could be used to manipulate and exploit people. Ensuring ethical practices and regulations is essential to protecting the rights and well-being of individuals.
- *Accuracy and representation*: Creating an accurate and comprehensive digital twin is difficult. Data must be collected and analyzed from a variety of sources, and data limitations and biases can affect twin accuracy and representation. Factors such as environmental influences, subjective experiences, and personal uniqueness can be difficult to capture in digital replicas.
- *Complexity and maintenance*: Developing and maintaining human digital twins requires significant computational resources and expertise. The process of creating and updating twin models, integrating new data, and refining algorithms can be complex and time-consuming. For example, Gemini requires constant effort to remain relevant and useful. Gemini is the first HIPAA compliant Digital Twin

platform available today, designed to ingest any type of information about the human body.
- *Limited contextual understanding*: Human digital twins can struggle to fully understand and simulate the complex social, cultural, and contextual factors that shape human behavior. While they can grasp certain physical and cognitive aspects, they may not fully understand the nuances and complexities of human experiences, emotions, or social interactions.
- *Dependence on data availability*: The accuracy and usefulness of human digital twins are highly dependent on the availability of relevant high-quality data. Limited, incomplete, or biased data can compromise the expressive and predictive abilities of twins.

Addressing these shortcomings and challenges is important to ensuring the responsible and ethical development and deployment of human digital twins, taking into account individual privacy, consent, and potential impacts on society as a whole.

Process of Human Digital Twins

Human digital twin technology (Fig. 2) is one of the key features of digital twin computing. Digital twin computing creates a digital twin that replicates not only the external aspects of humans, such as their physical and physiological characteristics, but also their internal characteristics, such as personality, emotions, thoughts, and abilities. This is because a human believes that by expressing the individuality of each person, humans can engage in interactions on the basis of the diversity that arises from each person's characteristics rather than interactions between average beings without individuality. Digital twins also incorporate social characteristics such as human conduct and interaction into the digital space.

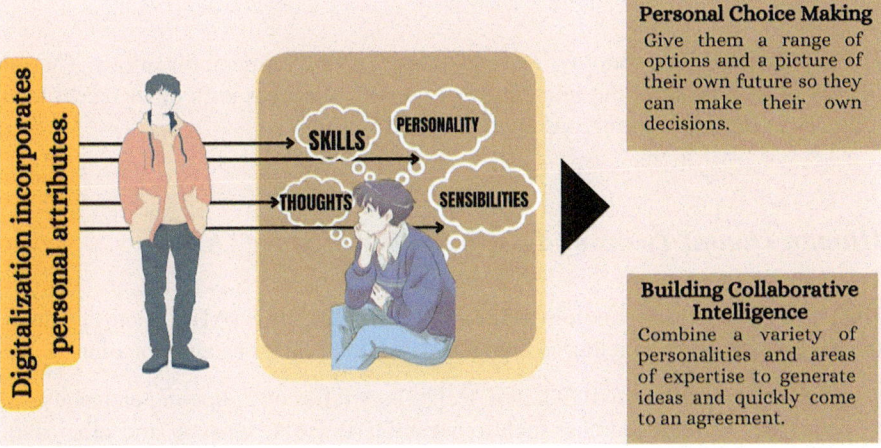

Fig. 2 Human digital twins

We have plenty of examples of what can be achieved with digital twin computing.

- *Proxy conferences between digital twins*: Digital twins with individual human personalities and traits can react to the approaches of others in cyberspace as if they were real humans. Conversely, digital twins can also approach other digital twins if they are allowed to act autonomously. For example, holding meetings between digital twins can help develop group intelligence by bringing together different personalities and fields of expertise to reach instant consensus without contacting people in the real world. For example, trends can be visualized, and groups can coordinate in real time.
- *Create a personal agent to work on your behalf*: Digital twin computing expands the scope of human activity from the physical world to cyberspace. In cyberspace, digital twins can act like humans and create powerful personal assistants that act on behalf of real humans. The fidelity of the digital twin also allows you to work in multiple locations simultaneously.
- *Have conversations that are impossible in the real world*: With the help of digital twins, nonexistent people, such as deceased people, can be communicated with to gather knowledge and experience. For example, people can communicate with their past and future personalities, which was unconceivable until recently. These interactions can be used to improve personal decision-making as well as self-understanding, self-discovery, idea generation, and more.
- *Using the digital twin as an interface*: People can also use their digital twin as an interface in cyberspace. For example, by exchanging or combining the skills of your digital twin with the skills of someone else's digital twin, you can create derivative content for your digital twin with skills (such as language skills) that you do not have. You can apply these newly acquired skills to a wide variety of real-world devices.

HDT Process with Various Recent Technologies

Human digital twin technology have extraordinary real-time applications. The following section lays out the integration of the HDT process with other recent technologies to introduce numerous applications (Fig. 3).

Human Digital Twins and Artificial Intelligence (AI)

Human digital twin technology and artificial intelligence (AI) are closely intertwined and can work together to enhance various applications and domains.

- *Data analysis*: Human digital twin technology relies on collecting and analyzing data from various sources, such as wearable devices, sensors, and other data-collection methods. AI techniques, such as machine learning and data analytics,

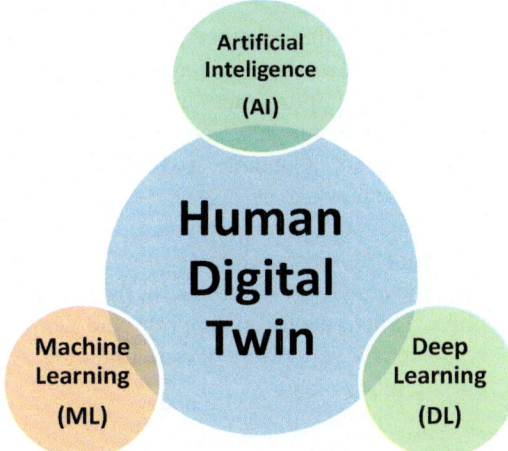

Fig. 3 Human digital twin with recent technologies

can be applied to process and analyze these data, uncover patterns, and extract valuable insights. AI algorithms can help identify correlations, make predictions, and provide personalized recommendations from the data collected from the human digital twin.
- *Simulation and modeling*: Human digital twins are created by using computational models that represent the behavior, characteristics, and interactions of an individual. AI techniques can be employed to develop and refine these models. Machine-learning algorithms can be trained on data from real individuals to improve the accuracy and fidelity of the digital twin's behavior and responses. This allows the digital twin to simulate different scenarios, predict outcomes, and provide valuable information for decision-making processes.
- *Personalization and adaptability*: AI plays a crucial role in enabling personalization in human digital twin technology. By leveraging AI algorithms, the digital twin can analyze individual data, preferences, and historical patterns to tailor recommendations, interventions, or simulations to the specific needs and goals of the individual. AI techniques can also enable the digital twin to adapt and learn over time, improving its understanding and responses by including feedback and new data.
- *Cognitive abilities*: AI techniques such as natural language processing (NLP) and computer vision can enhance the cognitive abilities of human digital twin technologies. By incorporating AI algorithms, digital twins can process and understand human speech, enabling natural, interactive communication. Computer vision techniques can be used to interpret visual information, enabling the digital twin to understand and respond to visual cues and stimuli.
- *Decision support*: AI can improve the decision-making capabilities of human digital twin technology. AI algorithms can analyze large numbers of data, identify patterns, and make predictions to support decision-making in a variety of applications. In healthcare, for example, AI can help digital twins analyze

medical data, suggest treatment options, and predict the effectiveness of interventions, enabling healthcare professionals to make informed decisions.
- *Continuous learning and improvement*: AI algorithms can facilitate continuous learning and improvement in human digital twin technology. By analyzing new data, updating models, and incorporating feedback, digital twins can become more accurate, adaptable, and effective over time. This enables digital twins to learn from new experiences and data while providing more-detailed and -personalized insights and recommendations. Overall, AI technology will play a key role in enhancing the capabilities of human digital twin technology. Because digital twins leverage AI algorithms for data analysis, simulation, personalization, cognitive skills, decision support, and continuous learning, they are valuable resources for supporting decision-making and improving the overall user experience in many areas.

Human Digital Twins and Machine Learning (ML)

Human digital twin technology and machine learning are closely related and can be viewed as complementary approaches. Machine learning is a subset of artificial intelligence (AI) that focuses on developing algorithms and models that enable computers to learn from data and make predictions and decisions without being explicitly programmed. Human digital twin technology and machine learning intersect in many ways [3].

- *Data-driven insights*: Human digital twin technology is based on collecting data from various sources, such as wearable devices, sensors, and other monitoring tools. Machine-learning techniques can be applied to analyze these data, reveal patterns, and generate meaningful insights. Digital twins technology can learn from the data and provide valuable information about behavior, health, performance, and other related aspects of human life by training machine-learning models on data collected from human individuals.
- *Predictive analytics*: Machine-learning algorithms can be used with human digital twin technology to make predictions by using historical data. By analyzing patterns and correlations in collected data, digital twins can use machine-learning techniques to predict future outcomes and trends. For example, in healthcare, machine-learning models can be used to predict the courses of diseases, identify potential risks, and assess the effectiveness of different treatment options.
- *Personalization*: Machine-learning algorithms enable personalization in human digital twin technology. By analyzing an individual's data and preferences, machine-learning models can identify personalized patterns, preferences, and needs. This will enable digital twins to develop customized recommendations, interventions, or simulations on the basis of an individual's specific characteristics and goals.

- *Adaptive models*: Machine learning can help develop adaptive models in human digital twin technology. Machine-learning algorithms can update and refine the models used in digital twins by continuously learning from new data. This allows the digital twin to adapt to changing conditions, integrate new information, and improve accuracy and effectiveness over time.
- *Decision support*: Machine-learning algorithms can provide decision support in human digital twin technology. By analyzing data and identifying patterns, machine-learning models can support the decision-making process. For example, machine learning in manufacturing can analyze production data to optimize processes or detect anomalies. In the medical field, machine-learning models help medical professionals make informed decisions that are based on patient data and medical records.
- *Anomaly detection*: Machine-learning algorithms can be deployed in human digital twin technology to detect anomalies and deviations from normal behavior. By training a model with normal patterns and data, the digital twin can use machine-learning techniques to identify abnormal behavior and identify potential risks.

Machine learning is especially useful in medical monitoring and performance-optimization scenarios. Using machine-learning techniques, human digital twin technology can improve skills in data analysis, prediction, personalization, adaptability, decision support, and anomaly detection. Machine learning plays a key role in enabling digital twins to learn from data, make accurate predictions, and provide personalized insights and recommendations on the basis of individual characteristics and goals.

Human Digital Twins and Deep Learning

Human digital twin technology and deep learning are related. Deep learning is a subset of machine learning that focuses on training multilayer artificial neural networks (ANNs) to learn and extract high-level representations from data. Human digital twin technology and deep learning intersect in several ways.

- *Data representation*: Deep-learning algorithms excel at developing complex representations from raw data. In the context of human digital twin technology, deep learning can be used to extract meaningful traits and representations from collected data. This enables digital twins to capture and understand complex patterns and relationships from data, enabling more-accurate and more-insightful analyses.
- *Feature extraction*: Deep-learning models can automatically learn hierarchical features from data. For human digital twin technology, deep learning can be used to extract relevant traits from various data sources, such as physiological measurements, sensor values, and behavioral data. By learning these capabilities,

digital twins can capture key traits and nuances that are important for understanding human behavior, health, and performance.
- *Complex pattern recognition*: Deep learning is characterized by recognizing and understanding complex patterns in data. This capability is beneficial for human digital twin technology in that the digital twin can identify complex relationships, correlations, and anomalies in the collected data. For example, deep-learning models can detect patterns in medical data that indicate specific diseases or conditions to aid in diagnosis and treatment planning.
- *Natural language processing*: Deep-learning techniques such as recurrent neural networks (RNNs) and transformers are commonly used for natural language processing tasks. In the context of human digital twin technology, deep learning enables digital twins to process and understand human language. This enables more-natural and -interactive communication between individuals and their digital twins, improving user experience and enabling personalized recommendations and explanations.
- *Time-series analysis*: Deep-learning models such as recurrent neural networks (RNNs) and convolutional neural networks (CNNs) are well suited for analyzing time-series data. Human digital twin technology uses deep learning to continuously analyze new data such as physiological signals, activity patterns, and behavioral sequences to uncover temporal dependencies and make predictions about future states and behaviors.
- *Transfer learning*: Deep-learning models trained on large-scale datasets can capture generic representations that can be transferred to related tasks or domains. In the context of human digital twin technology, transfer learning can be integrated with deep learning to use pretrained models for relevant tasks, such as image recognition or natural language understanding. Adapt these pretrained models to specific aspects of the digital twin's data analysis or prediction tasks.

Deep learning can help accelerate the training process for HDTs and improve their performance, especially when data availability is limited. By utilizing deep-learning techniques, human digital twin technology can benefit from the capabilities of deep neural networks when developing complex representations, recognizing patterns, processing natural language, analyzing time-series data, and leveraging transfer learning. Deep learning enables the digital twin to capture and understand intricate aspects of human behavior, health, and performance, enhancing the accuracy and effectiveness of its insights and recommendations.

Conceptual Paradigm for DT and HDT

Few keep in mind the complete human life cycle, and few understand the benefits of having a virtual human twin. Digital twins are especially useful in healthcare, particularly at identifying illnesses such as heart disease. The human digital twin is a duplicate of a human but set in a virtual world. It includes a database featuring information such as your age, weight, gender, family, and other facts. This is a virtual illustration that is digitally represented on a computer or server on the cloud. The synchronization of data depends on online communication between technologies by using the Internet, 4G, 5G, and Wi-Fi [4]. Smart wearables, sensors, smartphones, hospitals, and others constantly collect and store your data and wirelessly transmit them to a database [5]. With such data, it performs complicated tasks. By using cloud computing, deep learning, and other technologies, the human digital twin analyzes real-time and historical data to yield insights for diagnostics, predictions, and recommendations [6]. The human digital twin includes data from your surroundings to create a more accurate digital world. The nearby environment is included to factor in the effects that it has on the human whose digital twin is being developed. Other data include family history and genetics.

Digital twins can be used to compare seemingly identical products that are manufactured with different facilities. Making digital twins of each batch of products with different facilities allows for comparing them to determine whether the defects in one batch are also in another.

Similarity Between DT and HDT

While digital twins and human digital twins have different focuses and uses, they also have some similarities.

- *Representation*: Both digital twins and human digital twins involve creating a virtual representation. Their goal is to capture and model the behavior, characteristics, and interactions of each entity. Digital twins represent physical objects, systems, or processes, whereas human digital twins represent individuals or groups of people [7].
- *Data integration*: Both types of twins rely on data integration from various sources to improve understanding and simulation skills. Digital twins collect data from sensors, IoT-enabled devices, and other sources to monitor and analyze the behavior of physical systems. Human digital twins collect data from wearable devices, medical sensors, behavioral monitoring, and other sources to model and understand human behavior, health, and performance.
- *Real-time monitoring*: Both digital twins and human digital twins can monitor real-time data to provide insights and support decision-making. Digital twins monitor data from physical systems to assess performance, predict maintenance needs, and optimize operations [8]. Human digital twins monitor real-time data from individuals to track health metrics, analyze behavioral patterns, and provide personalized interventions and recommendations.
- *Simulation and prediction*: Both types of twins use simulation and prediction skills. Digital twins use simulations to test scenarios, predict outcomes, and optimize system performance. Human digital twins use simulation technology to model and simulate human behavior, health, and performance and to predict

outcomes, simulate interventions, and provide personalized recommendations that are based on modeled data.
- *Optimization and improvement*: Both digital twins and human digital twins aim to optimize and improve their respective entities. Digital twins focus on optimizing the operational efficiency, maintenance, and performance of physical systems. Human digital twins aim to optimize an individual's health, behavior, performance, and decision-making process by providing personalized insights, interventions, and recommendations.

Although DTs and HDTs have similarities, human digital twins have a specific focus on human traits, well-being, and personalized applications, whereas digital twins have a broader scope and focus primarily on industry- or system-level contexts.

Differences Between DT and HDT

The main differences between a digital twin and a human digital twin lie in their respective focuses and applications [9] (Table 2):

Characteristics of HDTs

Each HDT corresponds to a human represented in digital form. And each HDT has a unique ID tied to an account.

- When someone is born, their HDT is created with the permission of a doctor or other medical expert and their parent(s) at a medical institution. The next generation of humans inherit a number of genetic tendencies from their ancestors. Their HDTs also inherit those genetics and genetic tendencies. By having these traits and tendences, an HDT can be analyzed for the same or related medical problems and genetic traits for diagnoses and prognoses.
- Each human digital twin makes adjustments to include their human model's moods, treatments, vaccination status, injuries, and emotional responses to stimuli.
- The examinations, treatments (medication, injections, etc.), immunizations (vaccinations), fitness checkup data (diagnostic protocols, etc.), images taken at clinics (e.g., electrocardiograms), and facts such as blood pressure, weight, etc. of a person are transferred to the human digital twin at a clinic with consent, and the HDT updates its data in real time. As the human that the HDT is based on changes, so does the HDT, by incorporating real-world data. Height and weight, for example, can change over time, and whenever data on the corresponding human are taken, they are collected and transmitted to the HDT.

Human Digital Twin Processes and their Future

Table 2 Differences between digital twins and human digital twins

Elements	Digital twin	Human digital twin
Focus	Digital twins focus primarily on replicating and optimizing the behavior and performance of physical objects and systems.	Human digital twins represent and enhance human traits, behavior, and well-being.
Specific designs	Digital twins are commonly used in industrial settings to improve operational efficiencies and streamline physical processes.	Human digital twins are used in areas such as healthcare, sports, and personalized interventions to help individuals and to improve performance and experiences.
Data sources	Digital twins rely on data from sensors, IoT-enabled devices, and simulation models to capture and reflect physical systems' behaviors.	Human digital twins collect data from various sources, such as wearable devices, medical sensors, physiological measurements, and behavioral data, to model and understand human characteristics and activities.
Applications	Digital twins are used for tasks such as predictive maintenance, operation optimization, and simulation-based testing.	Human digital twins are used in applications such as personalized healthcare, performance optimization, decision-making support, and behavioral analytics.
Complexity of representation	Digital twins often focus on replicating the physical properties and operational aspects of objects and systems.	Human digital twins replicate physiological processes, cognitive abilities, emotions, and behavior. They aim to capture and model various human attributes.
Summary	Digital twins focus on replicating and optimizing physical systems.	Human digital twins are designed to express and enhance human traits, behaviors, and well-being in various domains.

- All wearable sensor data, such as weight, blood pressure, heart rate, respiration, blood sugar level, amount of exercise, and emotional reactions, are conveyed to and incorporated into the human digital twin [10]. Data on food intake and education are also incorporated into the human digital twin. In addition, environmental data are transmitted to the HDT, and it adjusts its environment accordingly. Data must remain coherent, and by analyzing new data and comparing it with historical data, inconsistences can be removed.
- A human digital twin incorporates large numbers of data and analyzes health information and diagnostics to make medical recommendations. Human digital twins rely on technologies such as computing, the Internet of Things, data fusion, smart diagnostics, clustering analytics, artificial neural networks, selection trees, and more. Health evaluations are communicated to their corresponding humans and those humans' physicians. HDTs are used in hospitals to inform predictions, suggestions, guides, and treatment plans. HDT suggestions make recommendations on exercise, dietary composition, treatment options, etc. In the case of an emergency, the HDT can dispatch an ambulance to supply first aid.
- A certified individual can log in to the human digital twin to analyze its content. VR/AR can offer vibrant three-dimensional (3D) snapshots of actual humans

and dynamically and visually display their health conditions. Even organs are reproduced thanks to the usage of 3D reconstructions.
- Passwords, fingerprints, iris recognition, encryption, authentication, IDs, and cryptographic strategies may be used to protect data privacy. Each HDT requires an account and password to log in and gain access. As an adult, you can control your account and password yourself. For children, payments and passwords are controlled by a parent or other guardian. A health practitioner or professional can be set as a legal alternate if necessary.
- Other security mechanisms include limiting access entry to healthcare facilities, requiring identification verification, and setting up vulnerability detection to log in to the HDT. A stable Internet connection is required for authentication and transmission. This approach should also include logging Internet protocol (IP) addresses and limited access to only those that are static and registered. Communication paths and routing paths are generally fixed. Healthcare facilities, families, and private information hubs now have the computing power, storage capacity, and bandwidth to run protection algorithms to secure protection and verification. To guarantee authenticity, any person or device that is granted access, such as relatives and private computers, should be registered, and the identification of every log-in should be traceable.

Human Digital Twin Technology and Human Beings

A human digital twin system (HDTS) comprises a virtual twin of a real human, including the traits, dispositions, behaviors, and other data needed to accurately represent a human. A human digital twin (HDT) is an alphanumeric representation of a real human as a twin. HDTs also appear on devices that can render them. Digital representations incorporate both real-time and historical data on a person's traits.

- *Physical*: anthropometric attributes, biomechanical attributes, time required to finish a task, eye movements, injuries, etc.
- *Physiological*: measurements of heart rate, heart rate variability, pores and skin responses, muscle tone, blood oxygen levels, electrophysiological signals, pluviography, blink rate and duration, peripheral blood flow, gastronomic activity, etc. Other physiological measures include fatigue, circadian rhythm, the degree of arousal, and the degree of engagement.
- *Perceptual*: auditory sensitivity, capability to decipher speech, visible sensitivity, sedation sensitivity, evaluation sensitivity, stress sensitivity, ache threshold, temperature sensitivity, etc.
- *Performance*: knowledge, abilities, aptitudes, workloads, situational awareness, decision-making, intuitive/analytical biases, etc.
- *Personality*: character types, mistrustfulness, and distrustfulness.
- *Emotional*: despair or anxiety.

- *Ethical*: values, beliefs, and practices.
- *Behavioral*: interactions with the HDT system.

Human digital twin technology and humans are different concepts with different characteristics and purposes.

- *Representations*: Human digital twin technology aims to create virtual representations or simulations of human characteristics, behaviors, and interactions. It uses data collection, modeling, and simulation techniques to create a digital version of a human. Humans, on the other hand, are creatures with bodies, consciousness, and subjective experiences [11].
- *Purpose*: The purpose of human digital twin technology is to provide insights, analytics, and support for various applications, such as healthcare, training, and decision-making. It is designed to support and enhance human capabilities in specific areas. Humans, on the other hand, have inherent purposes, goals, and aspirations that go beyond representation and simulation.
- *Data and modeling*: Human digital twins are created from data collected from various sources, such as wearable devices, sensors, and others. These data are used to create computational models and algorithms that simulate individual behavior and characteristics. Humans, on the other hand, have rich and complex natures that go beyond data and models. They have subjective experiences, emotions, perceptions, and the ability to make decisions on the basis of a variety of factors.
- *Autonomy and consciousness*: Human digital twins lack autonomy and consciousness because they are programmed systems that operate according to predefined rules and algorithms. They have no self-awareness or subjective awareness. However, humans have self-awareness, consciousness, and the ability to make decisions autonomously on the basis of their thoughts, beliefs, and experiences.
- *Physical presence*: A human's digital twin exists as a virtual representation within a digital platform or computer system. It has no physical presence and no sensory experience. Humans, on the other hand, have bodies, senses, and the ability to interact with the physical world.
- *Complexity and uniqueness*: Human digital twins aim to capture specific aspects of individual traits and behavior, but they may be limited in representing the complexity and uniqueness of the human as a whole. Humans are diverse and incredibly complex, physically, psychologically, and socially. Everyone has their own experiences, traits, and perspectives. Importantly, human digital twin technology is not intended to replace or duplicate humans. Instead, it serves as a tool that supports and enhances human capabilities in specific applications. Humans are endowed with qualities and characteristics beyond what can be simulated or represented by digital twin technology, making them irreplaceable in terms of consciousness, creativity, and subjective experiences [12].

Are Human Digital Twins Against Humans?

Human digital twin technology is not designed to "counter" humans. The purpose of human digital twin technology is to support and enhance human capabilities in various areas, such as healthcare, training, decision-making, and performance optimization [13]. The goal is to provide valuable insights, simulations, and predictions that help individuals to achieve their goals and improve their well-being. Human digital twin technology can revolutionize many aspects of our lives, but it is not intended to replace or diminish the importance of humans [1]. It aims to be a tool that complements and enhances human capabilities, providing personalized support and insight by using data analysis, modeling, and simulation techniques. Human digital twin technology augurs promising advances in personalized medicine, optimizing performance in various fields, enabling telemedicine delivery, and monitoring and improving decision-making processes. However, to ensure that this technology respects individual autonomy, privacy, and general well-being, the ethical and social implications of this technology must be considered. Ultimately, human digital twin technology should be seen as a tool to empower and support humans, not as an adversary. It is a complementary approach that enables more-personalized and more-effective solutions for the benefit of individuals and society as a whole.

Wearable Devices and HDTs

As technology continues to evolve and push the boundaries of what is possible, the concept of a human digital twin is rapidly becoming a reality. A human digital twin, a virtual replica of an individual, can be used to simulate and predict human behavior, health, and performance [1]. This cutting-edge technology can revolutionize industries as diverse as healthcare, sports, and entertainment. One area where the human digital twin is expected to have a major impact is wearable technology.

Wearable technology has come a long way since the arrival of fitness trackers and smartwatches. Wearables are now used not only to track physical activity but also to monitor vital signs, sleep patterns, and even mental health. As the human digital twin becomes more sophisticated, it will play a key role in shaping the future of wearable technology, providing personalized insights and recommendations on the basis of an individual's unique physiological and psychological profile.

One of the most important benefits of human digital twins in wearable technology is their potential to improve health [2]. By creating a virtual replica of a person, medical professionals can simulate different scenarios and predict how that person's body would respond to different treatments and procedures. This allows doctors to make more-informed decisions about the best course of action for their patients, ultimately leading to better health outcomes and lower healthcare costs. For example, a digital twin of a human can be used to predict how a patient would respond to a particular drug or treatment regimen, allowing physicians to tailor their approach to each individual's unique needs. This personalized medicine approach could improve patient outcomes and alleviate the trial-and-error process associated with conventional therapies [14].

In addition to healthcare, the human digital twin could also play an important role in the sports industry. Athletes can use digital replicas of themselves to analyze performance, identify opportunities for improvement and develop personalized training programs. By simulating different scenarios and analyzing their impact on the digital twin, athletes can optimize their training and performance, ultimately leading to better results on the field and court. Furthermore, human digital twins can also be used in the entertainment industry, especially in the areas of virtual reality (VR) and augmented reality (AR) [15]. By creating a digital replica of a person, VR and AR experiences can be tailored to the user's individual preferences and characteristics, resulting in a more immersive and personalized experience [16].

As the human digital twin continues to evolve, the possibilities for integration into wearable technology are almost limitless. For example, imagine a future where smartwatches monitor not only your physical activity but also your mental health and provide personalized recommendations on the basis of your unique psychological profile. This level of personalization could greatly improve the user experience and lead to the wider adoption of wearable technology. Wearable devices play important roles in the context of human digital twin technology by collecting real-time data on an individual's physical and physiological characteristics that can be used to create and update digital twin models [17, 18]. Here are some examples of human digital twin applications that use wearable devices (Fig. 4).

Data Collection

Wearable devices such as fitness trackers, smartwatches, biosensors, and other health-monitoring devices can collect a variety of data related to a person's physical activity, vital signs, sleep patterns, and more. These devices continuously or periodically collect data such as heart rate, the number of steps taken, hours asleep, calories burned, blood pressure, skin temperature, and other relevant metrics. These data provide valuable inputs for the development and refinement of digital twin models.

Real-Time Monitoring

Wearable devices can monitor a person's physiological and behavioral data in real time. Collected data can be wirelessly transmitted to a central system or cloud platform for analysis and integration into a digital twin model. Real-time monitoring

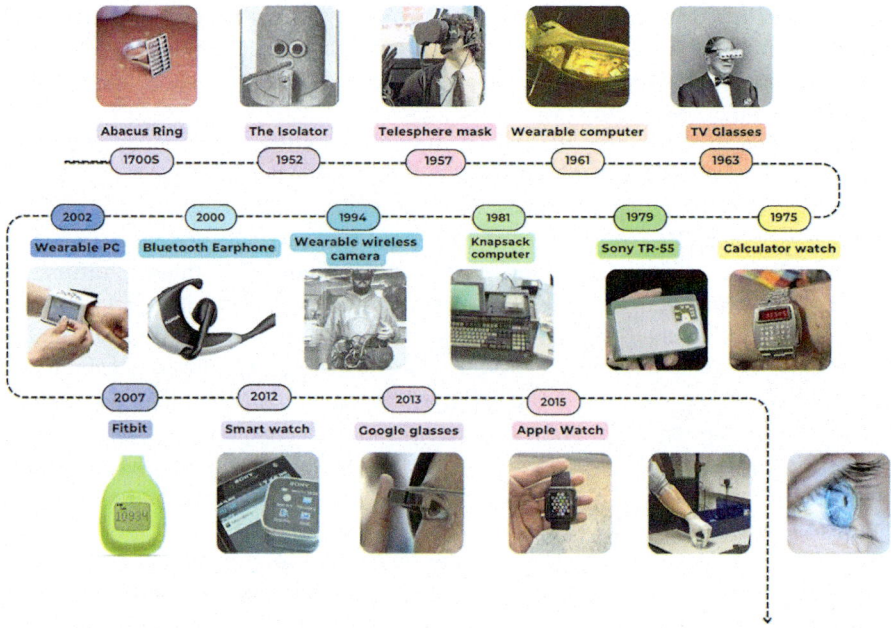

Fig. 4 Various wearable devices [19]

enables the timely detection of changes and anomalies, supporting proactive intervention and personalized insights.

Biomechanics and Motion Analysis

Wearable devices with motion sensors, accelerometers, or gyroscopes capture motion data to provide insights into a person's biomechanics and physical activity. This information can be utilized to simulate and analyze movement patterns, posture, gait, and motor performance in the digital twin. This allows you to optimize your technique, identify your risk of injury, and improve your physical performance.

Personalized Health Insights

Wearable devices can provide personalized health insights and recommendations. Combining data from wearable devices with other health-related information, digital twins can generate personalized recommendations on physical activity, exercise habits, sleep patterns, and stress management. This encourages healthier behaviors and lifestyle choices that are based on individual needs and goals.

Telemonitoring and Telemedicine

Integrating wearable devices with human digital twin technology enables telemonitoring and telemedicine applications. Real-time data collected from wearables can be sent to healthcare providers to monitor an individual's health status, track treatment effectiveness, and intervene if necessary. This allows for proactive health management, especially for those with chronic illnesses or who require regular monitoring. Wearable devices allow for the long-term tracking of personal health and lifestyle data. By continuously collecting data over time, trends and patterns in the digital twin can be identified and analyzed. These historical data provide a holistic view of an individual's health, enabling individualized predictions, the early detection of health problems, and the tracking of the effectiveness of interventions and lifestyle changes.

User Engagement and Behavioral Changes

Wearable devices can increase user engagement and drive behavioral changes. By providing real-time feedback, alerts, and data visualizations, individuals can become more aware of their behavior and make informed decisions to improve their health and well-being. Digital twins can use these data to simulate the possible consequences of changing behavior and provide personalized recommendations to encourage healthier habits. In summary, wearable devices play key roles in collecting real-time data on a person's physical and physiological characteristics, contributing to the development and optimizing the operation of the human digital twin. The integration of wearable technology enables personalized insights, remote monitoring, behavioral change support, and long-term tracking in the digital twin ecosystem.

Performance Optimization

Wearable devices are used to improve performance in many areas, including sports, fitness, and workplace productivity. Integrating portable data into a digital twin enables individuals to analyze their performance indicators, identify opportunities for improvement, and fine-tune routines and techniques accordingly.

Personalized Recommendations

By accessing data from wearable devices, human digital twins can provide personalized recommendations and insights. For example, the digital twin can suggest making adjustments to a personal exercise program on the basis of activity levels, sleep patterns, and heart rate data, or they can remind users to take breaks to avoid sitting for long periods of time.

Health and Wellness Tracking

Wearables help track health and wellness by continuously monitoring and recording data about physical activity, sleep quality, stress levels, and more. A human digital twin can use this information to create a comprehensive profile of an individual's general health, identify patterns and anomalies, and provide personalized advice to maintain or improve health.

Rehabilitation and Physical Therapy

Wearable devices can help monitor and control rehabilitation and physical therapy programs. By tracking an individual's movement, range of motion, and muscle activity, wearables provide data that can be integrated into their digital twin to monitor progress, track adherence to treatment plans, and adjust treatment approaches accordingly [1].

Interaction with Virtual and Augmented Reality

Wearables such as virtual reality (VR) headsets and augmented reality (AR) glasses can capture user movements and gestures. This information can be integrated into a digital twin, enabling a more immersive and interactive experience in virtual environments. This technology has applications in games, training simulations, and virtual collaboration.

Overall, wearable devices are transforming the human digital world by providing real-time data, enabling personalized insights and recommendations, and finding diverse applications in healthcare, wellness, performance optimization, and virtual environments.

Development of Human Digital Twins in Healthcare

Digital twin technology is gaining a lot of attention in the medical field and offers a variety of uses and benefits. Two main concepts should be considered when discussing digital twins in healthcare. The digital twins of physical systems or devices and the digital twins of humans [20].

Digital Twins in Healthcare

In healthcare, digital twins can represent physical systems or devices such as medical equipment, other equipment, and even hospital rooms. A digital twin acts as a virtual replica, reflecting its real-world counterpart in real time [21]. Data from sensors, devices, and other sources can be integrated to gain a comprehensive understanding of a physical system's behavior, performance, and status. Digital twins of physical healthcare systems can be used for a variety of purposes [22].

Predictive Maintenance The continuous monitoring of medical devices and the performance of medical equipment via digital twins allows healthcare providers to predict and correct maintenance issues before failures or disruptions occur.

Simulation and Testing Digital twins allow medical professionals to simulate scenarios and test different strategies and interventions before implementing them in the real world. For example, medical professionals can simulate surgical procedures to optimize outcomes and minimize risks [23].

Remote Monitoring and Optimization Digital twins enable the remote monitoring and optimization of medical device performance to ensure that medical devices are operating efficiently and meeting desired goals. This improves patient safety and improves resource utilization.

Human Digital Twins in Healthcare

A human digital twin represents a comprehensive digital model of an individual's physiological, genetic, and behavioral characteristics. The goal is to combine factors such as genetics, lifestyle, medical history, and real-time data to capture and simulate the complexities of individual health. Wearable devices and other sensors should be considered [1].

Personalized Medicine By analyzing an individual's digital twin, medical professionals can gain insights into the individual's unique characteristics and adjust treatment plans accordingly. This approach helps assess disease risk, optimize drug dosing, and identify the most effective interventions for individual patients.

Disease Management Human digital twins enable the continuous monitoring of health parameters and can provide real-time feedback to individuals and healthcare providers. This facilitates the early detection of health problems, better disease control, and preventive care.

Drug Development and Clinical Trials Human digital twins can simulate the effects of new drugs and treatments on virtual models before conducting costly and time-consuming clinical trials. This approach helps identify potential risks, optimize treatment protocols, and accelerate the development of new therapies.

Collectively, both the digital twins of physical systems and devices and the digital twins of humans offer valuable opportunities in healthcare by improving efficiency, personalization, and patient outcomes.

The Human Digital Twin: The Future of Personalized Drugs

In recent times, the conception of digital twins has gained traction in industries such as manufacturing, automotive, and aerospace. A digital twin is a virtual replica of a physical object or system that enables real-time monitoring, analysis, and optimization. Still, the operation of this technology isn't limited to such particulars [17]. A fairly new conception, the human digital twin, is poised to revolutionize the field of pharmaceuticals. A human digital twin incorporates data from sources such as electronic health records, wearable devices, and genetics to produce a virtual representation of an individual's health. This comprehensive digital model enables medical professionals to predict the onset of a complaint, cover the progressions of conditions, and customize treatment plans to optimize patient health.

One of the most promising applications of human digital twins is in the field of precision medicine. Precision medicine aims to tailor treatment to the unique characteristics of each patient, taking into account factors such as genetics, environment, and lifestyle [2]. By harnessing the power of human digital twins, healthcare providers can better understand the unique factors affecting patient health and design targeted interventions to address them.

For example, a human digital twin could help identify patients at risk of developing chronic conditions such as diabetes and cardiovascular disease. By analyzing data from digital twins, healthcare providers can identify specific lifestyle factors that may contribute to such risk and recommend targeted interventions to mitigate them. This proactive approach to healthcare could not only improve patient outcomes but also reduce the burden on healthcare systems by preventing the development of chronic diseases.

One more area where the human digital twin could have a huge influence is the growth of innovative medicines and treatments. Historically, drug development has been a slow, expensive process with a high failure rate. However, using human digital twins to simulate the effects of new drugs on virtual patients can give researchers valuable insights into their safety and efficacy before proceeding to clinical trials [18]. This may save time and resources and reduce the risks associated with drug development.

Additionally, human digital twins may play key roles in advancing the field of personalized medicine by allowing treatments to be customized according to an individual's unique genetic makeup. For example, certain cancer treatments are known to be more effective in patients with certain genetic mutations. By analyzing the genetic data of a patient's digital twin, healthcare providers can identify the most appropriate treatment options and increase the likelihood of positive outcomes.

Despite the immense potential of human digital twins, several challenges must be overcome before they can become a reality. Two of the main concerns are privacy and security issues. Given the sensitivity of medical data, ensuring that information that is stored in digital twins is protected from unauthorized access and misuse is imperative. In addition, integrating data from various sources requires the development of standardized protocols and data formats, creating significant technical challenges.

The Growth of the Human Digital Twin: A Revolution in Healthcare and Beyond

The rise of the human digital twin is revolutionizing now not only healthcare but also digital technology. A digital twin is a virtual duplicate of a physical object, system, or device that can be used to predict and optimize the general overall performance of its real-world counterpart [24]. Human digital twins are created with the aid of integrating an individual's biological, physiological, and behavioral data, permitting customized healthcare and a deeper understanding of human health. The idea of digital twins has been round for many years and is used in the aerospace industry to optimize the layout and overall performance of complicated systems. However, current advances in artificial intelligence (AI), machine learning, and data analytics are paving the way for improving HDTs in other fields. One of the benefits of human digital twins is their capacity to customize healthcare. By developing a digital duplicate of the human body, clinical specialists can simulate special remedies and strategies to predict and optimize consequences [2]. This helps customize therapy, reduces the risk of adverse side effects, and improves outcomes.

In addition to personalized healthcare, human digital twins can also play important roles in drug development and clinical trials. By simulating the effects of drugs on a virtual human body, researchers can gain valuable insights into their efficacy and safety, speeding up the drug development process and reducing the need for animal testing. Additionally, digital twins can be used to identify patient populations who are most likely to benefit from a particular drug, enabling more-targeted and more-efficient clinical trials.

The potential uses of human digital twins transcend healthcare, such as sports, nutrition, or even fashion. For example, digital twins can be used to optimize an athlete's training and overall performance with the aid of simulating the outcomes of various training programs, diets, and strategies. Similarly, digital twins can be used to create customized meal plans that are based primarily on an individual's physical traits and dietary needs. In the fashion industry, digital twins may be used to create customized clothing and add-ons for a person's frame, form, and size [25].

Despite the several benefits of human digital twins, there also are a few drawbacks and ethical issues that need to be addressed. Two of the primary concerns are the safety and privacy of the private data required to create digital twins. Storing these data securely and using them responsibly are crucial to preserving privacy and preventing the abuse of such technology. As with any new technology, the benefits of digital twins can fall disproportionately on those who can afford access to them, widening the divide between those who have access and those who do not. Addressing these concerns will require a combination of policy measures, industry standards, and public education to ensure that the benefits of human digital twins are fairly shared across society.

The Authority of Human Digital Twins in Disease Prevention and Management

The power of human digital twins in disease prevention and management is an emerging area for the future of healthcare. As technology continues to evolve, researchers and medical professionals are finding new ways to use digital twins to improve patient outcomes and reduce the strain on healthcare systems. A human digital twin could revolutionize how disease prevention, diagnosis, and treatment are approached by creating a virtual replica of an individual's physical and biological systems. A human digital twin is a virtual representation of a person's physical, physiological, and behavioral characteristics. This digital model can be used to simulate and predict the outcome of different interventions, allowing healthcare providers to make more-informed decisions about the best course of action for a given patient. In the context of disease prevention and management, human digital twins can be used to identify risk factors, monitor disease progression, and assess the efficacy of different treatment options.

One of the major benefits of using human digital twins for disease prevention and treatment is the ability to identify risk factors and predict the likelihood of developing certain diseases. By analyzing data from a patient's digital twin, healthcare providers can identify patterns and trends that may indicate an increased risk of developing a particular disease. This information can be used to develop personalized prevention strategies, such as lifestyle changes and targeted medical interventions, that can reduce the risk of developing the disease. Human digital twins can be used not only to identify risk factors but also to monitor disease progression and assess the efficacy of different treatment options. By simulating the impact of different interventions on a patient's digital twin, healthcare providers can gain valuable insights into how specific treatments affect an individual's health over time. This helps them make informed decisions about the most appropriate course of action and ensure that their patients receive the best possible care.

Additionally, human digital twins may play important roles in the development of new treatments and therapies. By providing virtual testing grounds for new interventions, digital twins can help accelerate the research and development process, ultimately leading to more-effective treatments for a wide range of diseases. This is especially valuable for rare or complex diseases where traditional clinical trials are difficult or time-consuming to conduct.

The use of human digital twins in disease prevention and management also has the potential to ease the strain on healthcare systems. By enabling disease onset and progression to be more accurately predicted, digital twins can allocate resources more efficiently and reduce the need for costly and invasive diagnostic procedures. Furthermore, by facilitating the development of personalized prevention strategies, digital twins can help reduce the prevalence of chronic diseases, ultimately improving public health and reducing healthcare costs.

Despite the many potential benefits of human digital twins in disease prevention and management, there are also some challenges that need to be addressed to reach their full potential [26]. One of the main concerns is the need to ensure the privacy and security of patient data given that creating a digital twin requires collecting and analyzing large amounts of sensitive information. Moreover, developing an accurate and reliable digital twin requires significant investment in R&D and the collaboration of a wide range of subject matter experts. The power of human digital twins in disease prevention and management represents a promising avenue for the future of healthcare. By harnessing the potential of this transformative technology, medical professionals can improve patient outcomes, reduce strain on health systems, and ultimately build a healthier future for all.

The Imminence of the Human Digital Twin: Challenges, Opportunities, and Estimations

The concept of digital twins has been around for several years, but recent technological advances have made digital twins more accessible and applicable across a wide range of industries. In the human context, digital twins can be thought of as virtual representations of people's respective physical, mental, and emotional states that can be used to monitor, predict, and improve their health and well-being. As we move toward a more connected, data-driven world, the idea of a human digital twin is gaining traction [27]. Researchers and companies are looking for ways to create digital replicas of individuals so that they can make better decisions about their health, lifestyles, and even career choices. This new technology could revolutionize healthcare, personalized medicine, and human performance optimization. However, the development and implementation of human digital twins also comes with some challenges and ethical considerations.

One of the biggest challenges in creating a human digital twin is the collection and integration of large numbers of data from various sources. Creating a comprehensive digital replica of an individual requires collecting data from wearable devices, medical records, social media, and other sources. These data must then be integrated and analyzed to consistently and accurately represent an individual's physical, mental, and emotional state. Ensuring the accuracy and reliability of these data is important because inaccuracies can lead to inaccurate predictions and recommendations. Another challenge is the potential violation of privacy and ethical concerns related to the use of personal data. To create a human digital twin, sensitive information such as medical records and personal habits must be collected and analyzed. This raises the question of who has access to these data, how these data are stored and how they are used [28]. Addressing these concerns is critical to gaining public trust and increasing the public's acceptance of human digital twins. In healthcare, digital twins could enable more-personalized and more-effective treatment as well as the early detection and prevention of disease [23]. For example, digital twins can help doctors predict how a patient would respond to a particular treatment on the basis of their unique genetic makeup and medical history. This may allow for more-targeted therapy and improve patient outcomes.

In the field of human performance optimization, digital twins can be used to help individuals make better decisions about their eating, exercise, and sleep habits. By analyzing data from wearable devices and other sources, digital twins can provide personalized recommendations to help individuals reach their health and fitness goals. This can improve overall health and reduce lifestyle diseases [25].

As technology continues to advance, various industries will likely develop and implement human digital twins. Predictions about the future of human digital twins include the integration of artificial intelligence and machine learning to improve the accuracy and predictability of digital replicas and the development of new wearable devices and sensors to collect richer data [29].

Privacy and Security Issues

Human digital twins need access to personal data, including sensitive health information, behavioral data, and other personal attributes. Ensuring the confidentiality and security of these data is essential to protecting individual rights and preventing unauthorized access and misuse. Implementing robust data encryption, secure data storage, and strong access controls are critical to addressing privacy and security concerns.

- *Data quality and integration*: Human digital twins require accurate, high-quality data to work effectively. However, data quality issues such as noise, missing values, and mismatches can affect the reliability and accuracy of digital twins. Integrating data from disparate sources and formats can also be challenging, requiring data cleansing, standardization, and synchronization processes to ensure that consistent and reliable information is incorporated into the digital twin.
- *Model accuracy and validation*: Developing an accurate computational model of a human digital twin is a complex task. Building models that fully capture the complexities of human behavior, health, and performance is challenging because of the inherent diversity and uniqueness of individuals. Validating models against and adjusting them to reflect real-world data to ensure their accuracy and generalizability are ongoing challenges that require rigorous testing, evaluation, and refinement.
- *Ethical considerations*: Human digital twins raise ethical concerns around consent, autonomy, and the responsible use of personal data. Individuals should be able to control their digital twins and provide their informed consent to the collection and use of data. For human digital twins to respect individual rights, avoid bias, and promote equitable outcomes, addressing ethical concerns around data ownership, transparency in algorithmic decision-making, and fairness is important.
- *Interdisciplinary collaboration*: The development and implementation of technology for a human digital twin requires collaboration across disciplines such as computer science, medicine, psychology, and ethics. Bridging the gaps between these disciplines and fostering interdisciplinary collaboration is critical to the development of comprehensive and effective digital twin systems that can capture multiple dimensions of human behavior and health.
- *User experience and adoption*: The human digital twin should be intuitive, easy to use, and seamlessly integrated into an individual's life to drive adoption and engagement. To ensure a positive user experience, user interfaces, interaction modalities, and personalized feedback mechanisms need to be designed to accommodate individual preferences and needs. Educating and raising awareness among individuals and medical professionals about the benefits and applications of human digital twin technology is also important for widespread adoption.

Conclusion

Human digital twins hold immense potential for personalizing medicine. A comprehensive and dynamic representation of an individual's health has the potential to revolutionize the way that healthcare is delivered, paving the way for more-accurate diagnoses, targeted interventions, and better patient outcomes. As technology continues to evolve, human digitals twin may become integral parts of the medical field, ushering in a new era of precision medicine.

Human digital twins will have huge impacts on the future of wearable technology. Digital twins provide personalized insights and recommendations on the basis of an individual's unique physiological and psychological profile to improve health outcomes, optimize performance, and enhance user experiences in VR and AR applications. As technology continues to evolve, human digital twins will become an integral part of our daily lives in a matter of time, shaping the way that people communicate with wearable technology and the world around us.

The rise of human digital twins has the potential to revolutionize healthcare and beyond, with many potential applications healthcare, drug development, sports, science, and fashion. However, the challenges and ethical considerations associated with this technology must be addressed if its benefits are to be shared responsibly and fairly. As digital technology and human biology continue to integrate, human digital twins will play key roles in shaping the future of healthcare and beyond.

The future of human digital twins presents both challenges and opportunities. Despite confidentiality and privacy concerns, the benefits of this technology to well-being, personalized drugs, and human performance optimization are enormous. As researchers and companies continue to explore the potential of human digital twins, they are committed to addressing these challenges and creating a future where digital replicas can be used to improve people's lives.

References

1. Coşgun, A. E. (2023). Digital twin and human digital twin for practical implementation in Industry 5.0. In *Global perspectives on robotics and autonomous systems: Development and applications* (pp. 168–183). IGI Global.
2. Wang, E., Tayebi, P., & Song, Y.-T. (2023). Cloud-based digital twins' storage in emergency healthcare. *International Journal of Networked and Distributed Computing, 11*, 75–87.
3. Javaid, M., Haleem, A., & Suman, R. (2023). Digital twin applications toward Industry 4.0: A review. *Cognitive Robotics, 3*, 71–92.
4. Yang, D., Sun, M., Zhou, J., Lu, Y., et al. (2023). Expert consensus on the "Digital Human" of metaverse in medicine. *Clinical eHealth, 6*, 159–163.
5. Manickam, S., Yarlagadda, L., Shynu, P. G., & Chowdhary, C. L. (2023). Unlocking the potential of digital twins: A comprehensive review of concepts, frameworks, and industrial applications. *IEEE Access, 11*, 135147.
6. Shengli, W. (2021). Is human digital twin possible? *Computer Methods and Programs in Biomedicine Update, 1*, 100014.
7. Miller, M. E., & Spatz, E. (2022). A unified view of a human digital twin. *Human-Intelligent Systems Integration, 4*, 23–33.
8. Jahromi, M. J. H., Bamakan, S. M. H., Qiang, Q., & Tabbakhian, H. (2023). The capability of distributed generation in digital twin platform. *Procedia Computer Science, 221*, 1208–1215.
9. Inamura, T. (2023). *Digital twin of experience for human–robot collaboration through virtual reality.*", International Journal of Automation Technology (Vol. 17, pp. 284–291).
10. Appelboom, G., Camacho, E., Abraham, M. E., Bruce, S. S., Dumont, E. L. P., Zacharia, B. E., D'Amico, R., Slomian, J., Reginster, J. Y., Bruyère, O., & Connolly, E. S. (2014). Smart

wearable body sensors for patient self-assessment and monitoring. *Archives of Public Health, 72*(1), 28.
11. Grabler, I., Steffen, E., Maier, G. W., & Roesmann, D. (2023). Chapter 1: Introduction— The digital twin of humans. In *The digital twin of humans* (pp. 3–10). Springer Science and Business Media LLC.
12. Ashvini, G., Salamzadeh, Y., & Abdul Rahim, N. F. (2023). Which E-leadership skills are needed to deploy digital strategies? A study on multinational companies in northern Malaysia. In *Multidimensional and strategic outlook in digital business transformation: Human resource and management recommendations for performance improvement; contributions to management science* (pp. 217–230). Springer International Publishing.
13. Breuer, S., Braun, M., Tigard, D., Buyx, A., & Müller, R. (2022). How engineers' imaginaries of healthcare shape design and user engagement: A case study of a robotics initiative for geriatric healthcare AI applications. *ACM Transactions on Computer-Human Interaction, 30*, 1–33.
14. Manjula Devi, C., Dharani, I., & Srinivasan, A. (2023). Machine learning and healthcare. In *Predicting pregnancy complications through artificial intelligence and machine learning* (pp. 14–33). IGI Global.
15. Tropmann-Frick, M., Jaakkola, H., Thalheim, B., Kiyoki, Y., & Yoshida, N. (2022). Information modelling and knowledge bases XXXIII. In *Frontiers in artificial intelligence and applications*. IOS Press.
16. Ahram, T. Z., Karwowski, W., Vergnano, A., Leali, F., & Taïar, R. (2020). *Intelligent human systems integration 2020*. Springer Science and Business Media LLC.
17. Sampedro, G. A. R., Putra, M. A. P., & Abisado, M. (2023). 3D-AmplifAI: An ensemble machine learning approach to digital twin fault monitoring for additive manufacturing in smart factories. *IEEE Access, 11*, 64128–64140.
18. Dlamini, Z. (2023). *Society 5.0 and next generation healthcare*. Springer Science and Business Media LLC.
19. Ling, Y., An, T., Yap, L. W., & Zhu, B. (2019). Disruptive, soft, wearable sensors. *Advanced Materials, 32*(18), e1904664.
20. Haleem, A., Javaid, M., Singh, R. P., & Suman, R. (2023). Exploring the revolution in healthcare systems through the applications of digital twin technology. *Biomedical Technology, 4*, 28–38.
21. Soori, M., Arezoo, B., & Dastres, R. (2023). *Digital twin for smart manufacturing, a review.*", Sustainable Manufacturing and Service Economics (Vol. 2, p. 100017).
22. Sirigu, G., Carminati, B., & Ferrari, E. (2022). Privacy and security issues for human digital twins. In *2022 IEEE 4th international conference on trust, privacy and security in intelligent systems, and applications (TPS-ISA)*. IEEE.
23. Vasiliu-Feltes, I. (2022). Chapter 4: Impact of digital twins on smart cities. In *Impact of digital twins in smart cities development* (pp. 104–126). IGI Global.
24. Wang, Z. (2022). *Mobility digital twin: Concept, architecture, case study, and future challenges*. Institute of Electrical and Electronics Engineers (IEEE).
25. Mourtzis, D., Angelopoulos, J., & Panopoulos, N. (2023). The future of the human–machine interface (HMI) in Society 5.0. *Future Internet, 15*, 162.
26. Asad, U., Khan, M., Khalid, A., & Lughmani, W. A. (2023). Human-centric digital twins in industry: A comprehensive review of enabling technologies and implementation strategies. *Sensors, 23*, 3938.
27. Pilz, S., Hellweg, T., Harteis, C., Rückert, U., & Schneider, M. (2023). Chapter 2; Who will own our global digital twin: The power of genetic and biographic information to shape our lives. In *The digital twin of humans* (pp. 11–35). Springer Science and Business Media LLC.
28. Farbiz, L. (2023). *Exploring the ethical and societal implications of incorporating user data into the ML workflow lifecycle*. Institute of Electrical and Electronics Engineers (IEEE).
29. Loaiza, J. H., Cloutier, R. J., & Lippert, K. (2023). Proposing a small-scale digital twin implementation framework for manufacturing from a systems perspective. *Systems, 11*, 41.

Digital Twin Application in Various Sectors

M. Mythily, Beaulah David, and J. Antony Vijay

A digital twin refers to a virtual representation or replica of a physical object, process, or system. It is a digital counterpart that simulates the behavior, characteristics, and performance of its real-world counterpart. The concept of a digital twin has gained popularity in various industries, including manufacturing, healthcare, transportation, and smart cities.

Main Components of Digital Twin

As seen in Fig. 1, a digital twin is a virtual version of a real system, process, or item. The development and operation of a digital twin encompass a number of significant elements that work together to enhance its precision, efficiency, and usefulness. Generally speaking, a digital twin's main parts are:

Physical Object: The physical object can be anything from a simple device to a complex system. It could be a machine, a building, a vehicle, an infrastructure network, or even a human body.

Sensors and Data: Sensors embedded in the physical object collect real-time data and send it to the digital twin. This data could include information about temperature, pressure, location, performance metrics, or any other relevant parameters.

M. Mythily (✉)
Karunya Institute of Technology and Sciences, Coimbatore, India

B. David
Hindusthan College of Engineering and Technology, Coimbatore, India

J. A. Vijay
Karpagam College of Engineering, Coimbatore, India

© The Author(s), under exclusive license to Springer Nature Switzerland AG 2024
A. Mishra et al. (eds.), *Transforming Industry using Digital Twin Technology*,
https://doi.org/10.1007/978-3-031-58523-4_11

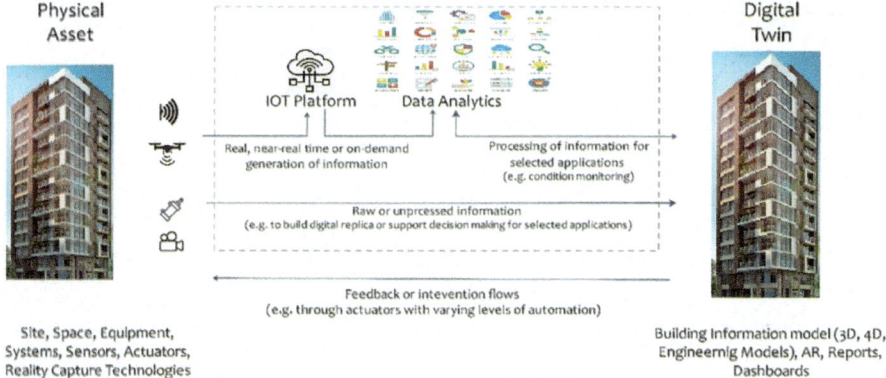

Fig. 1 Major components involved to simulate architecture

Virtual Model: The digital twin uses the collected data to create a virtual model that replicates the physical object. This model can be a 3D representation, a mathematical simulation, or a combination of various data-driven algorithms. The virtual model can be visualized, analyzed, and interacted with to gain insights and make informed decisions.

The purpose of a digital twin is to provide a deeper understanding of the physical object or system, enabling better monitoring, control, optimization, and predictive capabilities. It allows for real-time analysis, predictive maintenance, performance optimization, and scenario testing without directly impacting the physical object [1]. Digital twins can be used throughout the lifecycle of a product or system, from design and development to operation and maintenance.

Focal Areas of Digital Twin

Manufacturing: Digital twins can simulate production processes, optimize manufacturing operations, and improve quality control.

Healthcare: Digital twins of human organs or body systems can assist in personalized medicine, surgical planning, and remote patient monitoring.

Transportation: Digital twins of vehicles or transportation networks can aid in traffic management, route optimization, and predictive maintenance.

Smart Cities: Digital twins of entire cities can help urban planners optimize energy usage, manage resources, and enhance sustainability.

Digital twins offer a powerful tool for better understanding, monitoring, and optimizing physical objects and systems, leading to improved efficiency, productivity, and decision-making.

Application Areas of Digital Twin

Digital twins have numerous applications across various industries to simulate various real-time aspects as shown in Fig. 2. Some of the key applications of digital twins include:

Production and Logistics Operations: Digital twins can be used to model and enhance production lines, track machinery performance, forecast maintenance requirements, and increase overall productivity. Manufacturers can find bottlenecks, streamline workflows, and reduce downtime by building a virtual replica of the actual manufacturing system [2]. The manufacturing sector has benefited greatly from the introduction of digital twins, which have revolutionized several facets of the production process. Digital twins have made the following significant advances in manufacturing.

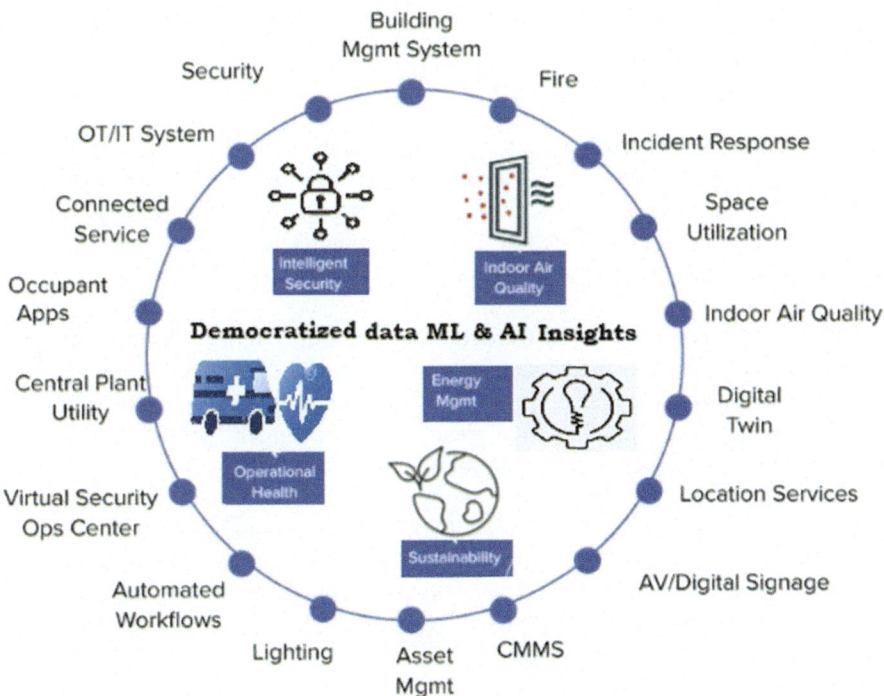

Fig. 2 Simulated features by DTT

Virtual Design and Simulation: Digital twins enable manufacturers to create virtual replicas of their products and production systems, allowing for design optimization and simulation before physical prototypes are built. This helps reduce costs and time-to-market by identifying and resolving issues early in the design phase.

Process Optimization: Digital twins can simulate and optimize manufacturing processes, helping to identify bottlenecks, improve efficiency, and reduce waste. By analyzing real-time data from sensors and integrating it with the digital twin, manufacturers can make informed decisions to optimize production parameters and increase overall productivity.

Predictive Maintenance: Digital twins facilitate predictive maintenance by continuously monitoring the performance of equipment and machinery. By analyzing data from sensors embedded in the physical assets and comparing it with the digital twin's model, manufacturers can predict maintenance needs, schedule repairs or replacements, and avoid unplanned downtime [7].

Quality Control: Digital twins enable real-time monitoring and analysis of production data, helping to identify deviations, defects, or quality issues. By comparing the digital twin's model with the actual production data, manufacturers can quickly detect anomalies, make adjustments, and ensure consistent product quality.

Training and Simulation: Digital twins can be used for training purposes, allowing operators and technicians to practice operating complex machinery or production systems in a virtual environment. This reduces the risk of accidents, improves skills, and enhances overall safety in the manufacturing facility.

Supply Chain Optimization: Digital twins can be extended to encompass the entire supply chain, enabling manufacturers to optimize inventory levels, track shipments, and improve logistics operations. By integrating data from suppliers, transportation systems, and production facilities, manufacturers can streamline the supply chain, reduce lead times, and enhance overall efficiency.

Continuous Improvement and Iterative Innovation: Digital twins provide manufacturers with valuable insights into the performance of their products and processes. By analyzing data from the physical system and leveraging the digital twin's model, manufacturers can identify areas for improvement, iterate designs, and implement continuous process and product innovation.

Internet of Things (IoT) and Smart Cities: Digital twins can be employed to manage and optimize complex IoT systems and smart city infrastructure. By creating virtual representations of physical assets, such as buildings, transportation networks, or energy grids, digital twins enable real-time monitoring, predictive analytics, and efficient resource allocation.

Overall, digital twins have had a transformative impact on the manufacturing industry, enabling manufacturers to achieve higher efficiency, improved product quality, reduced costs, and faster time-to-market. The ability to virtually model and simulate the entire production ecosystem brings enhanced visibility, control, and optimization opportunities to manufacturers, leading to significant competitive advantages.

Manufacturing Industry

Digital twins have made significant contributions to the manufacturing industry, revolutionizing various aspects of the production process. Here are some key contributions of digital twins in manufacturing:

Virtual Design and Simulation: Digital twins enable manufacturers to create virtual replicas of their products and production systems, allowing for design optimization and simulation before physical prototypes are built. This helps reduce costs and time-to-market by identifying and resolving issues early in the design phase [11, 12].

Process Optimization: Digital twins can simulate and optimize manufacturing processes, helping to identify bottlenecks, improve efficiency, and reduce waste. By analyzing real-time data from sensors and integrating it with the digital twin, manufacturers can make informed decisions to optimize production parameters and increase overall productivity [18].

Predictive Maintenance: Digital twins facilitate predictive maintenance by continuously monitoring the performance of equipment and machinery. By analyzing data from sensors embedded in the physical assets and comparing it with the digital twin's model, manufacturers can predict maintenance needs, schedule repairs or replacements, and avoid unplanned downtime.

Quality Control: Digital twins enable real-time monitoring and analysis of production data, helping to identify deviations, defects, or quality issues. By comparing the digital twin's model with the actual production data, manufacturers can quickly detect anomalies, make adjustments, and ensure consistent product quality.

Training and Simulation: Digital twins can be used for training purposes, allowing operators and technicians to practice operating complex machinery or production systems in a virtual environment. This reduces the risk of accidents, improves skills, and enhances overall safety in the manufacturing facility [8].

Supply Chain Optimization: Digital twins can be extended to encompass the entire supply chain, enabling manufacturers to optimize inventory levels, track shipments, and improve logistics operations. By integrating data from suppliers, transportation systems, and production facilities, manufacturers can streamline the supply chain, reduce lead times, and enhance overall efficiency.

Continuous Improvement and Iterative Innovation: Digital twins provide manufacturers with valuable insights into the performance of their products and processes. By analyzing data from the physical system and leveraging the digital twin's model, manufacturers can identify areas for improvement, iterate designs, and implement continuous process and product innovation.

Overall, digital twins have had a transformative impact on the manufacturing industry, enabling manufacturers to achieve higher efficiency, improved product quality, reduced costs, and faster time-to-market. The ability to virtually model and simulate the entire production ecosystem brings enhanced visibility, control, and optimization opportunities to manufacturers, leading to significant competitive advantages.

Internet of Things

Digital twins play a crucial role in the Internet of Things (IoT) and Smart Cities initiatives, bringing advanced capabilities for monitoring, managing, and optimizing complex systems. Here are some key contributions of digital twins in the context of IoT and Smart Cities:

System Monitoring and Control: Digital twins create virtual replicas of physical assets, such as buildings, infrastructure, or transportation networks, allowing real-time monitoring and control. By integrating data from IoT sensors and devices, digital twins provide a comprehensive view of the system's status, enabling proactive maintenance, efficient resource allocation, and effective incident response.

Predictive Analytics: Digital twins leverage historical and real-time data to predict future events and trends. In IoT-enabled systems, digital twins analyze sensor data and apply machine learning algorithms to identify patterns, detect anomalies, and forecast potential issues. This enables predictive maintenance, optimized energy consumption, and enhanced decision-making for Smart Cities.

Simulation and Optimization: Digital twins simulate and model the behavior of physical systems, enabling the testing of various scenarios and optimization of operations. For Smart Cities, digital twins can help optimize traffic flow, energy usage, waste management, or water distribution. By running simulations on the digital twin, city planners can make informed decisions to enhance efficiency, sustainability, and citizen well-being.

Urban Planning and Design: Digital twins support urban planning by creating virtual representations of cities. Planners can simulate and visualize the impact of infrastructure changes, zoning regulations, or transportation projects on the city's overall functionality. Digital twins enable data-driven decision-making, allowing planners to optimize land use, improve transportation networks, and enhance overall urban resilience.

Citizen Engagement and Services: Digital twins enhance citizen engagement by providing personalized services and information. By integrating data from IoT devices and citizen interactions, digital twins can deliver tailored recommendations, real-time updates, and interactive experiences. For example, citizens can receive personalized transportation routes, energy consumption insights, or safety alerts based on their specific needs and preferences [19].

Collaborative Planning and Coordination: Digital twins facilitate collaboration and coordination among stakeholders involved in Smart City initiatives. By providing a shared virtual environment, different entities, such as government agencies, utility providers, or transportation authorities, can collaborate, exchange information, and jointly plan for the city's development. Digital twins foster data sharing, interoperability, and holistic decision-making.

Emergency Response and Resilience: Digital twins contribute to improving emergency response and city resilience. By simulating emergency scenarios and integrating real-time data, digital twins can help emergency services respond effectively to incidents, optimize resource allocation, and enhance situational awareness. Digital twins also aid in designing resilient infrastructure that can withstand and recover from disasters.

The combination of digital twins and IoT technologies brings a comprehensive understanding of the city's functioning, enabling data-driven decision-making, efficient resource utilization, and improved citizen services in Smart Cities.

Healthcare and Medical Systems

Digital twins can be used in healthcare to model and simulate patient physiology, treatment plans, and medical devices. They enable personalized medicine, optimize treatment strategies, and support surgical planning. Digital twins can also assist in medical training and education. Digital twins have significant contributions to make in healthcare and medical systems, transforming various aspects of patient care, treatment planning, and medical research. Here are some key contributions of digital twins in the healthcare industry:

Personalized Medicine: Digital twins enable the creation of virtual models that represent individual patients, taking into account their unique physiological characteristics, genetics, and medical history. These virtual models can be used to simulate the effects of different treatments or interventions, allowing physicians to make personalized treatment decisions based on specific patient profiles.

Treatment Optimization: Digital twins help optimize treatment strategies by simulating and predicting the response of patients to different therapies. By integrating data from various sources, such as electronic health records, medical imaging, and genomic data, digital twins can provide insights into treatment outcomes, allowing physicians to tailor treatment plans and optimize medication dosages.

Surgical Planning and Training: Digital twins can assist in surgical planning by creating virtual replicas of patient anatomy. Surgeons can practice complex procedures, simulate different surgical approaches, and anticipate potential challenges or complications. Digital twins also support surgical training, allowing surgeons-in-training to gain experience in a virtual environment before performing procedures on actual patients.

Medical Device Design and Testing: Digital twins facilitate the design and testing of medical devices and implants. By creating virtual representations of devices and integrating them with patient-specific data, manufacturers can optimize device designs, evaluate performance, and ensure compatibility with individual patients. Digital twins also aid in predicting the long-term effects of implants and optimizing their lifespan.

Remote Patient Monitoring: Digital twins enable remote patient monitoring by integrating real-time data from wearable devices, sensors, and patient records. By comparing the real-time data with the patient's digital twin, healthcare providers can monitor vital signs, detect anomalies, and provide timely interventions. Digital twins support telemedicine and remote healthcare delivery, enabling continuous monitoring and personalized care outside traditional healthcare settings.

Drug Discovery and Development: Digital twins contribute to the drug discovery and development process. By simulating the effects of potential drugs on virtual models of human physiology, digital twins can predict drug efficacy, optimize dosage regimens, and identify potential side effects or interactions. This can help accelerate the development of new drugs and reduce the costs associated with clinical trials.

Medical Education and Training: Digital twins serve as valuable educational tools for medical students and healthcare professionals. They can be used to create interactive learning experiences, simulate patient cases, and provide virtual training environments for various medical procedures. Digital twins enhance medical education by offering realistic and immersive training opportunities [13].

By leveraging digital twins in healthcare and medical systems, personalized and data-driven approaches can be adopted, leading to improved patient outcomes, enhanced treatment efficacy, and more efficient healthcare delivery. Digital twins enable better understanding of individual patients, support informed decision-making, and contribute to advancing medical research and innovation.

Energy and Utilities

Digital twins can help monitor and optimize energy generation, distribution, and consumption. By modeling power plants, renewable energy systems, or smart grids, digital twins can improve energy efficiency, predict maintenance needs, and enhance overall grid reliability.

Digital twins have significant contributions to make in the energy and utilities sector, revolutionizing the way energy is generated, distributed, and consumed. Here are some key contributions of digital twins in the energy and utilities industry:

Power Generation Optimization: Digital twins can be used to model and simulate power plants, allowing operators to optimize generation parameters, improve efficiency, and reduce emissions. By analyzing real-time data from sensors and integrating it with the digital twin, operators can make informed decisions to maximize power output, minimize downtime, and enhance overall plant performance [17].

Renewable Energy Integration: Digital twins aid in the integration and management of renewable energy sources, such as solar and wind power, into the grid. By creating virtual replicas of renewable energy systems, digital twins can optimize their performance, predict power generation, and enable better grid integration. This helps ensure grid stability, improve forecasting accuracy, and enhance the overall reliability of renewable energy sources [10].

Grid Monitoring and Management: Digital twins enable real-time monitoring and control of energy distribution networks. By integrating data from sensors, smart meters, and other IoT devices, digital twins provide a comprehensive view of the grid's status, allowing operators to detect faults, balance loads, and optimize power flow. This results in more efficient energy distribution, reduced losses, and improved grid resilience.

Predictive Maintenance: Digital twins facilitate predictive maintenance in energy and utility systems. By continuously monitoring equipment and analyzing data from sensors embedded in the physical assets, digital twins can predict maintenance needs, schedule repairs, and minimize unplanned downtime. This leads to improved asset reliability, reduced maintenance costs, and optimized maintenance schedules.

Energy Efficiency and Demand Response: Digital twins help optimize energy consumption and enable demand response programs. By creating virtual replicas of buildings, facilities, or individual energy-consuming devices, digital twins provide insights into energy usage patterns, identify areas of inefficiency, and enable targeted energy-saving measures. Digital twins also support demand response initiatives by simulating load management strategies and optimizing demand-side flexibility.

Grid Planning and Expansion: Digital twins aid in grid planning and expansion by simulating and analyzing different scenarios. They can help optimize the placement of new infrastructure, evaluate the impact of new generation sources or loads, and forecast future demand. Digital twins enable data-driven decision-making, ensuring that the grid infrastructure is designed and expanded in a cost-effective and resilient manner.

Energy Trading and Market Analysis: Digital twins support energy trading and market analysis by providing real-time data and predictive insights. By integrating market data, weather information, and demand forecasts, digital twins enable better energy trading decisions, optimize portfolio management, and enhance market competitiveness. Digital twins facilitate the analysis of complex market dynamics, enabling energy companies to respond effectively to changing market conditions.

Overall, digital twins bring advanced monitoring, analysis, and optimization capabilities to the energy and utilities sector. They enable better resource management, improved energy efficiency, enhanced grid reliability, and support the transition toward a more sustainable and resilient energy future.

Aerospace and Defense

Digital twins find applications in the aerospace and defense sectors, where they can be used to simulate and analyze the performance of aircraft, satellites, or military equipment. They support design optimization, predictive maintenance, and training simulations.

Digital twins have significant contributions to make in the aerospace and defense industries, bringing advanced capabilities for design, testing, maintenance, and operational optimization. Here are some key contributions of digital twins in the aerospace and defense sectors:

Design and Development: Digital twins aid in the design and development of aircraft, spacecraft, and military equipment. By creating virtual replicas of these systems, digital twins enable engineers to simulate and analyze their performance, identify design optimizations, and enhance overall efficiency. Digital twins support iterative design processes and reduce the need for physical prototypes, saving time and costs.

Testing and Validation: Digital twins facilitate testing and validation of aerospace and defense systems. By simulating operational conditions and environmental factors, digital twins help engineers assess system behavior, identify potential issues, and validate performance before physical implementation. Digital twins support rigorous testing and verification processes, ensuring safety and reliability.

Predictive Maintenance: Digital twins enable predictive maintenance for aerospace and defense systems. By continuously monitoring real-time data from sensors embedded in aircraft, engines, or military equipment, digital twins can predict maintenance needs, schedule repairs, and minimize unscheduled downtime. This leads to increased operational availability, improved safety, and reduced maintenance costs.

Training and Simulation: Digital twins contribute to training and simulation in the aerospace and defense sectors. They create virtual environments for pilots, operators, and maintenance personnel to practice complex tasks, emergency procedures, and mission simulations. Digital twins enhance training effectiveness, improve decision-making under challenging conditions, and reduce the need for costly physical training exercises.

Mission Planning and Optimization: Digital twins aid in mission planning and optimization for defense applications. By creating virtual replicas of mission scenarios and integrating real-time data, digital twins enable commanders to assess various strategies, evaluate risks, and optimize resource allocation. Digital twins support mission readiness and enhance situational awareness.

Supply Chain Management: Digital twins play a role in optimizing supply chain management in aerospace and defense. By creating virtual replicas of the supply chain network, digital twins provide visibility into inventory levels, track components, and optimize logistics operations [9]. Digital twins enable efficient procurement, reduce lead times, and enhance supply chain resilience.

Performance Monitoring and Analytics: Digital twins provide advanced performance monitoring and analytics capabilities for aerospace and defense systems. By integrating real-time data from sensors, flight data recorders, and other sources, digital twins enable operators and analysts to monitor system performance, detect anomalies, and perform data-driven analysis for performance optimization and troubleshooting.

Digital twins bring enhanced efficiency, safety, and cost-effectiveness to the aerospace and defense industries. They enable iterative design processes, support predictive maintenance, enhance training and simulation, optimize mission planning, streamline supply chains, and provide valuable insights for performance optimization. The application of digital twins in aerospace and defense contributes to improved operational capabilities, reduced risks, and enhanced mission success.

Urban Planning and Architecture

Digital twins can aid in urban planning by creating virtual replicas of cities, enabling the assessment of different scenarios, optimizing resource allocation, and enhancing sustainability. They can also help architects and engineers in designing and visualizing buildings or infrastructure projects. Digital twins have significant contributions to make in the fields of urban planning and architecture, transforming the way cities are designed, developed, and managed. Here are some key contributions of digital twins in urban planning and architecture:

Visualization and Simulation: Digital twins enable the creation of virtual representations of cities and architectural designs, providing visualizations and simulations of planned projects. Urban planners and architects can use digital twins to explore different design options, assess their visual impact, and simulate how the proposed projects will interact with the existing urban environment. This helps stakeholders and decision-makers gain a comprehensive understanding of the project before implementation.

Design Optimization: Digital twins support design optimization by analyzing various aspects of urban planning and architectural projects. By integrating data from multiple sources, such as geographical data, environmental factors, and infrastructure networks, digital twins can assess the feasibility and performance of different design alternatives. This enables architects and urban planners to make informed decisions and optimize designs for factors like energy efficiency, traffic flow, environmental impact, and social integration [16].

Data-Driven Decision-Making: Digital twins provide a data-driven approach to urban planning and architecture. By integrating real-time data from IoT sensors, satellite imagery, and other sources, digital twins offer insights into factors like population density, traffic patterns, energy consumption, and environmental conditions. This data enables evidence-based decision-making and supports the development of sustainable, efficient, and resilient cities.

Stakeholder Collaboration: Digital twins facilitate collaboration and engagement among various stakeholders involved in urban planning and architecture. By providing a shared virtual environment, digital twins allow architects, urban planners, government officials, and community members to collaborate, visualize design proposals, and provide feedback. This improves transparency, inclusivity, and fosters a participatory approach to urban development.

Infrastructure Management: Digital twins help manage and monitor urban infrastructure. By integrating data from sensors and IoT devices, digital twins enable real-time monitoring of utilities, transportation systems, and public services. This allows for efficient infrastructure management, predictive maintenance, and better allocation of resources. Digital twins also aid in assessing the impact of infrastructure changes on the overall urban environment.

Urban Resilience and Sustainability: Digital twins contribute to enhancing urban resilience and sustainability. By modeling the interactions between different urban systems, such as energy, water, waste management, and transportation, digital twins can identify vulnerabilities, simulate disaster scenarios, and optimize resource allocation for resilience planning. Digital twins also assist in evaluating the environmental impact of urban projects, facilitating the development of sustainable and green cities.

Post-Construction Monitoring: Digital twins continue to provide value beyond the design and construction phase. They can be used for post-construction monitoring, enabling real-time tracking of building performance, energy usage, and environmental impact. This information helps identify areas for improvement, optimize operations, and enhance the long-term sustainability of urban infrastructure [14].

Digital twins revolutionize the way urban planning and architecture are approached, allowing for data-driven decision-making, stakeholder collaboration, and the creation of sustainable, resilient cities. They enable designers, planners, and policymakers to envision and optimize urban environments, leading to more efficient infrastructure, improved quality of life, and better integration of buildings with their surroundings.

Transportation and Logistics

Digital twins can optimize transportation networks, analyze traffic patterns, and improve logistics operations. They can simulate and predict the movement of vehicles, manage fleet operations, and enhance supply chain management.

Digital twins have significant contributions to make in the transportation and logistics sector, revolutionizing the way transportation systems are managed, optimized, and operated. Here are some key contributions of digital twins in transportation and logistics:

Traffic Management and Optimization: Digital twins enable real-time monitoring and optimization of traffic flow. By integrating data from various sources, such as traffic sensors, GPS devices, and weather information, digital twins provide a comprehensive view of the transportation network. This allows for proactive traffic management, congestion prediction, and optimization of signal timings, leading to improved traffic flow and reduced travel times [3].

Fleet Management and Optimization: Digital twins aid in the management and optimization of transportation fleets. By creating virtual replicas of vehicles and integrating real-time data from sensors and telematics devices, digital twins enable efficient fleet tracking, route planning, and optimization of delivery schedules. This leads to reduced fuel consumption, enhanced logistics operations, and improved customer satisfaction.

Supply Chain Optimization: Digital twins contribute to optimizing supply chain operations. By creating virtual representations of the supply chain network, digital twins provide visibility into inventory levels, track shipments, and optimize logistics operations. This helps improve delivery efficiency, reduce costs, and enhance overall supply chain performance.

Predictive Maintenance: Digital twins enable predictive maintenance for transportation assets. By continuously monitoring real-time data from sensors embedded in vehicles, trains, or aircraft, digital twins can predict maintenance needs, schedule repairs, and minimize unscheduled downtime. This leads to increased asset availability, reduced maintenance costs, and improved operational reliability.

Passenger Experience and Safety: Digital twins enhance the passenger experience and safety in transportation systems. By integrating data from various sources, such as passenger feedback, ticketing systems, and IoT sensors, digital twins can provide personalized travel recommendations, real-time updates, and optimize passenger flow. Digital twins also support safety monitoring by analyzing data from surveillance cameras, sensors, and predictive analytics, helping identify potential hazards and improve overall transportation security.

Infrastructure Planning and Design: Digital twins aid in the planning and design of transportation infrastructure. By creating virtual replicas of roads, railways, or airports, digital twins allow for simulations and analysis of different scenarios. This helps in assessing the impact of infrastructure changes, optimizing capacity, and improving overall transportation network design.

Autonomous Vehicle Development and Testing: Digital twins play a critical role in the development and testing of autonomous vehicles. By creating virtual models that represent autonomous vehicles and simulating their behavior in various scenarios, digital twins enable safe and efficient testing before physical implementation. This helps accelerate the development of autonomous vehicle technologies and ensures their readiness for real-world deployment.

Predictive Maintenance: Digital twins facilitate predictive maintenance strategies. By analyzing real-time data from sensors embedded in assets and comparing it with the digital twin model, potential failures or performance degradation can be predicted. This helps in scheduling maintenance activities, optimizing resource allocation, and minimizing unscheduled downtime. Predictive maintenance based on digital twins reduces costs associated with reactive repairs and enhances asset availability [5, 15].

Performance Optimization: Digital twins aid in optimizing asset performance. By simulating different operating conditions and scenarios, digital twins help identify optimal settings, configurations, or operational strategies. This enables asset managers to make data-driven decisions for performance improvements, energy efficiency, and reliability enhancements. Digital twins provide insights into how changes or upgrades to assets can impact overall system performance.

Lifecycle Management: Digital twins support asset lifecycle management from planning to decommissioning. By integrating data and information across different phases, digital twins enable effective asset tracking, maintenance history, and documentation. This helps in decision-making related to asset upgrades, replacements, or retirement, ensuring efficient asset utilization throughout their lifespan.

Remote Asset Management: Digital twins facilitate remote asset management. By connecting to assets through IoT devices, digital twins enable asset managers to remotely monitor and control assets. This reduces the need for physical inspections and enables timely responses to alarms or maintenance needs. Remote asset management through digital twins improves operational efficiency, reduces costs, and enhances safety for assets in remote or hazardous locations.

Data-Driven Insights and Analytics: Digital twins provide rich data-driven insights and analytics. By analyzing historical and real-time data from assets, digital twins help identify patterns, trends, and correlations. These insights support decision-making processes related to asset performance, maintenance strategies, and optimization initiatives. Digital twins enable data-driven asset management, leading to improved operational efficiency and cost-effectiveness.

Asset Planning and Design: Digital twins contribute to asset planning and design. By simulating different asset configurations and scenarios, digital twins help optimize asset placement, design, and layout. This aids in minimizing risks, optimizing resources, and ensuring that assets are designed to meet operational requirements and future needs.

Digital twins revolutionize asset management by providing a digital representation of physical assets, integrating real-time data, and enabling data-driven decision-making. They facilitate proactive maintenance, optimize asset performance, support remote management, and enhance the overall efficiency of asset management processes.

These are just a few examples of the applications of digital twins. As the technology evolves, its potential applications are likely to expand further into new domains and industries [4, 6].

Case Study

Title: Enhancing Architectural Design and Performance through Digital Twins: A Case Study

Abstract This case study explores the application of artificial intelligence (AI) in architectural design to optimize the creation of innovative and efficient building designs. By utilizing AI techniques such as generative design and machine learning, architects can leverage computational power to generate and evaluate numerous design options, leading to more efficient and aesthetically pleasing structures.

This case study examines the integration of digital twin technology in architectural design to enhance the efficiency, sustainability, and performance of buildings. Digital twins, virtual replicas of physical structures, enable architects and stakeholders to simulate, monitor, and optimize various aspects of a building's lifecycle. Through a specific case study, this research explores the benefits and challenges of utilizing digital twins in architectural design.

Introduction

Digital twin technology has revolutionized various industries, and its application in architecture offers significant advantages. This case study focuses on the implementation of digital twins in a large-scale commercial building project called "Green Towers." The goal was to create a sustainable, energy-efficient, and technologically advanced architectural design.

Conceptualization and Design

During the conceptualization phase, architects used digital twin technology to develop virtual models of Green Towers. These digital twins included detailed information about the building's structure, materials, systems, and environmental factors. Architects could simulate different design options and evaluate their performance in terms of energy efficiency, natural lighting, ventilation, and occupant comfort.

Simulation and Analysis

Digital twins enabled architects to conduct simulations and analyses to optimize various aspects of the building's design. Computational fluid dynamics (CFD) simulations helped assess airflow, thermal comfort, and energy consumption. Daylighting simulations evaluated the distribution of natural light, reducing the need for artificial lighting. Structural analysis assessed load-bearing capacities, stress points, and safety factors.

Energy Performance Optimization

By integrating energy modeling with the digital twin, architects could optimize the building's energy performance. Through machine learning algorithms, the digital twin analyzed historical and real-time data to identify energy consumption patterns. This information was used to adjust the design parameters, such as insulation, HVAC systems, and renewable energy integration, to achieve optimal energy efficiency.

Lifecycle Management

Digital twins provided architects and facility managers with real-time monitoring and control capabilities throughout the building's lifecycle. Sensor data integrated with the digital twin allowed for continuous monitoring of energy usage, indoor air quality, occupancy patterns, and equipment performance. This data-driven approach facilitated predictive maintenance, early fault detection, and proactive energy management.

Collaboration and Communication

Digital twins served as a collaborative platform, enabling architects, engineers, contractors, and stakeholders to communicate and visualize design decisions effectively. Real-time updates and interactive 3D models facilitated better coordination and decision-making, reducing construction errors and minimizing rework.

Challenges and Limitations

Implementing digital twins in architecture requires robust data integration, complex modeling techniques, and expertise in various domains. Challenges may arise in terms of data accuracy, privacy, interoperability, and managing large-scale models. Furthermore, the initial investment in developing digital twin infrastructure and training stakeholders on its usage should be considered.

The current contribution of digital twin technology in industry can help gather valuable insights and opinions. Figure 3 depicts the contribution of digital twins in various domains. This analysis aims to provide insights into the distribution of twin technology in current trends for their improvement.

Conclusion

The case study of Green Towers demonstrates the significant benefits of utilizing digital twins in architectural design. By leveraging real-time data, simulations, and optimization algorithms, architects can create sustainable and high-performance buildings. Digital twins provide a platform for iterative design improvements, energy optimization, lifecycle management, and effective collaboration among stakeholders. Addressing the challenges associated with data integration and model complexity is crucial for successful implementation.

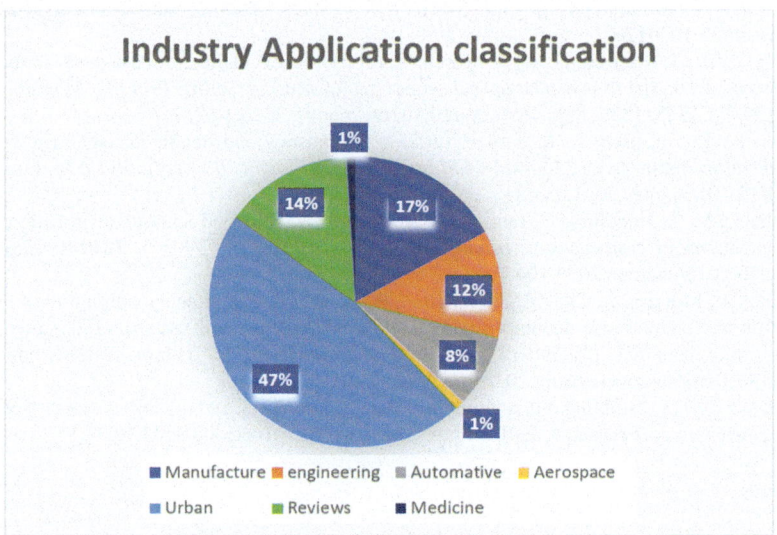

Fig. 3 Digital twin's contribution to supported industries

References

1. Attaran, M., & Gokhan, B. (2023). Digital Twin: Benefits, use cases, challenges, and opportunities. *Decision Analytics Journal, 6*(November 2022), 100165. https://doi.org/10.1016/j.dajour.2023.100165
2. Dubarry, M., Howey, D., & Wu, B. (2023). Perspective. Enabling battery digital twins at the industrial scale. *Joule, 7*(6), 1134–1144. https://doi.org/10.1016/j.joule.2023.05.005
3. Feng, H., Lv, H., & Lv, Z. (2023). Resilience towarded Digital Twins to improve the adaptability of transportation systems. *Transportation Research Part A, 173*(April), 103686. https://doi.org/10.1016/j.tra.2023.103686
4. He, B., Mao, H., Li, T., & Xiao, J. (2023). A closed-loop digital twin modeling method integrated with carbon footprint analysis. *Computers & Industrial Engineering, 182*(June), 109389. https://doi.org/10.1016/j.cie.2023.109389
5. Lai, X., Yang, L., He, X., Pang, Y., & Song, X. (2023). Digital twin-based structural health monitoring by combining measurement and computational data: An aircraft wing example. *Journal of Manufacturing Systems, 69*(2), 76–90. https://doi.org/10.1016/j.jmsy.2023.06.006
6. Liu, W., He, S., Mou, J., Xue, T., Chen, H., & Xiong, W. (2023a). Digital twins-based process monitoring for wastewater treatment processes ☆. *Reliability Engineering and System Safety, 238*(61773182), 109416. https://doi.org/10.1016/j.ress.2023.109416
7. Liu, S., Ren, S., & Jiang, H. (2023b). ScienceDirect. Predictive maintenance of wind turbines based on digital twin technology. *Energy Reports, 9*, 1344–1352. https://doi.org/10.1016/j.egyr.2023.05.052
8. Ogunsakin, R., Mehandjiev, N., & Marin, C. A. (2023). Towards adaptive digital twins architecture. *Computers in Industry, 149*(March), 103920. https://doi.org/10.1016/j.compind.2023.103920
9. Paiva, F., Trevisan, R., Santana, G., & Abel, M. (2023). A study on cloud and edge computing for the implementation of digital twins in the oil & gas industries. *Computers & Industrial Engineering, 182*(June), 109363. https://doi.org/10.1016/j.cie.2023.109363
10. Rantala, T., Ukko, J., Nasiri, M., & Saunila, M. (2023). Shifting focus of value creation through industrial digital twins – From internal application to ecosystem-level utilization. *Technovation, 125*(January 2022), 102795. https://doi.org/10.1016/j.technovation.2023.102795
11. Soori, M., Arezoo, B., & Dastres, R. (2023). Digital Twin for smart manufacturing. A review. *Sustainable Manufacturing and Service Economics, 100017*. https://doi.org/10.1016/j.smse.2023.100017
12. Sullivan, J. O., Sullivan, D. O., & Bruton, K. (2021). ScienceDirect. A case-study in the introduction of a digital twin in a large-scale smart manufacturing facility. *Procedia Manufacturing, 51*(2019), 1523–1530. https://doi.org/10.1016/j.promfg.2020.10.212
13. Sun, Y., Lu, Y., Bao, J., & Tao, F. (2023). Prognostics and health management via long short-term digital twins. *Journal of Manufacturing Systems, 68*(May), 560–575. https://doi.org/10.1016/j.jmsy.2023.05.023
14. Tuhaise, V. V., Handibry, J., Tah, M., & Abanda, F. H. (2023). Technologies for digital twin applications in construction. *Automation in Construction, 152*(May), 104931. https://doi.org/10.1016/j.autcon.2023.104931
15. Wang, J., Moreira, J., Cao, Y., & Gopaluni, R. B. (2023). Simultaneous digital twin identification and signal-noise decomposition through modified generalized sparse identification of nonlinear dynamics. *Computers and Chemical Engineering, 177*(May), 108294. https://doi.org/10.1016/j.compchemeng.2023.108294
16. Yoon, S. (2023). Building digital twinning: Data, information, and models. *Journal of Building Engineering, 76*(February), 107021. https://doi.org/10.1016/j.jobe.2023.107021

17. Yu, P., Ma, L., Fu, R., Liang, Y., Qin, D., Yu, J., & Liao, S. (2023). ScienceDirect. Framework design and application perspectives of digital twin microgrid. *Energy Reports, 9*, 669–678. https://doi.org/10.1016/j.egyr.2023.04.253
18. Zhang, J., Cui, H., Yang, A. L., Gu, F., Shi, C., Zhang, W., & Niu, S. (2023). An intelligent digital twin system for paper manufacturing in the paper industry. *Expert Systems with Applications, 230*(June), 120614. https://doi.org/10.1016/j.eswa.2023.120614
19. Zhu, Y., Cheng, J., Liu, Z., Cheng, Q., Zou, X., & Xu, H. (2023). Production logistics digital twins: Research profiling, application, challenges and opportunities. *Robotics and Computer-Integrated Manufacturing, 84*(February), 102592. https://doi.org/10.1016/j.rcim.2023.102592

A Review of Digital Twin Applications in Various Sectors

P. Kanaga Priya and A. Reethika

Introduction to Digital Twins: An Overview of the Concept and Its Potential

The technology of digital twins has emerged as a game-changing concept with the potential to revolutionize a wide range of industries. The definition of a digital twin is a virtual replica or simulation of a physical object, procedure, or scheme. This virtual representation mimics the physical and behavioral properties of its real-world counterpart in real time, documenting its interactions with the environment. Digital twins provide better knowledge, monitoring, and optimization of complex systems by linking the physical and digital realms. Digital twins have progressed from a theoretical concept to a real application, with significant usage across many industries. Because of developments in connectivity, sensors, data analytics, and processing capacity, the technology has acquired substantial traction. Within a virtual setting, digital twins can be used to acquire insights, forecast behavior, optimize performance, and simulate scenarios. Digital twins have several potential uses in a variety of industries. Figure 1 depicts a few areas where digital twin technology is being used to boost innovation and efficiency: manufacturing, energy, healthcare, transportation, construction, agriculture, aerospace, and smart cities.

In manufacturing, digital twins are transforming the way products are produced. By generating virtual replicas of the manufacturing process, companies can simulate different scenarios and identify potential issues early on. This enables improved product quality, optimized production processes, and predictive maintenance.

P. Kanaga Priya (✉)
Department of Computer Science and Engineering, KPR Institute of Engineering and Technology, Coimbatore, Tamil Nadu, India

A. Reethika
Department of Electronics and Communication Engineering, Sri Ramakrishna Engineering College, Coimbatore, Tamil Nadu, India

Fig. 1 Digital twins across various sectors

Manufacturers can use digital twins to make data-driven decisions, resulting in more efficient and cost-effective production operations. The energy sector is one more area where digital twins are making substantial strides. By creating digital replicas of energy systems, power plants, and grids, operators can simulate operations and maintenance procedures. This allows for the optimization of energy systems, prediction of maintenance needs, and reduction of energy consumption. Digital twins enable energy providers to enhance efficiency, improve plant reliability, and reduce downtime, ultimately making an impact on a more resilient and sustainable energy infrastructure.

In healthcare, digital twins could completely transform patient care and medical research. By creating personalized patient models based on real-time data, healthcare professionals can simulate surgical procedures, optimize treatment plans, and gain accurate insights into patient conditions. This technology holds the promise of improving diagnosis accuracy, enhancing treatment outcomes, and advancing personalized medicine. Transportation and construction are two sectors that greatly benefit from digital twin technology. In transportation, digital twins are employed to boost general efficiency, expedite deliveries, and optimize logistics. By simulating traffic flows, vehicle performance, and maintenance needs, transportation systems can be better managed and streamlined. Similarly, in construction, digital twins help improve project management, reduce errors, and enhance safety. Virtual replicas of construction sites enable present monitoring, resource optimization, and the identification of potential risks, leading to cost and time savings. Agriculture is also embracing digital twins to optimize crop yields and resource management. Farmers can create virtual models of their crops and leverage data insights to optimize water usage, fertilizer application, and pest control. By simulating different scenarios and understanding the interactions between various factors, digital twins enable more efficient and sustainable agricultural practices. Digital twins are critical in the aerospace industry for monitoring aircraft performance, forecasting maintenance needs, and enhancing safety. Real-time monitoring and analysis of aircraft data enables preventative maintenance, which reduces costs and improves overall reliability. Digital twins empower the aerospace industry to enhance operational efficiency, ensure passenger safety, and drive advancements in aviation technology. Smart cities leverage digital twin technology to simulate and optimize various aspects of urban life. By creating virtual models of cities, planners can test different scenarios and maximize resources for more effective and environmentally friendly city living. Digital twins assist the creation of smart and sustainable cities by enabling traffic flow control, energy usage optimization, and public safety advancements. Even though the potential aids of digital twins are significant, their implementation poses certain challenges like infrastructure requirements and data management. A DT is a representation of a physical thing that is linked to the object's true counterpart's real-time data exchange. It can be used for remote access, optimization, immediate monitoring, and other purposes.

Literature Review

Complex industrial systems that generate useful data for enhancing performance and maintenance have been brought about by Industry 4.0. DT's rising popularity has caused misunderstanding and a lack of agreement over its definition and types, though. To make straightforward identification possible, this document seeks to compile multiple definitions and DT kinds. It is essential for researchers, companies, and industries considering investment in this technology to comprehend the features, benefits, and drawbacks of DT. DT has been around for decades, but its effect has grown in recent years, with the ability to decrease costs, increase productivity, improve safety, and open the door to new applications. Determining the value of DT, setting rules and laws, managing data security problems, and assuring a qualified workforce are all challenges. To realize DT's full potential, a comprehensive understanding of its qualifications, implementation, benefits, and problems is required [1]. Energy DT technology has huge potential to change energy management and optimization in the process and energy industries. This study aims to provide better understanding and application of energy digital twin technology. It offers a multidimensional classification framework for digital twins, summarizes their uses across the lifecycle of a site, and recommends their implementation for carbon and environmental footprint reduction. The assessment underlines the issues that energy digital twins confront and proposes a framework for their implementation. Energy DT research, despite being less researched than other sectors, presents tremendous promise for energy efficiency, decarbonization, and cost optimization in the energy and process sectors. Energy efficiency, decarbonization, profitability, throughput, and quality improvement are the primary focal areas for energy digital twin deployment [2].

As a next-generation manufacturing system, intelligent manufacturing provides advantages such as enhanced quality, productivity, cost reduction, and production flexibility. Sustainability is becoming more popular, and sustainable production is evolving. Intelligent manufacturing uses digital twin technology to provide predictive maintenance and actual monitoring. The usage of DT in sustainable intelligent manufacturing provides practical benefits. This study examines intelligent manufacturing equipment, systems, and services; explores intelligent manufacturing sustainability; and introduces digital twin technology and its application. The direction of intelligent manufacturing's future development is also addressed, emphasizing the need for sustainable intelligent manufacturing and the function of DT in achieving efficiency, intelligent sensing, and failure prediction. Future directions are considered as well as a framework for digital twin-driven sustainable intelligent manufacturing [3].

DT are currently being used in multiple sectors like healthcare to improve clinical decision support and deliver personalized care. Combining digital twins and artificial intelligence (AI) allows for the integration and analysis of disparate data sources, resulting in improved diagnosis and treatment decision support. It focuses on the usage of digital twins and artificial intelligence (AI) in endometrial cancer care, addressing current limits and emphasizing the relevance of AI techniques. Digital twins provide a superior answer to the issues of rapidly rising cancer information, effective decision-making, and care that is patient-centered. They are in line with a value-based healthcare approach, which aims to control costs while providing high-quality care to patients [4]. The use of DTs has the potential to transform transport systems by offering instantaneous monitoring throughout their life cycle. This work aims to explore the topic of tiny highway simulation using real-time data integration. The suggested paradigm entails developing a DT model of the Geneva road that is constantly synchronized with actual traffic data streams. The methodological procedure of constructing the digital twin model is described, with special emphasis on the calibration aspects of the microscopic simulator SUMO. The digital twin paradigm enables simulation-based control optimization while the system is running, paving the path for real-time predictive analytics in traffic management. The results illustrate the approach's effectiveness and its potential for real-time study of a broader highway network [5].

The utilization of digital twins (DTs) is anticipated to become prevalent in the manufacturing industry, particularly in the Architecture, Engineering, and Construction (AEC) sector. Nevertheless, the implementation of DTs in the AEC industry is now at a nascent level, and it is imperative to establish standardized methodologies and criteria. The main contribution of the study is the development of a flexible methodological framework that can be applied to civil infrastructure. This framework addresses the challenge of employing digital technologies in the construction industry and offers a valuable solution [6]. The digital twin concept, characterized by its accurate and detailed modeling and two-way data exchange, has the capacity to diminish the distinction between state sensing, entity comprehension, and physical automation in the field of agriculture. The digital twin has the ability to address existing limits in decision-making assistance and automation in several agricultural sectors by utilizing data, modeling, and simulation. This study provides an extensive examination of the existing literature on digital twins in agriculture, encompassing present patterns and unresolved inquiries to enhance understanding and expertise. Although the digital twin has demonstrated potential in agriculture, there remain unresolved inquiries and concerns that necessitate attention before its widespread use in diverse agricultural contexts. To facilitate the expansion and application of digital twins in agriculture, it is imperative to conduct additional research in fields like as simulation, biological system modeling, and business model development [7].

Aerospace product assembly demands effective quality management and problem tracking due to single- or small-batch production and frequent disruptions. Digital twin (DT) technology, which blends virtual and physical space, provides a feasible solution by using the benefits of digital space. A digital twin–based quality management technique for aerospace product assembly that forecasts, analyzes, and resolves atypical quality concerns using the Grey-Markov model, T-K control chart, and Apriori algorithm. The implementation of this technique is addressed, using a digital twin–based quality control system for aerospace product assembly as an example. Two future research goals are to improve real-time feedback and to use text mining algorithms for rapid quality problem processing and decision-making [8]. The use of building information modeling (BIM) big data processing in smart city digital twins (DTs) to accelerate construction and increase data processing accuracy is discussed. The results reveal that the proposed methods significantly increase data analysis and classification performance. The work demonstrates the significance of BIM and DTs in smart city buildings, as well as the possibility for seamless integration of digital models and physical devices for real-time monitoring and maintenance. Future studies will concentrate on dealing with multi-source heterogeneity and applying deep learning and multimodal learning techniques to improve data representation in various big data applications [9].

The digital twins (DTs) and their applications in a variety of areas, including manufacturing, healthcare, transportation, and aerospace assembly, are discussed above. Understanding and characterizing DTs is emphasized to facilitate investment and decision-making by researchers, businesses, and industries. Cost reduction, increased productivity, improved safety, energy management, personalized treatment, and real-time monitoring are some of the possible benefits of DTs. Data security, worker skills, standardization, and handling the complexity of big data are mentioned as challenges. The papers propose approaches, algorithms, and frameworks for improving the application and effectiveness of DTs, highlighting the need for additional study in areas such as simulation, model creation, and data fusion [10–21].

Digital Twins in Manufacturing: Enhancing Efficiency, Quality, and Predictive Maintenance

Within the manufacturing sector, digital twins have emerged as a transformative technology, fundamentally disrupting conventional methods of production operations. By integrating digital twins, manufacturers can maximize efficiency, increase product quality, and apply predictive maintenance techniques, resulting in considerable cost savings and competitive benefits. By building a virtual reproduction of the manufacturing process, organizations obtain insights into the performance of individual machines, production lines, and the overall system. Real-time data from sensors implanted in physical assets can be collected and sent into the digital twin, enabling continuous monitoring and analysis. Manufacturers can utilize this capability to discover areas of congestion, optimize the sequence of tasks, and simplify operations in order to attain elevated levels of productivity. Digital twins are also important in boosting product quality. Manufacturers can uncover any defects and optimize parameters for greater product results by simulating the production process and analyzing data from the virtual replica. They can simulate various circumstances, run virtual tests, and make data-driven decisions to remove defects, decrease waste, and improve overall quality control. Real-time data synchronization between the physical manufacturing environment and the digital twin enables rapid alterations and improvements, ensuring constant product quality.

Today, the majority of new vehicle development occurs online. The digital enterprise solution portfolio also supports in the production of the product's digital twin, which subsequently permits accurate simulations to optimize the car before it is manufactured. The digital twin of the car serves as a depiction of the full vehicle including its software, mechanics, electrics, and physical qualities. This makes it possible to mimic and validate each stage of development, allowing for the early detection of problems and potential failures. Thanks to the digital enterprise portfolio, this capability applies to both conventional and electric cars.

The product's 3D data, for instance, can be used to model its behavior physically while optimizing elements like material characteristics, ventilation, and heat generation. The design and modeling of mechatronics, electronics, system-on-chips, and embedded software are also made easier by the virtual environment. Digitalization considerably reduces the requirement for physical prototypes, saving time and money. Additionally, it makes it possible for experts from diverse fields to work together on a project at the same time, makes it easier to configure numerous product iterations, and supports cutting-edge production methods like additive manufacturing. Moreover, digital twins enable predictive maintenance strategies in manufacturing, which can drastically reduce downtime and increase equipment reliability. By continuously monitoring the performance of machinery and capturing data on elements such as temperature, vibration, and energy consumption, prospective equipment breakdowns can be detected before they occur. This proactive strategy allows for scheduled maintenance tasks to be planned, limiting unexpected breakdowns and optimizing maintenance resources. Predictive maintenance based on digital twins not only extends the lifespan of equipment but also improves operational efficiency by decreasing unscheduled downtime and preventing costly production interruptions.

Another advantage of digital twins in manufacturing is the capacity to simulate and test various situations before making changes to the real environment. This virtual testing capability reduces the need for physical prototypes, minimizing costs and time associated with traditional trial-and-error methods. Manufacturers can assess the impact of process modifications, new equipment configurations, or production line reconfigurations within the digital twin. By analyzing the results and fine-tuning parameters virtually, they can optimize production processes without disrupting ongoing operations. Once the optimal solution is identified in the digital twin, it can be seamlessly implemented in the physical manufacturing environment, ensuring a smooth transition and minimizing the risk of costly errors. Furthermore, digital twins facilitate effective communication and collaboration among cross-functional teams involved in the manufacturing process. By providing a shared platform for data visualization and analysis, stakeholders from engineering, operations, quality assurance, and maintenance can access real-time information and make informed decisions collaboratively. This enhances transparency, aligns objectives, and improves overall coordination, resulting in streamlined operations and faster response times to production challenges. While the implementation of DT in manufacturing offers immense benefits, it also requires careful consideration of infrastructure, data management, and security aspects. Robust connectivity, reliable data integration, and scalable computational capabilities are essential for maintaining synchronization between the physical and virtual environments. Additionally, securing data security and protecting intellectual property become essential considerations when employing digital twins in manufacturing. In conclusion, digital twins have emerged as a disruptive technology in the manufacturing business, enabling greater productivity, improved product quality, and predictive maintenance capabilities. By establishing virtual duplicates of production processes and real-time data integration, manufacturers may optimize operations, decrease costs, and increase competitiveness. Digital twins help businesses to make data-driven decisions, adapt to market demands, and stay ahead in today's fast dynamic production scenario.

Leveraging Digital Twins in Healthcare: Revolutionizing Patient Care and Medical Research

Digital twins have found a remarkable application in the healthcare sector, ushering in a new era of personalized medicine and transformative advancements. Healthcare practitioners may revolutionize patient care, optimize treatment programs, simulate surgical operations, and advance medical research by leveraging the power of digital twins. The possibility of digital twins to generate tailored patient models is one of the most significant benefits of digital twins in healthcare. By integrating patient data, including medical history, genetic information, imaging scans, and real-time physiological assessments, healthcare professionals can construct entire digital copies of individual patients. These digital twins capture the unique characteristics and dynamics of each patient's body, allowing for personalized treatment strategies tailored to their specific needs. With digital twins, healthcare professionals can simulate various scenarios, predict the response to treatments, and optimize therapeutic interventions, leading to improved patient outcomes. In the realm of medical research, digital twins offer immense potential for accelerating discoveries and enhancing understanding. Researchers can leverage digital twins to simulate disease progression, study the effects of medications, and identify underlying mechanisms. By combining clinical data with computational models, they can gain insights into complex diseases, explore new therapeutic approaches, and test hypotheses in a simulated environment. Digital twins enable researchers to conduct in silico experiments, reducing reliance on traditional trial-and-error approaches and potentially expediting the development of novel treatments and interventions.

Digital twins also play a pivotal role in surgical planning and optimization procedures. Surgeons can create virtual replicas of patient anatomy allowing them to preoperatively visualize and simulate complex surgical procedures. This virtual practice empowers surgeons to plan precise incisions, evaluate potential complications, and optimize surgical strategies. By performing virtual surgeries within the digital twin, surgeons can refine their skills, enhance efficiency, and minimize risks during the actual operation. Digital twins enable a higher level of surgical precision and personalized care, ultimately leading to improved surgical outcomes and patient satisfaction.

Furthermore, digital twins facilitate the monitoring and management of chronic diseases. Digital twins provide real-time insights into a patient's physiological parameters by continuously gathering data from wearable devices, implantable sensors, and other sources, medication adherence, and overall health status. Healthcare providers can leverage this information to intervene proactively, detect early warning signs, and optimize treatment regimens. Digital twins encourage patients to actively participate in their care by offering them with tailored feedback, education, and support, enabling self-management of chronic illnesses. In addition to patient care, digital twins have the potential to reshape medical education and training. Healthcare professionals can utilize digital twins as educational tools, allowing students and trainees to interact with realistic patient cases and scenarios. This immersive learning experience enhances knowledge acquisition, decision-making skills, and clinical competence. Digital twins enable trainees to practice in a safe, controlled environment, promoting error-free learning and minimizing risks associated with real-life patient encounters.

However, leveraging digital twins in healthcare also poses challenges that need to be addressed. It is vital to secure the privacy and security of patient data, mandating effective data protection methods and adherence to regulatory requirements.

The integration of diverse data sources, interoperability of systems, and standardization of data formats face further obstacles in establishing digital twins across healthcare settings. In conclusion, digital twins have emerged as a disruptive tool in healthcare, transforming patient care and medical research. By creating personalized patient models, healthcare professionals can optimize treatment strategies and improve outcomes. Digital twins enable the simulation of surgical procedures, enhancing surgical precision and patient safety. They also expedite medical research by facilitating in silico experiments and the exploration of novel therapies. Digital twins have the potential to reshape medical education and training, enabling immersive learning experiences. As the healthcare sector continues to embrace digital twins, their potential to revolutionize patient care, advance medical research, and drive personalized medicine is truly remarkable.

Digital Twins in Energy and Utilities: Optimizing Operations and Grid Management

The energy and utilities sector faces numerous challenges in managing complex systems, optimizing operations, and ensuring reliable and efficient power distribution. Digital twins have emerged as a significant tool in this market, giving novel ways to boost operational performance, optimize energy systems, and expedite grid management. One of the primary applications of digital twins in the energy and utilities business is in optimizing operations. By establishing virtual replicas of power plants, substations, and other infrastructure components, operators can receive real-time insights into system performance, evaluate data, and detect potential bottlenecks or inefficiencies. Digital twins let operators to simulate multiple operating situations, assess the impact of modifications, and make informed decisions to maximize energy generation, limit downtime, and increase overall operational efficiency.

Moreover, digital twins play a key role in grid management. By developing a digital model of the electricity grid, utility firms can monitor and control grid operations in real time. Digital twins enable full visualization of the grid, allowing operators to discover areas of congestion, balance load distribution, and optimize power flow. They also offer predictive maintenance by continuously monitoring equipment health and identifying potential breakdowns before they occur. This proactive approach helps decrease unplanned outages, increase grid reliability, and optimize maintenance schedules, resulting in cost savings and improved customer satisfaction. These digital twins are made by engineers using information from numerous sensors and cameras, for example. The information gathered is then entered into software that creates a virtual model correctly simulating the behavior and attributes of the physical thing. Digital twins have three levels of complexity in the renewable energy industry. At the most fundamental level, they collect data from actual physical objects, such as solar panels, which can then be analyzed to forecast how well solar farms would perform. Digital twins are capable of modeling many scenarios at the intermediate level. To find the ideal configuration, businesses might experiment with operational parameters. For instance, the digital twin of a wind turbine may model how it would function in a hurricane, allowing operators to make the appropriate adjustments and keep stability.

Digital twins incorporate AI skills at the advanced level. They can spot variations in asset behavior and decide how to fix faults. A digital twin of a hydroelectric plant, for example, can help operators determine whether the facility is pumping too much water or is nearing capacity. Monitoring and decision-making are more efficient with this advanced degree of digital twin. Digital twins also aid in the grid integration of renewable energy sources. The power grid has gotten increasingly intricate and dynamic as renewable energy technologies such as solar and wind have gained prominence. Operators can utilize digital twins to model the integration of renewable energy sources, evaluate their influence on grid stability, and optimize their deployment for maximum efficiency. By modeling renewable energy generation, storage systems, and demand patterns, digital twins promote the integration of intermittent renewable sources, assure grid stability, and support the move toward a more sustainable energy mix. Furthermore, digital twins enable demand-side management and client involvement in the energy and utilities sector. By providing consumers with access to their digital twin, they can monitor their energy use, understand the impact of their usage habits, and make informed decisions to enhance energy efficiency. Digital twins also enable utility companies to provide personalized energy consumption feedback, offer targeted energy-saving recommendations, and incentivize customers to participate in demand response programs. This engagement enhances customer satisfaction, promotes energy conservation, and contributes to the overall optimization of the energy system.

The use of digital twins in the energy and utility sectors, on the other hand, poses its own set of hurdles. It involves sophisticated data collecting and integration from many sources, including sensors, meters, and control systems. Data security and privacy considerations also need to be addressed to ensure the protection of sensitive information. Additionally, the scalability of digital twins to span large-scale energy systems and the integration with existing infrastructure and legacy systems face considerable problems. Finally, digital twins have the potential to alter the energy and utilities industries by streamlining operations, boosting grid management, and facilitating the incorporation of renewable energy sources. With its ability to generate virtual replicas of energy systems, digital twins empower operators to make data-driven decisions, boost operational efficiency, and ensure dependable and sustainable power distribution. As the energy landscape continues to evolve, digital twins will play a critical role in creating smarter, more efficient, and resilient energy systems.

Transforming Transportation with Digital Twins: From Smart Cities to Autonomous Vehicles

The transportation business is undergoing a tremendous transformation, driven by technical discoveries and the increased demand for efficient, sustainable, and safe transportation systems. Digital twins have emerged as a significant enabler in this shift, offering a wide range of applications that are transforming transportation, from smart cities to autonomous vehicles. One of the key uses of digital twins in transportation is in the development and operation of smart cities. By developing virtual reproductions of cities, planners and policymakers can simulate and optimize different aspects of transportation, including traffic flows, infrastructure development, and public transportation systems. Real-time data from sensors, traffic cameras, and other sources can be evaluated using digital twins to enhance traffic management, eliminate congestion, and increase overall mobility. Digital twins enable decision-makers to make informed choices, plan for future growth, and construct sustainable and livable urban settings by modeling and simulating various scenarios. Digital twins are also important in the advancement of self-driving cars. Researchers and engineers may conduct comprehensive simulations, test algorithms, and develop autonomous driving systems by generating virtual reproductions of automobiles and their surrounding environs.

Digital twins enable the evaluation of vehicle performance in various road and weather conditions, the analysis of potential safety risks, and the optimization of navigation and decision-making algorithms. These virtual simulations help accelerate the development of autonomous vehicles, ensure their safety, and pave the way for their widespread adoption. Furthermore, digital twins support the optimization of transportation logistics and supply chain management. By creating virtual models of logistics networks, companies can simulate and optimize the movement of goods, reduce delivery times, and minimize costs. Digital twins enable the analysis of different transportation routes, the optimization of fleet management, and the prediction of potential disruptions. Digital twins enable end-to-end insight and control over the supply chain by integrating real-time data from sensors, GPS systems, and inventory management systems. This results in higher efficiency, lower waste, and increased customer satisfaction. Furthermore, digital twins contribute to the design of intelligent transportation systems. Digital twins allow real-time monitoring and analysis of transport networks by integrating data from varied sources such as vehicles, infrastructure, and passengers. This data-driven approach allows for the optimization of traffic signals, the implementation of adaptive traffic management strategies, and the provision of real-time information to travelers. Digital twins also support the development of connected vehicles and vehicle-to-infrastructure communication, enabling safer and more efficient transportation experiences.

However, the application of digital twins in transportation comes with issues that need to be addressed. Integration of different data sources, standardization of data formats, and system interoperability are all key challenges. Data security and privacy concerns also need to be addressed to ensure the protection of sensitive information. Furthermore, the scalability of digital twins to encompass large-scale transportation networks and the development of accurate and reliable models require substantial computational resources and expertise. In conclusion, digital twins are transforming transportation by enabling the development of smart cities, advancing autonomous vehicles, optimizing logistics, and enhancing transportation systems. Digital twins will continue to play a significant part in defining the future of transportation as it becomes more connected, efficient, and sustainable. The transportation industry can construct safer, more efficient, and sustainable transportation networks that meet society's changing needs by utilizing the power of digital twins.

Exploring Digital Twin Applications in Agriculture and Food Production: Precision Farming and Supply Chain Optimization

The agriculture and food production industry face numerous challenges, including the need for increased productivity, resource optimization, as well as providing a long-term and resilient food supply chain. Digital twins have emerged as a disruptive technology in this area, bringing novel solutions to optimize farming techniques, enhance crop yields, and streamline the entire food supply chain from farm to fork. One of the primary uses of digital twins in agriculture is precision farming. By developing virtual models of agricultural systems, farmers may monitor and analyze many elements such as soil conditions, weather patterns, crop growth, and pest infestations in real time. Farmers may utilize this real-time data connection to make data-driven decisions about irrigation, fertilizer application, and pest management. By simulating different scenarios and analyzing historical data, digital twins enable farmers to predict crop growth and plan for optimized harvesting. Precision farming using digital twins results in improved crop yields, reduced resource wastage, and increased sustainability.

Furthermore, digital twins contribute in the optimization of supply chains in the agriculture and food manufacturing industries. Stakeholders can acquire extensive visibility and control over the flow of agricultural products by building digital representations of the whole supply chain, from fields to processing facilities, distribution centers, and retail outlets. Digital twins provide real-time inventory monitoring, quality inspections, and logistical planning. Digital twins improve traceability, decrease waste, and optimize resource allocation along the supply chain by merging data from sensors, RFID tags, and other sources. These optimizations contribute to enhanced efficiency, reduced costs, and improved product quality and safety. Furthermore, digital twins aid in the optimization of supply chains in the agriculture and food manufacturing industries. The insights provided by digital twins help farmers optimize their resources, mitigate risks, and align their farming practices with market demands. Additionally, digital twins enable collaboration and information sharing among different stakeholders in the agriculture ecosystem, facilitating better coordination and efficiency across the entire value chain.

However, there are hurdles to implementing digital twins in agriculture and food production. Farmers and stakeholders need access to reliable and accurate data, which requires investment in data collection infrastructure and sensors. Data management and integration also pose challenges, as data from various sources and formats need to be harmonized and analyzed effectively. Additionally, there is a need for training and education to ensure farmers and stakeholders can utilize the full potential of digital twins and interpret the insights provided by these virtual models. In conclusion, digital twins offer immense potential in transforming agriculture and food production by enabling precision farming, supply chain optimization, and informed decision-making. By leveraging digital twins, farmers may boost production, minimize resource usage, and contribute to a more sustainable and resilient food supply chain. As the agriculture business expands, digital twins will play a key part in managing the issues and opportunities involved with supplying the growing need for food safely and efficiently.

Digital Twins in Construction and Infrastructure: Streamlining Design, Construction, and Maintenance

The construction and infrastructure industry is known for its complex and challenging projects that require seamless coordination, efficient design, and effective maintenance. Digital twins have grown as an important technology in the construction industry, with a wide range of applications that speed several stages of the construction lifecycle, from initial design to continual maintenance. One of the key applications of digital twins in construction is in design optimization. By developing virtual reproductions of buildings, bridges, or other infrastructure projects, designers and engineers can see the structure in a digital environment. Digital twins enable the investigation of multiple design possibilities, evaluation of structural performance, and identification of potential collisions or conflicts before construction begins. This early diagnosis and resolution of design concerns result in enhanced productivity, decreased rework, and cost savings throughout the building phase.

Furthermore, digital twins enable for real-time monitoring and control across the building. By combining data from a number of sources, including sensors, drones, and building information modeling (BIM), construction managers may follow progress, monitor quality, and assure adherence to project schedules. Digital twins enable the depiction of the construction site in a virtual environment, allowing stakeholders to remotely watch construction activity and make informed decisions. Real-time data analysis and simulation capabilities of digital twins also improve safety management, identifying possible dangers and improving building processes to promote worker safety. Digital twins continue to give value beyond the construction phase by simplifying ongoing maintenance and asset management. By generating a digital replica of the built building, facility managers may monitor and assess the performance of various systems, such as HVAC, electrical, or plumbing. Digital twins enable the integration of data from sensors, equipment maintenance records, and energy consumption monitoring, providing insights into the health of the infrastructure and predicting repair needs. Proactive maintenance planning based on real-time data analysis helps reduce downtime, optimize resource allocation, and extend the lifespan of the assets.

Moreover, digital twins support the concept of smart buildings and infrastructure. By integrating digital twins with Internet of Things (IoT) devices and building management systems, operators can create intelligent environments that enhance energy efficiency, occupant comfort, and overall operational performance. Digital twins enable for real-time monitoring and administration of building services such as lighting, ventilation, and security. This integration enables data-driven decision-making, predictive analytics, and process automation, resulting in optimized energy usage, lower costs, and better user experiences. However, the adoption of digital twins in construction and infrastructure does come with challenges. Data acquisition and integration from multiple sources, interoperability with existing systems, and data security and privacy concerns are among the key hurdles. Furthermore, the scalability of digital twins to encompass large-scale projects and the need for collaboration and information sharing among stakeholders pose additional challenges. In conclusion, digital twins are revolutionizing the construction and infrastructure industry by streamlining design, construction, and maintenance processes. By leveraging digital twins, stakeholders can optimize design, improve project coordination, enhance safety, and enable proactive maintenance. As the building and infrastructure industries embrace digital transformation, digital twins will play a critical role in delivering sustainable, efficient, and resilient projects.

Enhancing Performance and Safety in Aerospace and Defense Through Digital Twins

The aerospace and defense industry operate in a highly complex and safety-critical environment where performance, reliability, and safety are of paramount importance. Digital twins have emerged as a powerful tool in this industry, offering a range of applications that enhance performance, optimize maintenance, and improve safety across various aerospace and defense systems. One of the primary uses of digital twins in aerospace is monitoring and optimizing the performance of airplanes. By developing virtual replicas of aircraft, manufacturers and operators can collect and evaluate real-time data from sensors, flight data recorders, and maintenance logs. Digital twins enable the monitoring of critical systems, such as engines, avionics, and flight controls, providing insights into their performance and detecting potential anomalies or failures. This real-time monitoring allows for predictive maintenance, proactive troubleshooting, and the optimization of aircraft performance to ensure efficient and safe operations.

Moreover, digital twins facilitate the simulation and analysis of various flight conditions and scenarios. By creating virtual models of aircraft and their operating environments, engineers can conduct extensive simulations, test new designs, and evaluate system performance. Digital twins enable the analysis of factors such as aerodynamics, structural integrity, and fuel efficiency, leading to the optimization of aircraft design and performance. These simulations also contribute to the development of advanced flight control systems, pilot training programs, and safety enhancements. Digital twins are critical in the aerospace sector for enhancing maintenance practices. By integrating data from sensors, onboard diagnostics, and historical maintenance records, digital twins enable condition-based maintenance and predictive analytics. Operators can monitor the health of important components, estimate maintenance needs, and schedule maintenance activities accordingly. This proactive strategy avoids unnecessary downtime, decreases maintenance costs, and assures efficient utilization of resources. Furthermore, digital twins support remote monitoring and troubleshooting, allowing for real-time collaboration between maintenance crews and experts, leading to faster resolution of issues and improved aircraft availability.

In the defense sector, digital twins have applications in the optimization of military systems and operations. By building virtual reproductions of military equipment, vehicles, or even entire battlefields, commanders can simulate and assess numerous scenarios, review mission plans, and enhance operational performance. Real-time monitoring and analysis of weapon systems, logistics, and situational awareness are made feasible by digital twins, which improve decision-making capabilities and mission success. Additionally, digital twins enable training and simulation exercises, providing a realistic and immersive environment for military personnel to develop their abilities and tactics. However, the introduction of digital twins in aerospace and military is not without problems. Data integration from various sources, including legacy systems, data security, and protection, and maintaining data correctness and dependability are all major challenges. The scalability of digital twins to span complex and interconnected systems, interoperability with existing infrastructure, and the necessity for specialized skills further pose barriers to general adoption. Finally, digital twins have the potential to alter the aerospace and military industries by boosting performance, optimizing maintenance, and increasing safety. Stakeholders can use digital twins to monitor and optimize aircraft performance, simulate and analyze various scenarios, and optimize maintenance methods. In the aerospace and defense industry, the deployment of digital twins facilitates proactive decision-making, enhances operational efficiency, and increases safety. Digital twins will continue to play a vital role in driving innovation and changing the aerospace and defense industries as technology and data management capabilities grow.

Unleashing the Potential of Digital Twins in Retail and Consumer Goods: Personalized Experiences and Inventory Management

With the advent of digital technologies, the retail and consumer goods business is undergoing a fundamental upheaval. Digital twins, in particular, have the potential to transform the retail business by providing personalized consumer experiences, optimizing inventory management, and improving overall operational efficiency. One of the primary applications of digital twins in retail is in creating individualized consumer experiences. By creating virtual replicas of customers and their preferences, retailers can analyze vast amounts of data to understand individual buying behaviors, preferences, and trends. This allows for the customization of marketing strategies, product recommendations, and personalized offers tailored to each customer. Digital twins enable retailers to create targeted and personalized campaigns, resulting in improved customer engagement, loyalty, and satisfaction.

Furthermore, digital twins support the optimization of inventory management in the retail industry. By creating virtual models of physical stores and warehouses, retailers can monitor and manage inventory levels in real time. Digital twins integrate data from numerous sources, like point-of-sale systems, supply chain information, and customer demand patterns, to deliver accurate and up-to-date insights into inventory levels, stock replenishment needs, and product availability. This real-time visibility allows merchants to optimize inventory levels, eliminate stockouts, and prevent overstocking, resulting in enhanced operational efficiency and cost savings. Digital twins also facilitate the simulation and analysis of store layouts and merchandising strategies. Retailers can create virtual replicas of physical store environments and test different layouts, product placements, and visual merchandising techniques. This enables retailers to optimize store layouts for better traffic flow, enhance product visibility, and create engaging shopping experiences for customers. By analyzing customer movement patterns and behavior within the virtual store environment, retailers can identify opportunities to increase sales and improve overall customer satisfaction.

Furthermore, digital twins contribute in the optimization of supply chains in the retail and consumer goods industries. Stakeholders can acquire extensive insight and control over the flow of products by building digital representations of the whole supply chain, from manufacturers and suppliers through distribution centers and retail locations. Digital twins enable real-time monitoring of inventory levels, production capacities, transportation logistics, and demand fluctuations. This integration of data and simulation capabilities helps retailers optimize their supply chain processes, reduce lead times, and respond quickly to changing customer demands, ensuring efficient and effective product delivery. However, the implementation of digital twins in retail and consumer goods does come with challenges. Data integration and interoperability among many systems and stakeholders, data security and privacy issues, and the necessity for sophisticated analytics capabilities are some of the significant barriers. Additionally, the adoption of digital twins needs organizational preparation, investment in IT infrastructure, and the development of data-driven competencies within retail enterprises. In conclusion, digital twins have the ability to open new possibilities in the retail and consumer goods business by providing tailored customer experiences and streamlining inventory management. By leveraging digital twins, retailers can deliver tailored marketing strategies, optimize store layouts, enhance supply chain efficiency, and provide seamless and engaging shopping experiences for customers. Digital twins will play a critical role in generating innovation, boosting operational efficiency, and fulfilling the evolving demands of the modern customer as retailers continue to embrace digital transformation.

Challenges and Future Directions of Digital Twin Adoption Across Sectors

While the concept of digital twins holds tremendous promise for different sectors, its widespread implementation does not come without hurdles. Addressing these difficulties and determining a route for the future is important to realizing the full potential of digital twin technology. Here, we analyze some of the significant hurdles and suggest potential future routes for digital twin adoption across sectors.

Data Integration and Interoperability: One of the primary challenges is integrating data from diverse sources and ensuring interoperability among different systems. Digital twins rely on data from various sensors, devices, and platforms, making it essential to establish seamless data integration frameworks. Future directions involve developing standardized protocols, data models, and interfaces to enable smooth data exchange and interoperability.

Scalability and Complexity: As digital twin applications expand, scalability becomes a critical challenge. Building and managing digital twins for large-scale systems and complex environments require robust computational infrastructure and data processing capabilities. Future directions involve developing scalable architectures, leveraging cloud computing, and exploring edge computing solutions to support the growth and complexity of digital twin deployments.

Data Security and Privacy: The sensitive nature of data used in digital twins raises worries regarding security and privacy. Protecting data from illegal access, guaranteeing data integrity, and complying with privacy requirements are crucial. Future directions involve implementing robust security measures, adopting encryption and authentication mechanisms, and establishing clear governance frameworks to protect data privacy and security.

Analytics and Decision-Making: Extracting relevant insights from the massive volumes of data generated by digital twins is a significant challenge. Machine learning and artificial intelligence, for example, are critical in analyzing data, recognizing trends, and making informed judgments. Future directions involve enhancing analytics capabilities, developing advanced algorithms, and leveraging real-time analytics to enable proactive decision-making based on digital twin insights.

Organizational Readiness and Culture: Adopting digital twins often requires organizational transformation and a cultural shift toward data-driven decision-making. Overcoming resistance to change, fostering a culture of innovation, and building digital literacy across the organization are critical challenges. Future directions involve investing in change management, providing training and education, and promoting a culture of collaboration and openness to embrace digital twin technologies.

Cost and Return on Investment: Implementing digital twins involves significant investments in technology infrastructure, data management, and expertise. Calculating the return on investment (ROI) and presenting the value proposition of digital twin adoption can be problematic. Future directions involve conducting cost-benefit analyses, developing business models, and showcasing successful case studies to justify the investment in digital twin technology.

Looking ahead, several future directions can shape the adoption of digital twins across sectors:

Advanced Simulation and Predictive Capabilities: Future digital twins will incorporate more sophisticated simulation and predictive capabilities, enabling stakeholders to anticipate and simulate various scenarios accurately. This will enhance decision-making, optimize operations, and improve resource utilization.

Collaboration and Ecosystem Integration: Digital twins will increasingly involve collaboration among multiple stakeholders, integrating diverse systems, and sharing data across ecosystems. This collaborative approach will foster innovation, create synergies, and enable holistic solutions to complex challenges.

Augmented and Virtual Reality Integration: Integrating digital twins with augmented reality (AR) and virtual reality (VR) technology increases visualization and interaction. This integration will allow stakeholders to interact with digital twins in immersive environments, improving comprehension and decision-making.

Ethical and Sustainable Digital Twins: Future digital twins will prioritize ethical and sustainable practices, incorporating principles of fairness, transparency, and environmental responsibility. This shift toward ethical and sustainable digital twins will align with broader societal and environmental goals.

Edge Computing and Internet of Things (IoT) Integration: Leveraging edge computing and IoT technology will enable real-time data processing, lowering latency and enabling speedier decision-making.

Conclusion

Digital twins have considerable promise for addressing environmental challenges. Digital twins can aid in the development of sustainable solutions and contribute to society by simulating and modeling certain scenarios. While digital twins have proved their revolutionary potential in a variety of domains, challenges exist.

Significant expenditures in infrastructure, data management, and understanding are necessary for deployment. Overcoming these hurdles and cultivating stakeholder participation is vital for the broad implementation of digital twin technology. Finally, the examination of digital twin applications across sectors emphasizes their significance in giving real-time insights, lowering costs, optimizing processes, and improving safety. Digital twins can revolutionize industries, drive innovation, and open up new avenues for growth and improvement as technology advances.

References

1. Singh, M., Fuenmayor, E., Hinchy, E. P., Qiao, Y., Murray, N., & Devine, D. (2021). Digital twin: Origin to future. *Applied System Innovation, 4*(2), 36.
2. Yu, W., Patros, P., Young, B., Klinac, E., & Walmsley, T. G. (2022). Energy digital twin technology for industrial energy management: Classification, challenges and future. *Renewable and Sustainable Energy Reviews, 161*, 112407.
3. He, B., & Bai, K. J. (2021). Digital twin-based sustainable intelligent manufacturing: A review. *Advances in Manufacturing, 9*, 1–21.
4. Kaul, R., Ossai, C., Forkan, A. R., Jayaraman, P. P., Zelcer, J., Vaughan, S., & Wickramasinghe, N. (2023). The role of AI for developing digital twins in healthcare: The case of cancer care. *Wiley Interdisciplinary Reviews: Data Mining and Knowledge Discovery, 13*(1), e1480.
5. Kobayashi, K., Daniell, J., & Alam, S. B. (2024). Improved generalization with deep neural operators for engineering systems: Path towards digital twin. *Engineering Applications of Artificial Intelligence, 131*, 107844.
6. Gao, C., Wang, J., Dong, S., Liu, Z., Cui, Z., Ma, N., & Zhao, X. (2022). Application of digital twins and building information modeling in the digitization of transportation: A bibliometric review. *Applied Sciences, 12*(21), 11203.
7. Pregnolato, M., Gunner, S., Voyagaki, E., De Risi, R., Carhart, N., Gavriel, G., Tully, P., Tryfonas, T., Macdonald, J., & Taylor, C. (2022). Towards civil engineering 4.0: Concept, workflow and application of digital twins for existing infrastructure. *Automation in Construction, 141*, 104421.
8. Purcell, W., & Neubauer, T. (2022). Digital twins in agriculture: A state-of-the-art review. *Smart Agricultural Technology, 13*, 100094.
9. Zhuang, C., Liu, Z., Liu, J., Ma, H., Zhai, S., & Wu, Y. (2022). Digital twin-based quality management method for the assembly process of aerospace products with the Grey-Markov Model and Apriori Algorithm. *Chinese Journal of Mechanical Engineering, 35*(1), 105.
10. Lv, Z., Chen, D., & Lv, H. (2022). Smart city construction and management by digital twins and BIM big data in COVID-19 scenario. *ACM Transactions on Multimedia Computing, Communications, and Applications, 18*(2s), 1–21.
11. Shoji, K., Schudel, S., Onwude, D., Shrivastava, C., & Defraeye, T. (2022). Mapping the postharvest life of imported fruits from packhouse to retail stores using physics-based digital twins. *Resources, Conservation and Recycling, 176*, 105914.
12. Singh, M., Srivastava, R., Fuenmayor, E., Kuts, V., Qiao, Y., Murray, N., & Devine, D. (2022). Applications of digital twin across industries: A review. *Applied Sciences, 12*(11), 5727.
13. Onwude, D., Bahrami, F., Shrivastava, C., Berry, T., Cronje, P., North, J., Kirsten, N., Schudel, S., Crenna, E., Shoji, K., & Defraeye, T. (2022). Physics-driven digital twins to quantify the impact of pre-and postharvest variability on the end quality evolution of orange fruit. *Resources, Conservation and Recycling, 186*, 106585.

14. Torzoni, M., Tezzele, M., Mariani, S., Manzoni, A., & Willcox, K. E. (2024). A digital twin framework for civil engineering structures. *Computer Methods in Applied Mechanics and Engineering, 418*, 116584.
15. Chen, Z. S., Chen, K. D., Xu, Y. Q., Pedrycz, W., & Skibniewski, M. J. (2024). Multiobjective optimization-based decision support for building digital twin maturity measurement. *Advanced Engineering Informatics, 59*, 102245.
16. Wang, J., Li, X., Wang, P., & Liu, Q. (2024). Bibliometric analysis of digital twin literature: A review of influencing factors and conceptual structure. *Technology Analysis & Strategic Management, 36*(1), 166–180.
17. Fei, T. A., Xuemin, S. U., Cheng, J., Yonghuai, Z. H., Weiran, L. I., Yong, W. A., Hui, X. U., Tianliang, H. U., Xiaojun, L. I., Tingyu, L. I., & Zheng, S. U. (2024). makeTwin: A reference architecture for digital twin software platform. *Chinese Journal of Aeronautics, 37*(1), 1–8.
18. Wang, B., Zhou, H., Li, X., Yang, G., Zheng, P., Song, C., Yuan, Y., Wuest, T., Yang, H., & Wang, L. (2024). Human digital twin in the context of industry 5.0. *Robotics and Computer-Integrated Manufacturing, 85*, 102626.
19. Arowoiya, V. A., Oke, A. E., Ojo, L. D., & Adelusi, A. O. (2024). Driving factors for the adoption of digital twin technology implementation for construction project performance in Nigeria. *Journal of Construction Engineering and Management, 150*(1), 05023014.
20. Opoku, D. G., Perera, S., Osei-Kyei, R., Rashidi, M., Bamdad, K., & Famakinwa, T. (2024). Digital twin for indoor condition monitoring in living labs: University library case study. *Automation in Construction, 157*, 105188.
21. Costa, E. A., Rebello, C. M., Schnitman, L., Loureiro, J. M., Ribeiro, A. M., & Nogueira, I. B. (2024). Adaptive digital twin for pressure swing adsorption systems: Integrating a novel feedback tracking system, online learning and uncertainty assessment for enhanced performance. *Engineering Applications of Artificial Intelligence, 127*, 107364.

Digital Twin-Enabled Solution for Smart City Applications

Pawan Whig, Balaram Yadav Kasula, Ashima Bhatnagar Bhatia, Rahul Reddy Nadikattu, and Pavika Sharma

Introduction

The concept of a "digital twin" is widely recognized as a simulation technique that maximizes the use of physical models, sensors, historical operational data, and various other inputs to integrate data across multiple disciplines, physical attributes, scales, and probabilities. It serves as a tool for virtually simulating physical products, reflecting the entire life cycle of the corresponding physical entity or product in a mirrored digital space, as noted by Whig, Velu, and Naddikatu in 2022 [1–3]. However, the interpretation of "digital twins" has evolved, and there is currently no universally accepted definition. Nevertheless, the core elements generally acknowledged in digital twins encompass physical entities, virtual models, data, connections, and services, as highlighted by Alkali et al. in 2022 and Whig, Velu, and Sharma in 2022 [4, 5].

At the heart of digital twins lies a two-way mapping interaction between real and virtual spaces. This interaction, in contrast to unidirectional mapping that transfers data only from physical items to digital objects, involves bidirectional mapping or a "digital shadow." This term signifies that a change in the state of the physical item leads to a corresponding change in the digital object, but not vice versa, as illustrated in Fig. 1, following the insights of Whig, Kouser, Velu, et al. in 2022 [6, 7].

Digital twins initially found their application and significance in product and manufacturing design. Subsequently, they expanded their presence across diverse

---

P. Whig (✉) · A. B. Bhatia
Vivekananda Institute of Professional Studies-TC, New Delhi, India

B. Y. Kasula · R. R. Nadikattu
University of the Cumberlands, Williamsburg, KY, USA

P. Sharma
Department of ECE, BPIT, New Delhi, India

© The Author(s), under exclusive license to Springer Nature Switzerland AG 2024
A. Mishra et al. (eds.), *Transforming Industry using Digital Twin Technology*,
https://doi.org/10.1007/978-3-031-58523-4_13

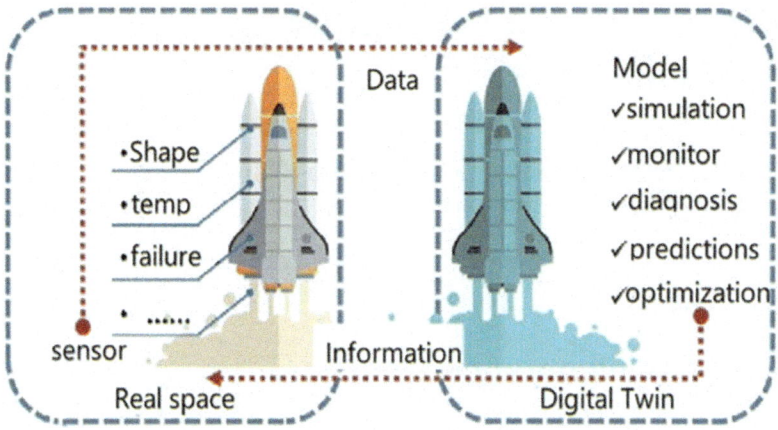

Fig. 1 Digital twin example

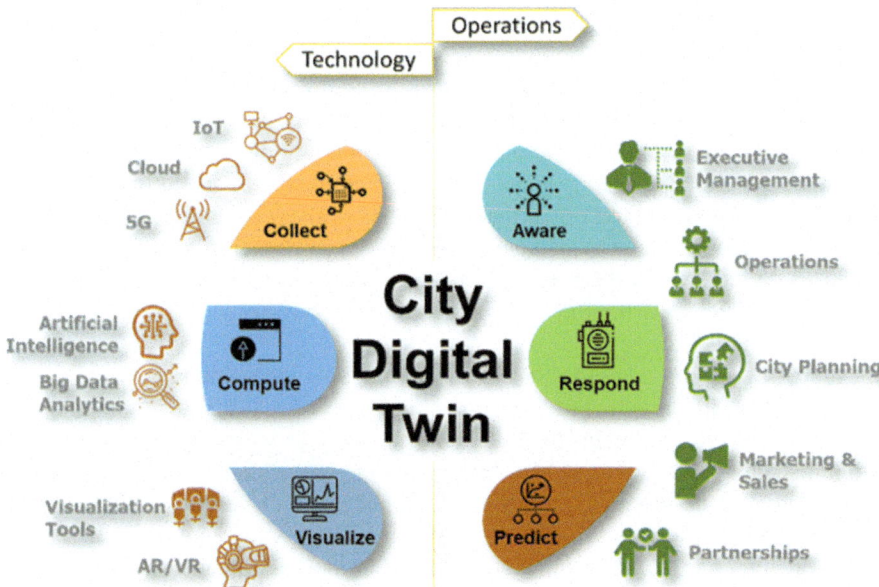

Fig. 2 Digital twin city

sectors, including aerospace, automation, shipbuilding, healthcare [8], and energy. The emergence of smart cities has been a gradual evolution fueled by advancements in technologies such as the Internet of Things (IoT), big data, cloud computing, and artificial intelligence (AI). The transition from static 3D modeling to digital twins, combining dynamic digital elements, has been notable [9, 10].

The concept of a digital twin city, as depicted in Fig. 2, represents a comprehensive application of the digital twin idea at the city level. It aims to establish a sophisticated, extensive system capable of mapping both the real world and the virtual environment, facilitating bidirectional communication [11, 12]. Building a digital twin city requires a technological foundation and a robust data infrastructure. The term "data foundation" encompasses the vast amounts of big data generated daily by numerous sensors and cameras throughout the city, along with the digital subsystems incrementally developed by municipal administration agencies [13–15]. The technical basis involves essential technologies such as 5G, big data, cloud computing, and AI.

In a digital twin city, information on infrastructure operational status, allocation of municipal resources, and the movement of people, goods, and vehicles is collected through sensors, cameras, and various digital subsystems. Leveraging technologies like 5G for data transmission to both the cloud and local government facilitates enhanced city efficiency [16]. The development of digital twin cities is poised to bring about significant advancements in intelligent urban planning, management, and services, serving as a transformative "new starting point" for building smart cities. This progress contributes to achieving goals related to intelligent city planning, management, services, and the comprehensive visualization of city information [8, 17].

The digital twin city holds a crucial role as a vital component within a smart city framework, representing the ultimate objective of a digital city. It is a foundational capacity that empowers cities to attain smartness. This transition in urban informatization, propelled by technology, marks a shift from qualitative to quantitative changes, providing greater opportunities for creativity in the creation of smart cities [7]. Digital twins, known for their affordability and usability, have gained widespread adoption across various industries with the growth of the Internet of Things (IoT). They play a pivotal role in showcasing the smart city concept, enabling efficient control from urban planning to land-use optimization. Digital twins allow for testing ideas before implementation, preemptively identifying issues. These technologies can be applied for planning and assessing architectural elements such as housing, solar panels, wireless network antennae, and public transportation [18]. Figure 3 illustrates the different levels within digital twin cities [19–23].

Even though Singapore is one of the most technologically advanced nations in the world, Thomas Pramotedham, CEO of Esri Singapore, thinks that every city starting its digital transformation path must first create a digital twin. Government agencies can only successfully examine what can be done with the data to better city life, provide economic opportunities, and rebuild a closer community with a digital twin in place, according to Pramotedham. Although the idea is currently novel in many nations, it is expected to catch on during the next 5–10 years [2]. Challenge Advisory is doing a lot of initiatives to reduce this duration. The 2019 event for digital twin smart city planning is one of them. The goal of this lecture is to raise awareness of both the concept itself and the notion of utilizing simulations to improve our communities and towns [12, 14, 18, 24–26]. Beyond our wildest dreams, the city would benefit from data-rich computer models that reproduce its physical characteristics and record its operations in real time. Digital twin implementation may help

Fig. 3 Different levels in digital twin cities

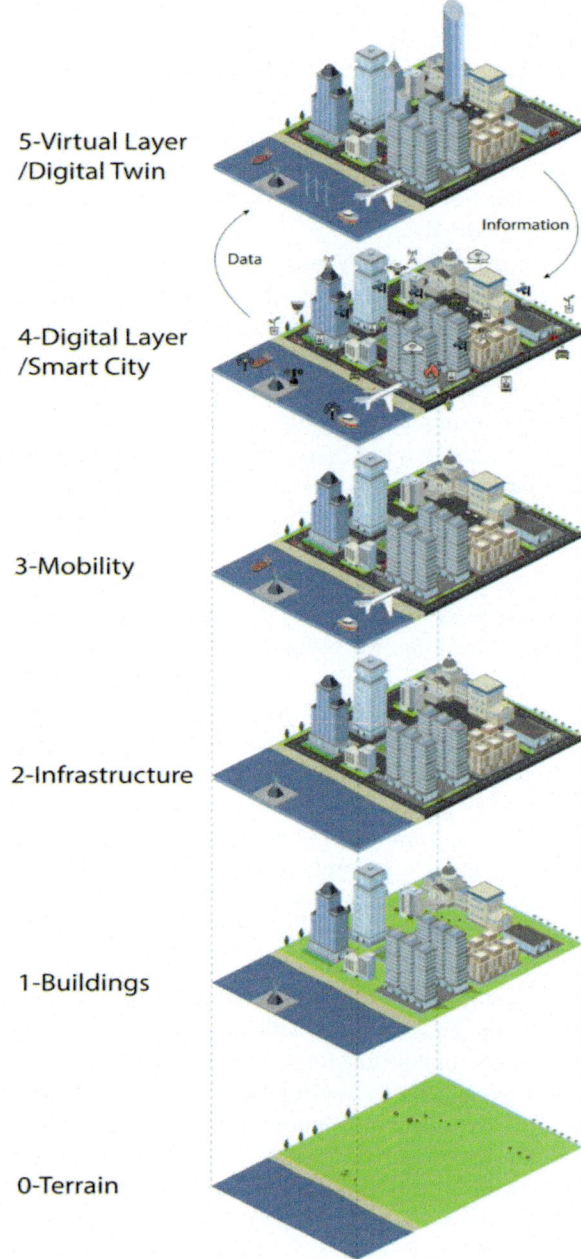

with areas including energy use, waste management, security monitoring, mobility enhancements, and infrastructure management [6].

The Evolution of Smart Cities

In the 1960s and 1970s, the Community Examination Agency utilized computer files, cluster examination, and infrared aerial photography to collect data, generate insights, and allocate resources strategically to areas most in need, aiming to prevent potential disasters and reduce deficiencies. This period marked the inception of the smart city concept [3]. Subsequently, the evolution of smart cities can be categorized into three distinct generations [15, 27, 28]. The stages of development in smart buildings are visually represented in Fig. 4.

During the era of Smart City 1.0, technology industry leaders took the forefront. Despite the city's limited understanding of the potential impacts of technology on daily life, this generation was focused on the widespread deployment of technology within urban spaces. Smart City 2.0, in contrast, witnessed a shift where cities themselves took the lead. Visionary municipal leaders played a key role in determining the future of the city and devising optimal ways to leverage advanced technology and other innovations [5]. In the third generation, neither technology companies nor local government officials assumed dominant leadership roles. Instead, a paradigm of citizen co-creation emerged. This recent evolution appears to be motivated by

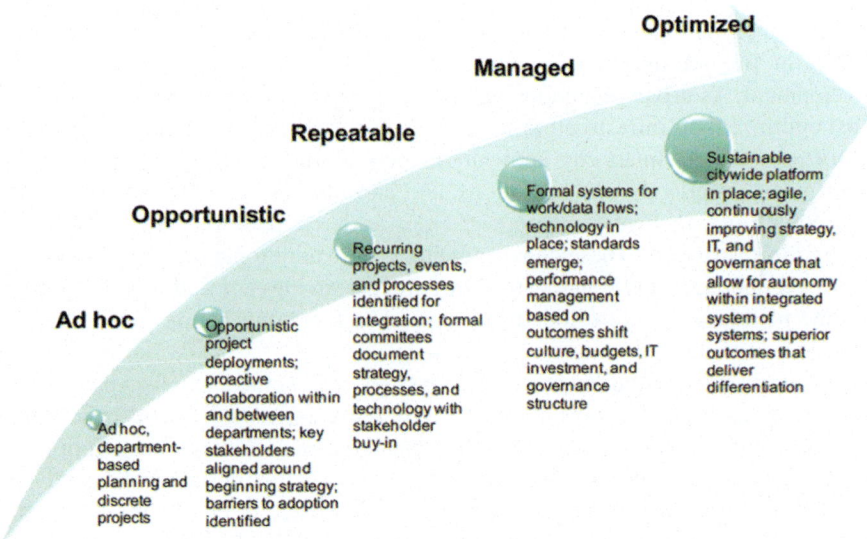

Fig. 4 Stages in smart cities development. (Source: IDC Government Insights, 2013)

Smart City 1.0	Smart City 2.0	Smart City 3.0
The creators of technological advancements encourage cities to implement their solutions, with the aim of improving the efficiency of urban management. Technology is the key element of the smart city 1.0 concept. Technological innovations are often implemented in cities that are not fully prepared for this process.	Local authorities play the key role in the development of smart cities 2.0. They focus on new technologies, to explore various options for improving the quality of life in cities. Cities introduce programs and projects which support the implementation of modern technologies in various areas of life. In a smart city 2.0, the significance of the quality of life and local governance is equated with that of modern technology.	This is the latest and the advanced generation of cities. Citizens play the k urban development. Loc residents consciously ch participate in the proces building modern cities, t on modern social partici tools, and are creative. In the smart city 3.0, urba created for users and wit involvement.

Fig. 5 Comparison between different versions

concerns about equity and a desire to cultivate a smart community that values communal presence. As part of this collaborative approach, Vienna, for instance, empowered locals to invest in solar power facilities and emphasized public participation in addressing issues like gender equality and affordable housing [10, 29–31]. Figure 5 provides a concise comparison between the different versions of smart cities.

Why Smart Cities Are Necessary

The primary goal of a smart city is to offer an urban environment that enhances the quality of life for its residents while concurrently promoting overall economic development. As urban populations continue to expand, metropolitan communities must optimize their infrastructure and resources to accommodate the growing number of inhabitants. Smart city applications play a crucial role in facilitating these adjustments, improving city operations, and elevating the residents' quality of life [11]. Smart city applications empower cities to derive additional value from their existing infrastructure. These enhancements lead to additional revenue sources and improved operational efficiency, resulting in cost savings for both the government and the populace. Figure 6 illustrates various factors contributing to the development of smart cities.

Singapore, often regarded as a benchmark for smart cities, employs devices and cameras to monitor the movement of registered cars, crowd density, and public space cleanliness. Businesses and individuals can track energy consumption and waste output. Singapore is also exploring geriatric monitoring technology and autonomous vehicles, including full-size robotic buses, to ensure the health and well-being of its senior residents [9].

In Kansas City, Missouri, smart city efforts encompass Wi-Fi connectivity along its two-mile streetcar track. The city's data visualization software provides public

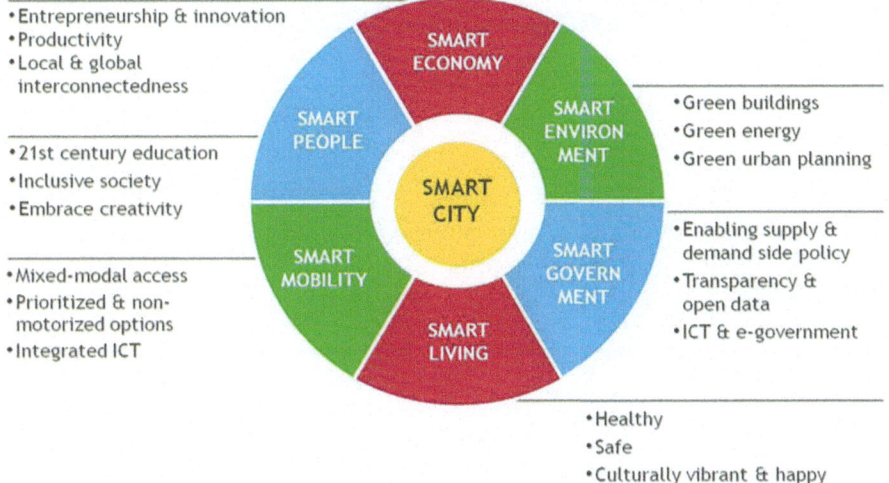

Fig. 6 Various factors for smart cities

access to information on available parking spaces, traffic flow, and pedestrian hotspots.

San Diego deployed 3200 smart sensors in 2017 to optimize parking and traffic, enhance environmental awareness, improve public safety, and enhance overall quality of life. The city features solar-to-electric charging stations to facilitate electric car usage, and networked cameras aid in traffic monitoring and criminal investigation.

Barcelona, Spain, implements smart transit and bus systems with features such as free Wi-Fi, USB charging ports, and real-time updates on bus schedules. The city also offers a bike-sharing scheme and smart parking software with online payment options. Sensors are utilized to monitor humidity, rainfall, pollution levels, noise, and temperature [13].

Challenges and Issues with Smart Cities: City officials face the challenge of promoting the use of open, democratized data among constituents and raising awareness of the benefits of adopted smart city technology. The key to cultivating engaged and empowered smart citizens lies in fostering collaboration between the public and private sectors and local citizens. Smart city initiatives should include plans to make data public and accessible to residents through open data sites or mobile apps, allowing residents to interact with and understand the data. Smart city apps can enable residents to perform personal tasks such as checking home energy usage, paying bills, and finding effective public transit. Figure 7 outlines various design challenges in smart cities.

Moreover, new smart metropolises that are being developed from the bottom up, such as Neom in Saudi Arabia and Buckeye in the desert region of Arizona, have the challenge of needing to attract citizens because they lack an existing population. These potential "smart cities" also lack any track record of accomplishment to

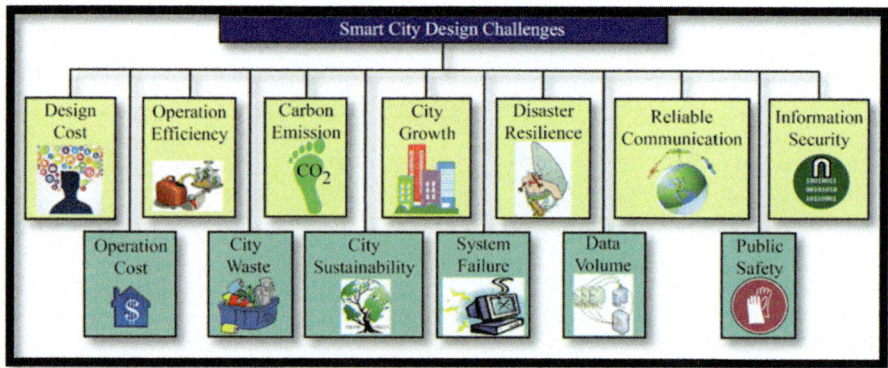

Fig. 7 Various smart city design challenges

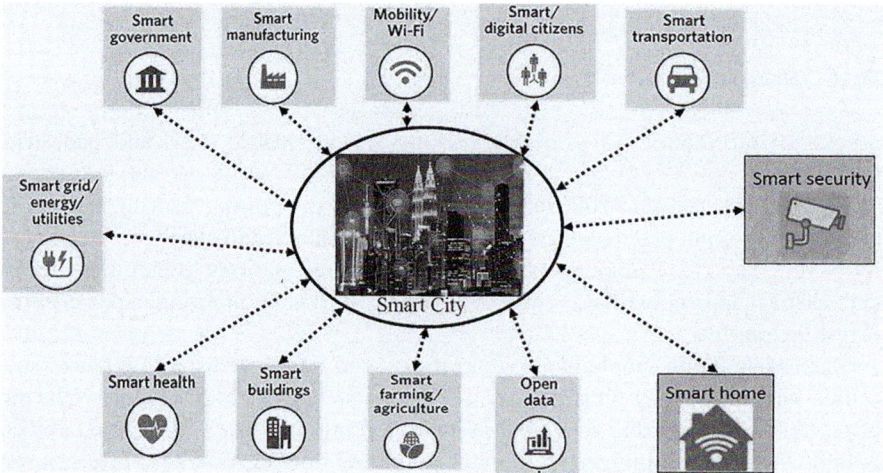

Fig. 8 Smart city components

inspire confidence [16]. There have been questions raised over whether a sustainable water source is even accessible as Neom and Buckeye have been constructed.

How Smart Cities Operate

Clever metropolises use their network of unified IoT plans and additional technology to improve livability and spur economic development as shown in Fig. 8. Four stages are followed by successful smart cities:

Data collection: Real-time data is gathered by smart sensors placed all around the city.

The analysis is the process of evaluating the data gathered by smart sensors to derive useful conclusions.
Communication: Through effective communication networks, the insights discovered during the analytical phase are shared with decision-makers.

Cities take action by utilizing the information gleaned from the information to develop answers, enhance operations and asset organization, and raise the resident's excellence of life.

Smart City Technology

Using tools like the Internet of Things (IoT) sensors and IoT analytics platforms, a smart city (or smart town) is a community that employs electronic means to collect data about its operations, processes, and infrastructure, including utilities, mobility, and infrastructure as shown in Fig. 9. Utilizing this information will enhance operations and enable improved resource, asset, and service management. In the end, the insights provided by this data assist cities in implementing certain projects and enhancing the quality of life for their residents [5].

In contrast to adopting a generic one-size-fits-all strategy, several towns are concentrating their smart city efforts on specific initiatives. The winners of Infrastructure

Fig. 9 IoT components for smart cites

Canada's Smart City Challenge are involved in a variety of activities, as can be seen below, each tailored to the unique requirements of the town.

Reduced energy poverty is a goal of the Town of Bridgewater in Nova Scotia.

Associated communities in Nunavut are attempting to put protective and preventative measures in place to lower the incidence of suicide. The winning project in Ontario's Wellington County and the City of Guelph aims to create a circular food economy that reduces waste and gives every citizen access to nutritious food. A smart city effort will be implemented in the City of Montreal, Quebec, to expand food access and enhance mobility.

Although digital twins may be employed in a variety of fields, including engineering, urban planning, and utilities, they all have a few important things in common. A digital twin is essentially a virtual depiction of actual things, procedures, actions, and connections, whether they be created naturally, artificially, or both. In this sense, a digital twin is more than just a 3D representation of a city, albeit it is.

Location's Function in Digital Twins

Location is crucial for digital twins since they are a key tool for creating smarter cities. A digital twin needs data to accurately reflect reality. In many cases, the problem is not a shortage of data but rather compiling massive amounts of various data kinds, such as those given by IoT devices, into context. This context is provided by location, which connects several data kinds into a single view.

Planners can design scenarios and assess implications in context and with more detail by utilizing data from smarter, more connected communities than they can with conventional techniques like physical models, spreadsheets, and straightforward digital 3D models. For instance, IoT sensors on automobiles assist planners in comprehending how and where traffic moves as shown in Fig. 10. This information may be utilized to enhance pedestrian security along specific routes, improve traffic flow, and promote alternate modes of transportation where they are most practical.

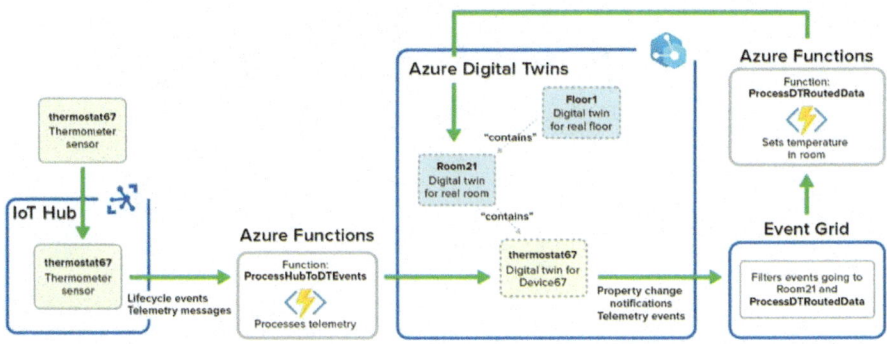

Fig. 10 Example to show planners design scenarios and assess implications

Planners may also use digital twins to investigate how proposed developments relate to and affect the surrounding environment using data from the actual world. Planners can assess whether a proposed development is beneficial by considering how it would change the mix of residential and retail properties in an area. Additionally, to better serve the city's underserved residents, planners may employ a digital twin to assess transportation times between different parts of the city in conjunction with socioeconomic data.

In each of these scenarios, the location offers a context that connects various data kinds, producing a comprehensive and accurate depiction of reality. All of these data are combined by a GIS into a visual, analytical system that models both the present situation and a desired one. The instances of how actual societies are utilizing digital twins are explored here.

Examples of Digital Twins

Boston: Potential Effects of Development

In this online picture as shown in Fig. 11, Boston's projected developments are depicted in the shadows they create over Boston Commons, a park in the city. Urban planners in Boston are utilizing ArcGIS Urban to explore several possibilities, including various pieces of data, such as building footprint, height, and location, and achieve their goals—in this case, reducing the number of shadows thrown on the park.

Fig. 11 Boston's projected developments

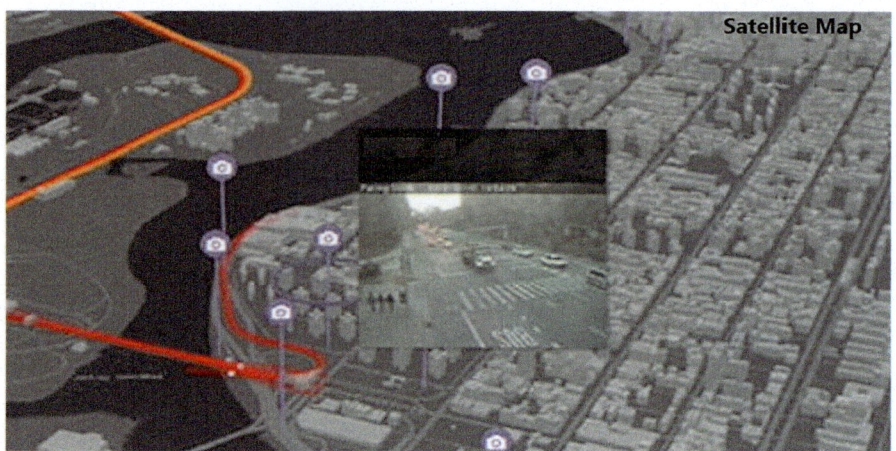

Fig. 12 Dashboard created by New York City

The City of New York has developed a dashboard for its Vision Zero initiative, as depicted in Fig. 12. This dashboard presents crucial indicators related to pedestrian safety, utilizing data from various sources and IoT sensors. These sensors include traffic sensors seamlessly integrated into a geographic context, intelligent cameras, and systems for tracking pedestrian traffic. Leveraging predictive analytics and real-time monitoring of traffic collisions, public safety agencies can enhance their ability to regulate traffic flow and improve pedestrian safety. The dashboard serves as a comprehensive tool for monitoring and addressing key aspects of the Vision Zero initiative in the city.

How Can Smart Cities Become Even Smarter?

As early as 2002, the term "smart communities" was used to describe locales where local decision-makers and stakeholders were collaborating and building alliances across electronic networks and the Internet to innovate and derive new economic and social value. Later, the concept of a "smart city" was more precisely defined and began to refer to a framework built on the Internet of Things and cloud computing that gathers, manages, and analyzes data to make cities more effective and responsive to inhabitants. Smart initiatives have been implemented in the transportation sector to improve parking, decrease traffic congestion, and make energy, water, and lighting systems more cost-effective and responsive to the requirements of the public.

According to the Center for Strategic and International Studies, there were 379 fully implemented smart city projects in 61 countries 2 years ago. Despite the pandemic, the landscape has not stopped changing as a result of the widespread interest in transforming static neighborhoods into intelligent, responsive, living urban areas. The smart city market, according to a market research firm, will grow to $2.75 trillion in value by 2023, and local government leaders throughout the world are on pace to adopt intelligent technology more and more.

Even the Epidemic Was Unable to Reverse the Situation

Although the coronavirus has irrevocably altered how we live, how we use our spaces, and how we see the future, it appears that the idea that urban existence necessitates clever technology used in smart ways will endure, especially when in some circumstances it may potentially aid in the pandemic-fighting effort. For instance, the city of Chicago used anonymized cellphone data to follow people's travels and analyze travel patterns to monitor the success of stay-at-home and self-isolation policies. Or in Helsinki, city hall had a Special Operations Group that made decisions using information from digital technology platforms and scenario analysis.

However, the tale is not without its flaws. When it comes to data collection technology, such as smart city initiatives, individuals frequently have serious concerns about data privacy and security, and these fears are valid. For instance, during the epidemic, Singapore, which was formerly known as the "smartest city in the world," used a QR code for identification anytime residents went into a public area. It was first said to be used for contact tracing, but it subsequently emerged that the police had been utilizing it for their inquiries. The incident greatly increased popular suspicion. Smart city initiatives in Zhengzhou came under fire after failing to stop fatalities caused by catastrophic floods.

Creating Urban Digital Twins

These examples highlight the vast potential of smart city technologies, indicating that there is still room for further development. Similar considerations apply to the concept of an "urban digital twin," which aims to digitally replicate various aspects of a city or neighborhood, spanning from social dynamics to environmental attributes. The goal is to empower policymakers, urban planners, and the general public to simulate the potential effects of any changes before implementing actual policies and construction projects. The foundation for these digital twins relies on large datasets that can be analyzed and processed by sophisticated algorithms and computer models, necessitating the integration of cloud computing and the Internet of Things (IoT).

While this technology is still in its early stages, remarkable instances demonstrate its immense potential. Its utility lies in the ability to conduct various tests without directly impacting people's lives, allowing municipal planners to propose optimal courses of action while minimizing potential harm.

One notable example is Singapore, recognized as the "smartest city in the world 2019." The city has been developing its "Virtual Singapore" model, serving as a digital twin that not only portrays the unique infrastructure of this city-state in the twenty-first century but also incorporates crucial information for future development. The simulation is utilized by the municipal council to identify areas for the removal of architectural impediments, benefiting individuals with decreased mobility. This showcases the practical applications and positive impacts of digital twin technology in urban planning.

Algorithms Evaluate the Effects of Becoming Green

Although making a digital twin may seem straightforward at first glance and even seem simple, the path to a digital world that accurately replicates reality is a long one that requires dogged, consistent labor. According to Pintér, it takes between 2 and 3 months to personalize the digital environment, depending on the intricacy and size of the region. Even though they created the biggest specific database in the world—Urban Nature Atlas—as part of the previously stated Naturvation project, which should ease some of their load, the availability of data further complicates the scenario.

Case Study

A Versatile Production System (VPS) as shown in Fig. 13 is a production facility that operates on a small scale. The flowchart below shows the general physical flow of the Versatile Production System. A series production batch flow is used in this demonstration, which has five stations from delivery to production, each of which is carrying out a separate subprocess. The post-processed end product is shown by the caption "Container" on the far left.

Data for Modules

The five steps of the Versatile Production System are described by nine pieces of data. Each column in a different data set represents a reading from a unique sensor that is connected to every stage of a subprocess of the Versatile Production System.

Delivery Module

Maize is delivered as raw material and transferred to the plant's storage in this module. The maize is moved to the storage module by a pressure conveyor after being moved via a conveyor belt in a stainless-steel funnel is shown in Fig. 14 and its heat map in Fig. 15.

Digital Twin-Enabled Solution for Smart City Applications

```
VPS <- as.data.frame(matrix(data = 0, nrow = 7, ncol = 7))
VPS[[1,7]] <- "Product[Processing]"

VPS[[4,3]] <- "physical-flow"
VPS[[5,4]] <- ""
VPS[[6,5]] <- "physical-flow"
VPS[[7,6]] <- ""

name <- c(expression(Container),
          expression(._.),
          expression(Delivery),
          expression(Storage),
          expression(Dosing),
          expression(Filling),
          expression(Production))
plotmat(A = VPS, pos = 7, curve = 0.7, name = name, box.type = "none",
        lwd = 0.65, arr.len = 0.295, my = -0.25,
        dtext = 0.5, main = "Versatile Production System - Industry 4.0 or IIoT")
```

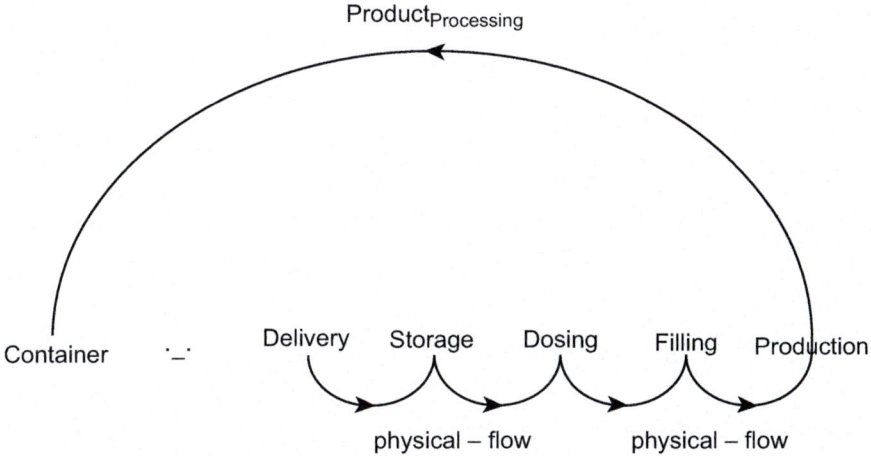

Fig. 13 Versatile Production System

Storage Module

A container is a storage module. The corn will be moved pneumatically through a pipe to the dosing module if there is a specific quantity of corn in the storage module, and a specific quantity of storage capacity in the dosing module (the module after that) is shown in Fig. 16 and its heat map in Fig. 17.

$$\ldots \text{features are binary} = \begin{cases} \text{Silo Full} & \text{i.e. 0 or 1} \\ \text{Silo Min Full} & \text{i.e. 0 or 1} \\ \text{Aspirator } O_{ON} & \text{i.e. 0 or 1} \\ \text{Belt Conveyor } O_{ON} & \text{i.e. 0 or 1} \\ \text{Funnel Blocked} & \text{i.e. 0 or 1} \end{cases}$$

Timestamp

Timestamp	SiloFull	SiloMinFull	Aspirator.O_xOn	BeltConveyor.O_xOn	FunnelBlocked
1.52E+12	0	1	0	0	0
1.52E+12	0	1	0	0	0
1.52E+12	0	1	0	0	0
1.52E+12	0	1	0	0	0
1.52E+12	0	1	0	0	0
1.52E+12	0	1	0	0	0

Fig. 14 Delivery module

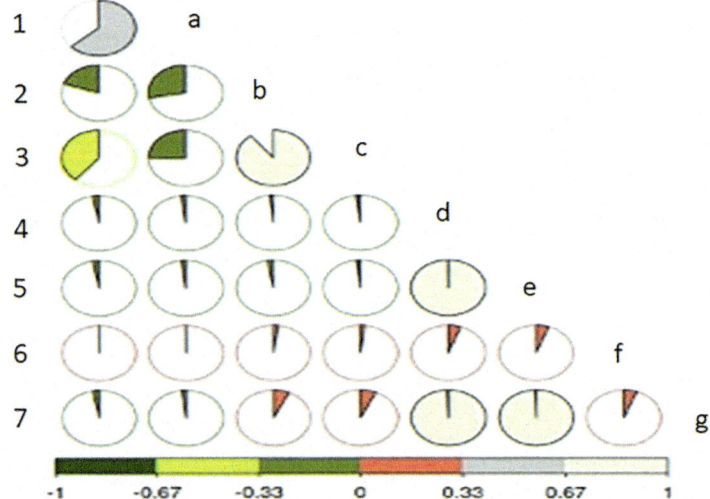

Fig. 15 Heat map to show delivery module

Supply Module

A predetermined quantity of maize is transported by the module to the supply module. Therefore, the right amount of corn is pneumatically transported to the filling module using a load cell and a dosing screw (Figs. 18 and 19).

$$\ldots \text{features are binary} = \begin{cases} \text{Timestamp} \\ \text{Funnel Blocked } I_{Signal} & \text{i.e. 0 or 1} \\ \text{Storage Silo Full } I_{Signal} & \text{i.e. 0 or 1} \\ \text{Storage Silo Min Full } I_{Signal} & \text{i.e. 0 or 1} \\ \text{Aspirator } O_{ON} & \text{i.e. 0 or 1} \\ \text{Muscle Trigger } O_{ON} & \text{i.e. 0 or 1} \\ \text{Muscle Trigger}_{ON} & \text{i.e. 0 or 1} \end{cases}$$

Timestamp	SiloFull	SiloMinFull	Aspirator.O_xOn	BeltConveyor.O_xOn	FunnelBlocked
1.52E+12	0	1	0	0	0
1.52E+12	0	1	0	0	0
1.52E+12	0	1	0	0	0
1.52E+12	0	1	0	0	0
1.52E+12	0	1	0	0	0
1.52E+12	0	1	0	0	0

Fig. 16 Storage module

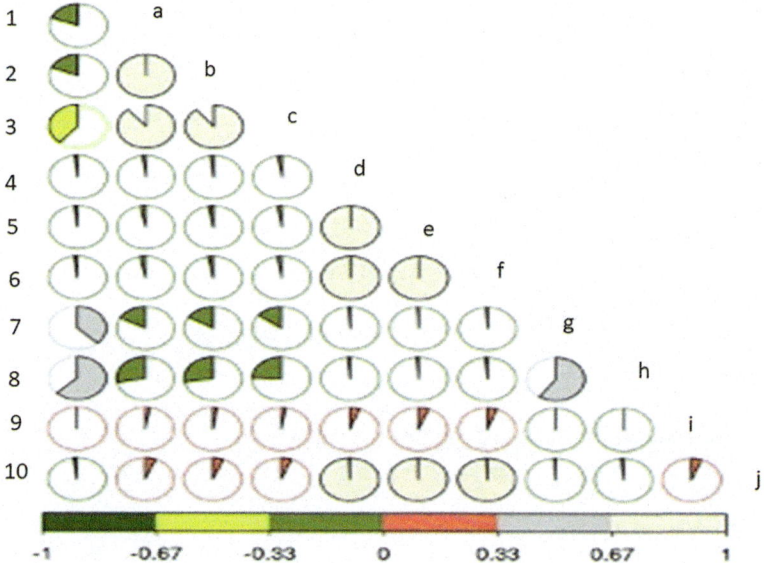

Fig. 17 Heat map for storage module

Production Module

Each subprocess was either conducted individually or (or) may have been run in parallel, i.e., concurrently, according to the time stamp of the data sets. There are mean and standard deviation figures for every data collection.

Timestamp

Weight Sensor $I_{uiValue}$

... features are binary = $\begin{cases} \text{Auger } O_{ON} & \text{i.e. 0 or 1} \\ \text{Aspirator } O_{ON} & \text{i.e. 0 or 1} \\ \text{Funnel Blocked } I_{Signal} & \text{i.e. 0 or 1} \end{cases}$

Timestamp	SiloFull	SiloMinFull	Aspirator.O_xOn	BeltConveyor.O_xOn	FunnelBlocked
1.52E+12	0	1	0	0	0
1.52E+12	0	1	0	0	0
1.52E+12	0	1	0	0	0
1.52E+12	0	1	0	0	0
1.52E+12	0	1	0	0	0
1.52E+12	0	1	0	0	0

Fig. 18 Supply module

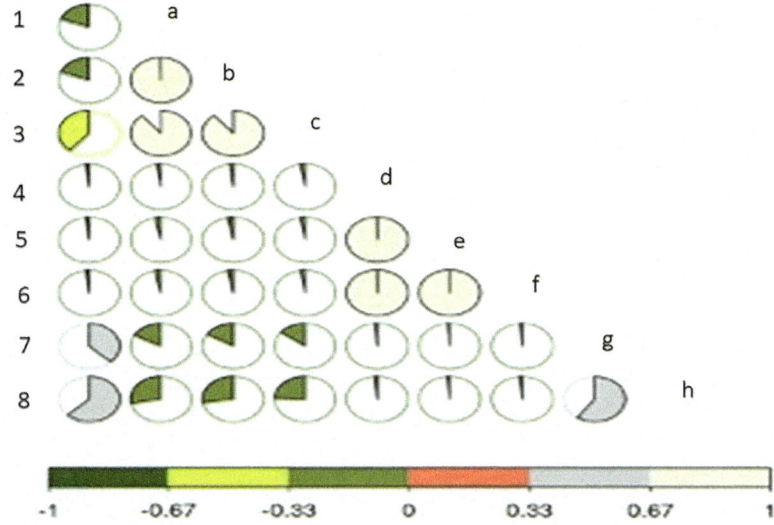

Fig. 19 Heat map of supply module

Let us now consider the "WHAT IF" scenario in which we execute each subprocess in series, i.e., one after the other, much like a scalable production process. The following inquiry is: What is the mean and standard deviation of its time record? These numbers may enable us to gauge the process's growing complexity as well as its overall systemic instability.

The contrast between the mean and standard deviation values of the Parallel Processes and a Series Process is seen in the WHAT IF simulation graph (below). The mean is on the x-axis, and the standard deviations for the Parallel and Series Processes are displayed on the left and right sides, respectively, of the y-axis (main and secondary axes). I have assumed that each data set's Time Records are normally distributed due to the large number of observations in each data set (Figs. 20 and 21).

{ Emergency Stop   i.e. 0 or 1
  Fan On   i.e. 0 or 1
  Heat Cover Closed i.e. 0 or 1
  Heater On   i.e. 0 or 1
  Pot Available   i.e. 0 or 1
  Produce   i.e. 0 or 1
  Scale Cup Down   i.e. 0 or 1
  Scale Cup Up   i.e. 0 or 1
  Switch to Pot   i.e. 0 or 1
  Track Swith at Cup   i.e. 0 or 1
  Track Switch at Pot i.e. 0 or 1
  Turn Scale Cup   i.e. 0 or 1 }

Timestamp	SiloFull	SiloMinFull	Aspirator.O_xOn	BeltConveyor.O_xOn	FunnelBlocked
1.52E+12	0	1	0	0	0
1.52E+12	0	1	0	0	0
1.52E+12	0	1	0	0	0
1.52E+12	0	1	0	0	0
1.52E+12	0	1	0	0	0
1.52E+12	0	1	0	0	0

Fig. 20 Production module

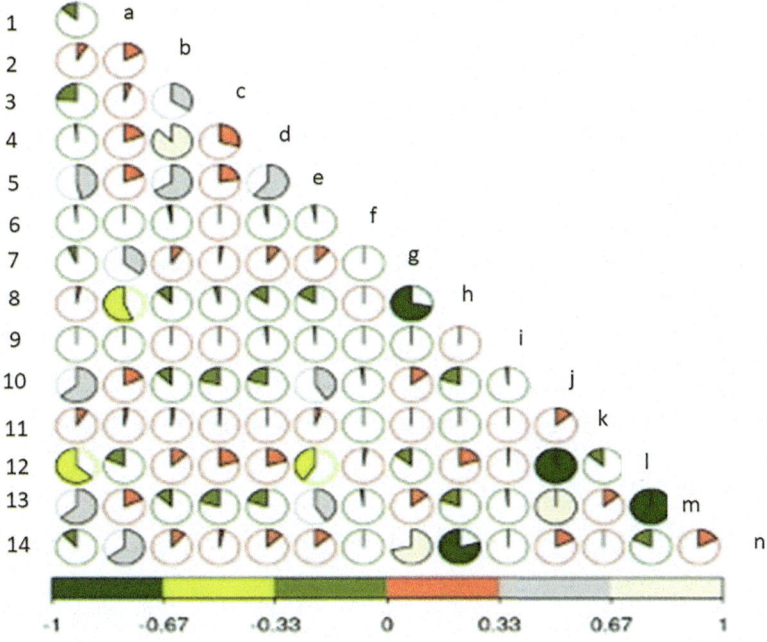

Fig. 21 Heat map for production module

Observations

Series Process and Parallel Process mean values are comparable. On the other hand, if processes are executed in series instead of parallel, the standard deviation of those processes is three times higher. Mathematically, we may state that the operation of a series production process is more complex and three times more volatile than the operation of a parallel production process. Hence, by designing a digital twin of a given system not only makes the system smart, but it can be optimized easily by taking all small parameters and tuning them in a systematic way.

Conclusion

A smart city is established by integrating the physical and digital environments through the incorporation of the real and digital biospheres, facilitated by digital urbanization, the Internet of Things (IoT), and fog computing. The digital twin urban serves as a unique starting point for the construction of modern smart cities, providing insights, control, and intelligent capabilities for both people and things. Smart cities based on digital twin technology offer numerous opportunities for economic transformation, efficient urban management, and enhanced public services, contributing to a more harmonious development of human life and the environment. To ensure the successful and cost-effective implementation of diverse smart city applications, the realization of smart cities necessitates the development of a comprehensive spatial information infrastructure. The vast data generated by smart cities presents both new opportunities and challenges. Advancing the growth of the digital services sector and better understanding the varied smart applications of the Internet + smart city require substantial efforts in technical innovation and research. The development of a smart city is a significant undertaking, requiring top-level design and comprehensive planning tailored to the unique features of each city. The establishment of an operation brain and operation center is crucial for the effective functioning of smart cities, contributing to their overall success and sustainability.

References

1. Alkali, Y., Routray, I., & Whig, P. (2022). Strategy for reliable, efficient and secure IoT using artificial intelligence. *IUP Journal of Computer Sciences, 16*(2), 16–25.
2. Anand, M., Velu, A., & Whig, P. (2022). Prediction of loan behaviour with machine learning models for secure banking. *Journal of Computer Science and Engineering (JCSE), 3*(1), 1–13.
3. Chopra, G., & Whig, P. (2022). A clustering approach based on support vectors. *International Journal of Machine Learning for Sustainable Development, 4*(1), 21–30.
4. Jupalle, H., Kouser, S., Bhatia, A. B., Alam, N., Nadikattu, R. R., & Whig, P. (2022). Automation of human behaviors and its prediction using machine learning. *Microsystem Technologies, 28*, 1–9.

5. Khera, Y., Whig, P., & Velu, A. (2021). Efficient effective and secured electronic billing system using AI. *Vivekananda Journal of Research, 10*, 53–60.
6. Mamza, E. S. (2021). Use of AIOT in health system. *International Journal of Sustainable Development in Computing Science, 3*(4), 21–30.
7. Tomar, U., Chakroborty, N., Sharma, H., & Whig, P. (2021). AI based smart agricuture system. *Transactions on Latest Trends in Artificial Intelligence, 2*(2) https://ijsdcs.com/index.php/TLAI/article/view/90
8. Jimenez, J. I., Jahankhani, H., & Kendzierskyj, S. (2020). Health care in the cyberspace: Medical cyber-physical system and digital twin challenges. In *Digital twin technologies and smart cities*. Springer.
9. Velu, A., & Whig, P. (2021). Protect personal privacy and wasting time using Nlp: A comparative approach using Ai. *Vivekananda Journal of Research, 10*, 42–52.
10. Whig, P., Velu, A., & Ready, R. (2022). Demystifying federated learning in artificial intelligence with human-computer interaction. In *Demystifying federated learning for blockchain and industrial Internet of Things* (pp. 94–122). IGI Global.
11. Velu, A., & Whig, P. (2022). Studying the impact of the COVID vaccination on the world using data analytics. *Vivekananda Journal of Research, 10*(1), 147–160.
12. Whig, P., Velu, A., & Bhatia, A. B. (2022). Protect nature and reduce the carbon footprint with an application of blockchain for IIoT. In *Demystifying federated learning for blockchain and industrial Internet of Things* (pp. 123–142). IGI Global.
13. Whig, P. (2019). Exploration of Viral Diseases mortality risk using machine learning. *International Journal of Machine Learning for Sustainable Development, 1*(1), 11–20.
14. Whig, P., Velu, A., & Naddikatu, R. R. (2022). The economic impact of AI-enabled blockchain in 6G-based industry. In *AI and blockchain technology in 6G wireless network* (pp. 205–224). Springer.
15. Whig, P., Velu, A., & Nadikattu, R. R. (2022). Blockchain platform to resolve security issues in IoT and smart networks. In *AI-enabled agile Internet of Things for sustainable FinTech ecosystems* (pp. 46–65). IGI Global.
16. Whig, P., & Ahmad, S. N. (2014). Simulation of linear dynamic macro model of photo catalytic sensor in SPICE. *COMPEL: The International Journal for Computation and Mathematics in Electrical and Electronic Engineering, 33*(1/2), 611–629.
17. Enders, M. R., & Hoßbach, N. (2019). Dimensions of digital twin applications: A literature review. In *25th Americas conference on information systems, AMCIS 2019* (pp. 2575–2584). Association for Information Systems.
18. Whig, P., Nadikattu, R. R., & Velu, A. (2022). COVID-19 pandemic analysis using application of AI. In *Healthcare monitoring and data analysis using IoT: Technologies and applications* (pp. 1–15). Institution of Engineering and Technology.
19. Kaur, M. J., Mishra, V. P., & Maheshwari, P. (2020). The convergence of digital twin, IoT, and machine learning: Transforming data into action. In *Digital twin technologies and smart cities*. Springer.
20. Kharchenko, V., Illiashenko, O., Morozova, O., et al. (2020). Combination of digital twin and artificial intelligence in manufacturing using industrial IoT. In *The 11th IEEE International Conference on Dependable Systems, Services and Technologies, DESSERT'2020*.
21. Whig, P., Kouser, S., Velu, A., & Nadikattu, R. R. (2022). Fog-IoT-assisted-based smart agriculture application. In *Demystifying federated learning for blockchain and industrial Internet of Things* (pp. 74–93). IGI Global.
22. Mondoro, A., & Grisso, B. (2019). On the integration of SHM and digital twin for the fatigue assessment of naval surface ships. In *Proceedings of the 12th International Workshop on Structural Health Monitoring 2019*.
23. O'Dwyer, E., Pan, I., Charlesworth, R., et al. (2020). Integration of an energy management tool and digital twin for coordination and control of multivector smart energy systems. *Sustainable Cities and Society, 62*, 102412.

24. Khatib, M. M. E., Al-Nakeeb, A., & Ahmed, G. (2019). Integration of cloud computing with artificial intelligence and its impact on telecom sector—A case study. *iBusiness, 11*(1), 1–10.
25. Kritzinger, W., Karner, M., Traar, G., Henjes, J., & Sihn, W. (2018). Digital twin in manufacturing: A categorical literature review and classification. *IFAC-PapersOnLine, 51*(11), 1016–1022.
26. Li, D., Shao, Z., Yu, W., Zhu, X., & Zhou, S. (2020). Public epidemic prevention and control services based on big data of space-time locations make cities smarter. *Journal of Wuhan University Information Science Edition, 45*(4), 475–487, 556. https://doi.org/10.13203/j.hugis20200145
27. Li, R., Zhao, Z., Zhou, X., et al. (2017). Intelligent 5G: When cellular networks meet artificial intelligence. *IEEE Wireless Communications, PP*(5), 2–10.
28. Liu, S., Bao, J., Lu, Y., et al. (2020). Digital twin modeling method based on biomimicry for machining aerospace components. *Journal of Manufacturing Systems, 58*, 180–195.
29. Yu, Y., Fan, S., Peng, G., et al. (2017). Application of digital twin model in product configuration management. *Aviation Manufacturing Technology, 60*(7), 41–45.
30. Zheng, Y., Yang, S., & Cheng, H. (2019). An application framework of digital twin and its case study. *Journal of Ambient Intelligence and Humanized Computing, 10*(3), 1141–1153.
31. Whig, P., Velu, A., & Sharma, P. (2022). Demystifying federated learning for blockchain: A case study. In *Demystifying federated learning for blockchain and industrial Internet of Things* (pp. 143–165). IGI Global.

Combining Digital Twin Technology and Intelligent Transportation Systems for Smart Mobility

Ajay Dureja, Aman Dureja, Varun Kumar, and Sachin Sabharwal

Introduction

One of the important parameters for measuring the development process of any nation is increasing the number of vehicle owners. But vehicle ownership also increases traffic congestion. In the past decade in India, vehicle ownership has massively increased and resulted in road blockages and traffic congestion in the cities. The primary reasons for the traffic congestion in any city are the road space and the nonavailability of adequate infrastructure. So there is a need for intelligent transportation systems in smart cities. With new means of transportation, in the earlier days, mobility allowed people to move freely, and thanks to the rapid increase in urbanization across the world, such mobility is driving progress and growth. Intelligent transportation systems (ITSs) with IoT-enabled devices can reduce traffic congestion. An ITS can be defined as a transportation system augmented by the use of technologies. The main objectives of ITSs are to evaluate, develop, analyze, integrate, and synchronize with new technologies to achieve traffic efficiency, improve environmental factors, save energy, conserve time, and ensure and enhance safety and comfort for drivers, pedestrians, and other traffic groups [13].

A. Dureja (✉)
Department of IT, Bharati Vidyapeeth's College of Engineering, Paschim Vihar, Delhi, India

A. Dureja
Department of IT, Bhagwan Parshuram Institute of Technology (BPIT), Rohini, Delhi, India

V. Kumar
Department of Management, Mangalmay Institute of Management and Technology, Greater Noida, India

S. Sabharwal
Department of Management, Jagan Institute of Management Studies, Rohini, Delhi, India

Digital twin technology and ITSs are two of the interconnected components of IoE-based smart cities. Digital twin technology is a virtual representation of a physical object or system that allows for the real-time monitoring, analysis, and simulation of its behavior. In the context of intelligent transportation systems, a digital twin could be used to model and monitor transportation infrastructure and vehicles, including traffic flow, road conditions, and vehicle performance. Intelligent transportation systems, on the other hand, use advanced technologies such as sensors, communication networks, and data analytics to improve the safety, efficiency, and sustainability of transportation systems. ITSs can leverage the data collected from digital twin models to optimize traffic flow, reduce congestion, and improve safety.

In an IoE-based smart city, digital twin technology and ITSs work together to create a more intelligent and responsive transportation system. By using real-time data from digital twins, ITSs can make more-informed decisions about traffic management, route planning, and public transportation services, ultimately improving the quality of life for residents and visitors alike. The intelligent transportation system in smart cities requires smart traffic planning and innovative ideas for public transportation system [11, 12]. In earlier days, transportation had fewer traffic jams and produced fewer emissions, which therefore had fewer negative effects on people's quality of life and the environment in smart cities, resulting in more positive habits among the population. Thanks to the development of IoT and IoE technologies, stainable ecosystems can be developed in smart cities. Advancements in IoT increase quality of life [1].

The foregoing technological situation provides a new, relevant structure for peoples and the development of urban areas, especially metropolitan cities. The idea of the Internet of Things (IoT) was introduced in parallel with wireless sensor networks (WSNs). The term *Internet of Things* was coined by Kevin Ashton. It refers to identifiable objects and their representations is an "Internet-like" structure. IoT objects include buildings, cars, planes, and traffic lights as well as humans, plants, and animals. IoT is a fast developing technology. Over 50 billion IoT gadgets were expected to exist before the end of 2021 [3, 4]. We can say that IoT is "a dynamic global network infrastructure with self-configuring capabilities based on standard and interoperable communication protocols where physical and virtual 'Things' have identities, physical attributes, and virtual personalities and use intelligent interfaces, and are seamlessly integrated into the information network" [5] or "a global infrastructure for the information society enabling advanced services by interconnecting (physical and virtual) things based on, existing and evolving, interoperable information and communication technology" [6].

Digital twin technology and ITS can work together to create a more intelligent and responsive transportation system. By using real-time data from digital twins, ITSs can make more-informed decisions about traffic management, route planning, and public transportation services. This can ultimately reduce traffic congestion, improve safety, and improve the overall transportation experience for commuters and travelers.

The Internet of Things and Digital Twin Technology

IoT makes a difference in people's lives, helps them to control and maintain their lives, and lets them work smarter than they could without IoT. Enhancing the digital market, including smartphones, could lead to automation and the development of robotized homes; in this way, IoT establishes a new basis for industry, commerce, and business. IoT presents opportunities to businesses by allowing them to view their entire frameworks and how they work, from the implementation of work and equipment to the supply chain and coordination among their operations [2].

IoT allows companies, firms, etc. to automate forms and reduce input costs. It additionally reduces costs spent on waste, reduces the cost of production, and enhances and reduces the cost of service delivery systems. Finally, it provides a straightforwardness to transactions for customers and buyers.

The Pros and Cons of IoT

Some of the merits of IoT include the following:
- It collects data from any place at any time on any electronic device.
- It enhances communication between electronic devices and enables synchronization.
- It transfers information data over a network to save time and money.
- It automates everyday jobs making a difference to move forward the quality of a business's administrations and decreasing the requirement of human efforts and intervention.

Some of the advantages and disadvantages of IoT include the following:

- As the number of connected and associated smart devices increases day by day and massive amounts of information in the form of data are shared and synchronized between electronic devices, the potential for hackers to steal private data increases [8, 9].
- Enterprises may in the long run need to deal with large numbers—perhaps indeed millions—of IoT smart devices, and collecting and overseeing the information from all those smart devices will be challenging.
- If viruses infect or bugs affect the framework, each associated gadget will likely be corrupted.
- The lack of a worldwide standard for IoT compatibility has stymied communication between gadgets from different manufacturers.

Digital Twin Technology

Digital twin technology creates an online replicated version of a tangible or physical object, system, or process. It uses real-time information from sensors, IoT devices, and other sources to model and simulate the behavior, performance, and characteristics of a physical counterpart. The following are some of the key aspects and applications of digital twin technology:

- *Virtual representation*: Digital twins create a detailed virtual model that mirrors a physical object or system, capturing its geometry, attributes, and functionalities. This virtual representation serves as a foundation for analysis, monitoring, and optimization.
- *Real-time data integration*: Digital twins are connected to their respective physical counterparts through IoT devices and sensors, continuously collecting data on its status, condition, and performance. These real-time data are integrated into the digital twin, ensuring that the virtual model reflects the current state of the physical object.
- *Monitoring and analysis*: Digital twins enable the real-time monitoring and analysis of physical assets or systems. By comparing the data collected from IoT devices with the virtual model, stakeholders can gain insights into the physical performance, identify anomalies or deviations, and make informed decisions on the basis of both real-time and historical data.
- *Predictive analytics*: With access to real-time and historical data, digital twins can leverage advanced analytics and machine-learning algorithms to make predictions. This enables proactive maintenance, the optimization of operations, and predictions about future behavior that are based on historical patterns and real-time conditions.
- *Simulation and what-if analysis*: Digital twins can simulate the behavior of physical assets or systems under different conditions and scenarios. This allows stakeholders to assess the effectiveness of its flexibility, evaluate performance, and control operations without making physical modifications.
- *Remote monitoring and control*: Digital twins facilitate the remote monitoring and control of physical assets or systems. By accessing the virtual model, stakeholders can remotely monitor performance, diagnose issues, and control operations, reducing the need for their physical presence and enabling efficient management.
- *Optimization and decision-making support*: Digital twins provide a platform for optimization and decision-making support. By analyzing data and simulating various scenarios, stakeholders can identify inefficiencies, optimize processes, and make data-driven decisions to improve performance, reliability, and sustainability.
- *Life-cycle management*: Digital twins can be used throughout the entire life cycle of an asset or system. From design and development to operation and maintenance, digital twins provide insights and support decision-making at every stage, increasing efficiency, reducing costs, and improving outcomes.

Digital Twin Technology and the Internet of Things (IoT)

Digital twin technology and the Internet of Things (IoT) are closely related and can complement each other to enhance various industries and domains. The following is an overview of how digital twin technology and IoT work together:

- *IoT devices and data collection*: The Internet of Things refers to a network of interconnected physical devices embedded with sensors, actuators, and connectivity capabilities. These devices collect and transmit data about their environment, operations, and performance. IoT devices include sensors in buildings, manufacturing equipment, vehicles, wearables, and more [7].
- *Integrating IoT data with digital twins*: Digital twin technology uses real-time data from IoT-enabled smart devices or gadgets as inputs to create a digital and virtual replica or simulation of the physical object or system. The data collected by IoT devices, such as temperature, pressure, location, and status information, are used to update and populate the digital twin model.
- *Real-time monitoring and control*: IoT devices continuously provide data to the digital twin, enabling the real-time monitoring and control of the physical counterpart. Changes in the physical object or system can be incorporated into the digital twin, allowing for immediate analysis, visualization, and decision-making.
- *Predictive analytics and optimization*: By combining IoT data with advanced analytics and machine-learning algorithms, digital twins can generate insights and predictions. These predictions can be used to optimize operations, identify potential issues, and help stakeholders to make informed decisions. For example, a digital twin of a manufacturing plant can use IoT data to predict equipment failures, optimize production processes, and reduce downtime.
- *Remote management and maintenance*: Digital twins integrated with IoT devices enable the remote management and maintenance of physical assets. Real-time data from IoT sensors allows for the remote monitoring of performance, conditions, and energy consumption. Maintenance activities can be scheduled on the basis of predictive analytics, reducing costs and improving efficiency [10].
- *Iterative improvement and simulation*: IoT devices continue to gather data from their corresponding physical assets or systems, providing feedback to digital twins. This iterative process continuously improves the digital twin model, enhancing its accuracy and effectiveness. Additionally, digital twins can simulate different scenarios by utilizing IoT data, helping to evaluate the impact of changes and optimize performance.
- *Collaboration and data sharing*: IoT devices and digital twins can facilitate collaboration and data sharing among stakeholders. The real-time data collected by IoT devices can be shared with relevant parties, improving decision-making and coordination. Multiple stakeholders, such as manufacturers, service providers, and maintenance teams, can access and contribute to the digital twin, fostering collaboration and knowledge exchange.

Thus, the integration of digital twin technology and IoT enables real-time monitoring, predictive analytics, remote management, and collaboration. By leveraging IoT data, digital twins provide comprehensive and dynamic representations of physical assets and systems, enhancing operational efficiency, maintenance practices, and decision-making processes across various industries.

Internet of Everything (IoE) Architecture

Our world is teeming with IoT devices such as mobile phones, pagers, smart watches, etc. These devices can communicate with one another though their built-in sensor devices. Some special IoT devices work as aggregators or distributors and thus form an IoT Edge. The data that are generated from these IoT devices are received by actuators which in turn produce IoT solutions. An IoT app acts as an interface between the IoT Edge and the IoT cloud infrastructure. IoT-enabled gadgets and devices provide more virtualization and digitalization and are available in individual or small single-board computers (i.e., Raspberry Pi). These types of devices and computers can be openly linked without a wired network by using a small quantity of additional energy, and they can be installed whenever electricity is available.

IoT Edge devices include smartphones, tablets, and other handheld devices like pagers, smart watches, etc. In an IoT cloud infrastructure, the main component is the cloud, which gathers all the onboard data and the real-time data from the outside world through a special application programming interface (API) known as a smart city API. A smart city API can receive data and carry out the services available on the IoT app. Figure 1 shows the framework or architecture of an IoT solution.

IoE– and Digital Twin Technology–Based Smart Cities: A Future Vision

The PLEEC project aims to improve energy efficiency, especially for urban development [16]. It focuses on developing technology, analyzing city structures, and evaluating people's behavior to assess the key elements that affect energy requirements and the emissions of cities and urban areas.

Consequently, the PLEEC program includes six normal cities (Eskilstuna, Sweden; Jyväskylä and Turku, Finland; Tartu, Estonia; Stoke-on-Trent, UK; Santiago de Compostela, Spain), and the major areas that would benefit from urban improvement were found by two web-based rounds of overviews (Fig. 2). The assessment yielded five major variables and the most imperative range domains.

To enhance the productive capacity of a city according to its classification, its shape can be redesigned to align with a smart city profile, benefiting both residents and residences.

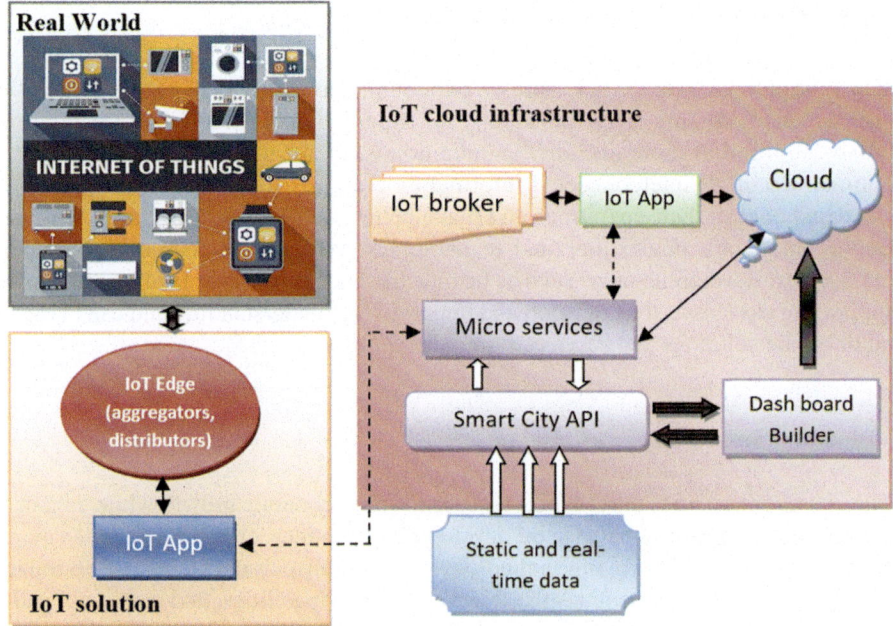

Fig. 1 Internet of Everything (IoE) architecture

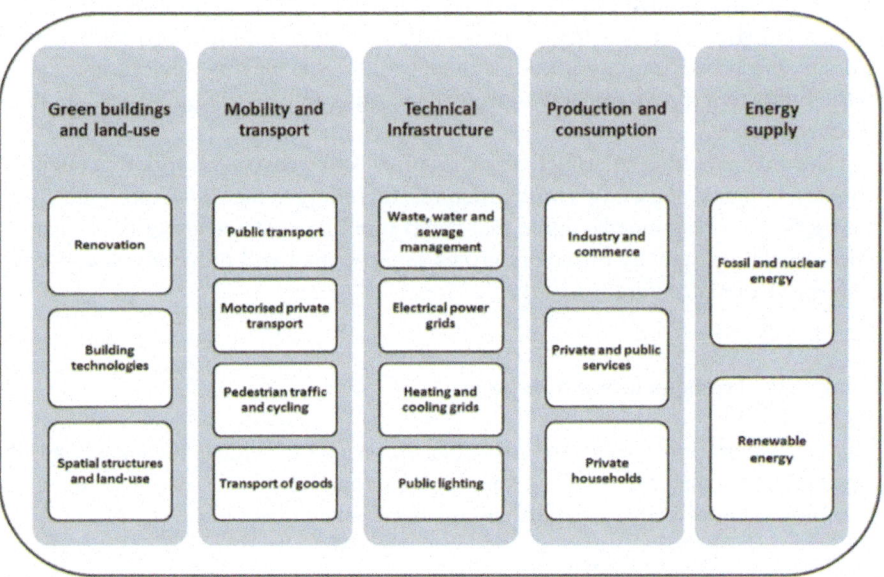

Fig. 2 Major areas and range domains in the PLEEC project. (*Source*: [16])

Smart Cities and Smart Transportation

The definition of smart cities varies from country to country, where "a city with digital technology integrated into all city functions," by the Smart City Council, is a broad definition. A more data-driven definition would be that "the IBM city makes optimal use of all interconnected information currently available to better understand and control their operations and make better use of limited resources." In the future, smart city innovations and technologies will have increased capacity to include new developing areas such as healthcare, education, and planning, all while supporting citizens as they develop and improve their understanding and use of digital or online solutions and services ("intelligent citizens") [15].

Vision of a Smart City

The aim of smart cities or AI-based cities is to spur economic growth while enhancing residents' work–life balance by developing local areas, creating smart transportation systems and improving technology. Smart transportation aims to optimize residents' mobility. It combines the Internet, Wi-Fi capabilities, and electronic gadgets to make city travel more sustainable and efficient. Smart transportation is a method that integrates cutting-edge technologies into transportation structures. This includes cloud computing, wireless communication, location-based offerings, computer vision, and different types of equipment to improve mobility.

One way to optimize transportation is to focus on traffic lights on the road. The most effective model uses timers and avoids collisions. AI equipped with sensors and suitable processing algorithms connect with other nearby devices and cycles traffic lights to improve the flow of transit and other transportation by making them faster and safer.

The concept behind smart delivery structures is to weave a network that connects vehicles, humans, and a city's infrastructure. According to the BIA Communications survey, 83% of respondents agree that smart mobility systems are critical to smart cities. As the complexity of factors grows, smart transit will propel the transformation of city infrastructure forward.

Smart City Transportation Solutions

Smart transportation systems have to work as a whole to achieve their goals. In this way, including every block of any smart transit community is essential for the overall achievement of the city's goals and has a massive effect on road conditions.

1. Connected (linked) vehicles

    Connected vehicles are crucial units in any smart city. They depend on a car-to-infrastructure (V2I) type of communication. Connected cars collect data on the surrounding congestion, including various motors, street incidents, and diverse devices within the region, such as parking sensors. This technology continuously gathers information on the surrounding road conditions. These vehicles have communication channels for drivers and alerts that warn drivers about nearby pedestrians and cars. Connected vehicles are so promising that experts predict there could be more than 233 million of them around the world by using the end of 2025.

2. Mobility as a service (MaaS)

    Mobility as a service (MaaS) considers transportation as a product. It intends to improve mobility by gathering data and assessing the critical elements of someone's journey all on one interface. Its tasks commonly include reserving vehicles, sharing journey details, paying bills, renting transportation such as bicycles or vehicles, and providing options for directions.

3. Self-driving vehicles

    Self-driving vehicles still lack human judgment, but they are quickly improving. Self-driving vehicles can seriously reduce congestion and the number collisions and injuries sustained on the roads. They can also offer faster and less-expensive transportation, encouraging people to take taxis or public transit instead of individual vehicles.

4. Digital services

    Digital services include contactless payment for public transport, dynamic navigation, real-time monitoring services, and more. Figure 3 depicts all above features for Smart City with Transportation Solutions.

Fig. 3 Smart cities and smart transportation [14]

The Main Benefits of Transportation Technology

Using AI to help regulate the transportation of a smart city has the following advantages:

- Smart transportation is more secure: Unlike human drivers, AI does not get distracted, exhausted, or emotional. Self-reliant transportation systems have fewer chances of injuries, accidents, and collisions because they lack these human vulnerabilities. Instead, they are based on machine learning with IoT and fifth generation (5G) wireless services, which develop self-reliant transportation systems.
- Smart transportation offers more-precise control: AI-based transportations system are better able to control traffic than traditional transportation systems are.
- Smart transportation is more ecofriendly: AI systems are also more ecofriendly than traditional transportation systems are.
- Smart transportation is more effective: AI and IoT-enabled systems optimize the use of available resources. They can reduce costs thanks to their conservation of fuel and lower electricity consumption. In other words, a smart public transportation system is more cost-effective than a system that relies on people who drive their own individual vehicles.
- Smart transportation quickly provides insights: AI transportation systems are better able to quickly share information on problem areas, accidental locations, etc. with the traffic control facilitator. They also help protect the public and provide better management systems for traffic control rooms.

AI-Based Transportation Systems and Digital Twin Technology: A Future Vision

AI Vision for Accident Prevention

According to the Centers for Disease Control and Prevention (CDC), approximately 1.35 million people die due to motor vehicle collision or other means of accidents globally every year, with approximately 3,700 daily deaths. Most of these annual deaths are incurred by pedestrians, cyclists, and motorcyclists. Such injuries arise because of a number of factors, such as poor visibility, driver fatigue, a lack of focus, and technical failures, among others. Smart transportation entails an IoT network that includes sensors mounted on highways and busy streets in addition to those on vehicles. These sensors, along with closed-circuit television (CCTV) cameras, can offer timely data to drivers that indicate how closer their vehicles are to pedestrians, static structures, and other vehicles at any given moment. In a well-connected smart city, such data are autonomously captured by smart cameras. Then these data can be sent to vehicles that are connected to the system. As soon as a vehicle receives such data, it can alter its course accordingly. The algorithms in computer-equipped vehicles can anticipate potential dangers earlier and thus help to avoid collisions and other accidents.

These days, dynamic image capturing and processing are crucial components in vehicles. Vehicles in smart cities use computerized vision tools for occupant and pedestrian safety. Moreover, the data accrued and analyzed by predictive applications may be used by municipalities and corporations to develop smart network tasks.

As with any AI-enabled system, smart transportation systems continue to improve over time. The sensors become better at identifying highway symbols and functions, lane markings, limitations, and other street-related information. Additionally, such structures end up markedly better at detecting other devices and potential collisions/accidents and at relaying information to drivers and pedestrians, through smartphone apps or wearable devices.

Safety technology which includes intuitive emergency braking, lane centering, blind-spot tracking, collision alert systems, adaptive cruise manage, smart velocity systems, nighttime sensors, pedestrian tracking, and street sign identification systems rely on digital image capturing and cognitive information processing. In the modern age of big data, AI vision enables analyzing raw data on the fly thanks to algorithms that carry out several million probability-based calculations and generate valuable insights and predictions to optimize road safety.

Smart transportation additionally boosts the sustainability of smart cities. For instance, data-gathering devices on delivery cars allows transport agencies to identify the best routes for vehicles and drivers. In this way, organizations can reduce the distances that deliver vehicles travel. As a result, gas and car usage can be better managed. A smart city such as Amsterdam actively prioritizes sustainability and citizen protection while making city improvement plans. In addition to safety and sustainability, computerized predictions can enhance mobility in smart cities through smart travel management.

AI Vision for Traffic Management

Smart transportation is not limited to vehicle-based devices and applications. The idea also entails the optimization of street networks in a smart city. Smooth mobility is an important factor for gauging the livability, workability, and sustainability of any smart city. For example, in the event of medical emergencies where people need to be rushed to the hospital, a smart transportation system that can prioritize mobility could save lives. Mobility is also useful for enhancing the everyday sustainability and workability of smart cities too.

So, how is AI vision useful for control and mobility in smart cities? Traffic-automation mechanisms that cannot usually be computerized tend to produce numerous errors. As the number of daily commuters grows and globalization intensifies, vehicle tracking becomes even more crucial. Smart cities feature some of the busiest highways and large numbers of vehicles, increasing traffic concerns.

Smart transportation tools and IoT-equipped devices enable autonomous monitoring and communication. Smart traffic lights and parking and traffic guidance structures use computer-processed predictions and IoT to support drivers in making decisions on selecting optimal routes to take to reduce commute times. Vehicle sensors and smartphone apps complement AI vision–based smart transportation applications for this purpose.

Smart roads and highways are also part of a smart city's transportation network. IoT and AI-based programs in smart cities can control the speed limit of vehicles on all sorts of roads. Such programs continue to autonomously monitor the flow of traffic and send indicators that vehicles are crossing the speed limit. This feature of smart transportation is beneficial mainly for preventing collisions with speeding vehicles.

A geospatial traffic steering program, which is also a part of AI vision–powered smart transportation networks, uses global positioning system (Gps), geographic information system (GIS), and radiofrequency devices for precise traffic tracking. Such equipment works in unison to provide three-dimensional (3D) spatial and geographical information, such as vehicle proximity, traffic density, upcoming impediments, and traffic inflows on specific routes, to self-driving vehicles and to the drivers of other vehicles.

One example of smart city–based traffic management can be found in the city of Darmstadt, Germany.

AI Vision for Autonomous Driving

Over the past few years, almost every carmaker has dipped their toes into the driverless-vehicle market. The number of self-driving vehicles is slated to grow over the next few years. Such vehicles enable smart transportation in smart cities. What is more, such vehicles' AI vision systems generate and gather a plethora of data to evaluate their mobility, occupant and pedestrian protection, and fuel efficiency.

Self-driving vehicles depend on several smart cameras for object recognition—for example, it identifies pedestrians, traffic lights, and other vehicles even in moderate- to low-visibility conditions—so that they can automate the deployment of protection features such as airbags and automatic brakes. Moreover, such vehicles also depend on AI prediction for 3D mapping to improve decision-making during course selection by using speed and parking data. This reduces the number of injuries sustained in smart cities.

Furthermore, AI vision and IoT make self-driving vehicles more connected in that they allow vehicles to autonomously communicate with other smart vehicles and with smart transportation devices and applications. For example, two self-driving vehicles in a slim lane can foresee a potential collision when another vehicle rushes toward them from another angle. The usage of IoT and AI prediction allows vehicles to either pull over for protection or use an alternate lane to move away from a potential collision. This is arguably the most important feature of what connected smart vehicles can do.

Smart transportation would not exist without AI, IoT, and a few other smart city technologies. AI prediction, in particular, is instrumental in studying dynamically captured data and communicating them across vehicles and smart transportation devices. In this way, AI vision plays a vital role in making smart cities livable, workable, and sustainable.

Smart cities promote three attributes: livability, workability, and sustainability. Livability focuses on promoting healthy balances between the environment, work, routines, etc. Also, healthy work–life balances can more easily develop in smart cities. Workability refers to healthy working environments with good opportunities for employment. Finally, sustainability is a major component in the development of any smart city in that it enables the proper usage of technology while decreasing energy consumption, pollutants, and accidents.

Already, AI vision is being used in countless unique programs in smart cities. One of the essential software applications is smart transportation. AI, IoT, and predictive software lend their unique capabilities to smart cities to transform transportation networks and vehicles in several ways.

AI-Based Transportation Systems Combined with Digital Twin Technology

AI-based transportation systems integrated with digital twin technology are revolutionizing the way we plan, manage, and optimize transportation networks. By harnessing the power of artificial intelligence and creating virtual replicas of physical assets and systems, these intelligent transportation systems are unlocking unprecedented capabilities. Through real-time data collection from IoT devices, AI algorithms can analyze and interpret information in digital twins, enabling predictive analytics, optimization, and informed decision-making. AI-based transportation systems with digital twin technology offer benefits such as the real-time monitoring of traffic conditions, proactive congestion management, optimized route planning, the predictive maintenance of infrastructure, and improved user experiences. These intelligent systems can enhance the efficiency, safety, and sustainability of transportation networks, providing smarter and more connected mobility ecosystems for the future.

Conclusion

Digital twin technology finds applications in various industries, such as manufacturing, healthcare, energy, transportation, and smart cities, among others. It enables improved asset management, predictive maintenance, process optimization, and informed decision-making, increasing operational efficiency, cost savings, and innovation. The major objectives of intelligent transportation systems (ITSs) are to evaluate, to create, to analyze, and to integrate within new technologies to try to achieve promote traffic efficiency, improve environmental factors, save energy, save time, and ensure safety and security. In short, the integration of digital twin technology and IoT enables real-time monitoring, predictive analytics, remote management, and collaboration. By leveraging IoT data, digital twins provide comprehensive and dynamic representations of physical assets and systems, enhancing operational efficiency, maintenance practices, and decision-making processes across various industries. The integration of digital twin technology with intelligent transportation systems (ITSs) holds immense potential for transforming smart mobility. By combining real-time data insights, predictive analytics, and enhanced decision-making capabilities, this integration can revolutionize the way that transportation networks operate and offer numerous benefits to users, cities, and transportation authorities.

The integration of digital twin technology with ITS enables the real-time monitoring and visualization of transportation systems. By collecting data from various sources, such as traffic sensors, GPS devices, and weather stations, digital twins provide comprehensive overviews of traffic conditions, congestion hotspots, and system performance values. This real-time information empowers stakeholders to make informed decisions and take proactive measures to optimize traffic flow, reroute vehicles, and effectively manage emergency situations.

Furthermore, digital twins integrated with ITSs facilitate predictive analytics and simulations. By leveraging historical and real-time data, digital twins can predict traffic patterns, optimize routes, and anticipate congestion. This enables transportation authorities to proactively manage traffic, optimize public transportation schedules, and improve overall system efficiency. The simulation capabilities offered by digital twins allow stakeholders to test different scenarios and evaluate the impact of potential changes or interventions, ensuring optimal decision-making.

Integrating digital twin technology with ITS also enhances infrastructure maintenance and optimization. Digital twins monitor the condition and performance of transportation infrastructure, such as bridges, tunnels, and roads, by analyzing data from sensors and incorporating environmental factors. This predictive maintenance approach enables transportation authorities to identify maintenance needs, prevent potential failures, and optimize infrastructure usage. By proactively managing infrastructure, cities can reduce costs, improve safety, and enhance the overall quality of their transportation networks.

Connected and autonomous vehicles (CAVs) can significantly benefit from the integration of digital twin technology with ITS. Real-time data from digital twins, such as those on traffic conditions, infrastructure status, and weather information, can enhance the decision-making capabilities of CAVs. This enables autonomous vehicles to navigate more efficiently, improve safety, and provide a seamless user experience. The integration of digital twins with ITS plays a crucial role in ensuring the successful integration and deployment of CAVs into smart mobility systems.

Moreover, integrating digital twin technology with ITS enhances the overall user experience. Real-time information on traffic conditions, public transportation schedules, parking availability, and alternative routes empowers users to make informed decisions and choose the most efficient and convenient transportation options. By providing up-to-date and personalized information, digital twins integrated with an ITS contribute to a seamless and user-centric mobility experience.

Collaborative decision-making is another significant advantage of integrating digital twin technology with an ITS. Digital twins can be shared across stakeholders, including transportation authorities, urban planners, and service providers. This promotes collaboration, data sharing, and joint decision-making for optimizing transportation systems, improving efficiency, and addressing challenges collectively. The shared digital representation allows stakeholders to align their efforts, share insights, and work toward the common goal of smart and sustainable mobility.

In conclusion, integrating digital twin technology with intelligent transportation systems inaugurates a new era of smart mobility. The combination of real-time data insights, predictive analytics, simulation capabilities, and collaborative decision-making empowers transportation authorities and stakeholders to optimize traffic flow, improve infrastructure maintenance, enhance user experiences, and facilitate the successful integration of connected and autonomous vehicles. By leveraging the power of digital twin technology within the framework of an ITS, we can create more-efficient, sustainable, and user-centric transportation systems for the future.

References

1. Dastjerdi, A. V., & Buyya, R. (2016). Fog computing: Helping the internet of things realize its potential, *49*(8). https://doi.org/10.1109/MC.2016.245
2. Yan, Z., Yan, Z., Zhangand, P., & Vasilakos, A. V. (2014). A survey on trust management for Internet of Things, *42*(42). https://doi.org/10.1016/J.JNCA.2014.01.014
3. Evans, D. (2011). The internet of things: How the next evolution of the internet is changing everything. *CISCO White Paper, 1*, 1–11.
4. Ray, S., Jinand, Y., & Raychowdhury, A. (2016). The changing computing paradigm with internet of things: A tutorial introduction, *33*(2). https://doi.org/10.1109/MDAT.2016.2526612
5. van Kranenburg, R. (2008). *The internet of things: A critique of ambient technology and the all-seeing network of RFID*. Institute of Network Cultures.
6. TSSO, I. (2012). Next Generation Networks-Frameworks and functional architecture models. *International telecommunication union technical report*.
7. Abomhara, M., & Køien, G. M. (2015). Cyber security and the internet of things: Vulnerabilities, threats, intruders and attacks, *4*(1). https://doi.org/10.13052/JCSM2245-1439.414
8. Serpanos, D. (2018). The cyber-physical systems revolution, *51*(3). https://doi.org/10.1109/MC.2018.1731058
9. Bertino, E., & Islam, N. (2017). Botnets and internet of things security. *Computer, 50*(2), 76–79. https://doi.org/10.1109/MC.2017.62
10. Zhang, L., Zhang, L., & Du, B. (2016). Deep learning for remote sensing data: A technical tutorial on the state of the art. *IEEE Geoscience and Remote Sensing Magazine, 4*(2), 22–40. https://doi.org/10.1109/MGRS.2016.2540798
11. Levine, J., & Underwood, S. E. (1996). A multiattribute analysis of goals for intelligent transportation system planning, *4*(2). https://doi.org/10.1016/0968-090X(96)00004-6
12. Chowdhury, M., & Sadek, A. W. (2003). *Fundamentals of intelligent transportation systems planning*. Artech House.

13. Report to the ITS Standards Community ITS Standards Testing Program By Battelle Memorial Institute for US, Department of Transportation (USDOT), Chapter 2.
14. An introduction to smart transportation: Benefits and examples. Accessed October 18, 2023. [Online]. Available: https://www.digi.com/blog/post/introduction-to-smart-transportation-benefits
15. Parnell, G. S., Specking, E., Pohland, E. A., & Buchanan, R. K. (2023). Smart cities—A structured literature review. https://doi.org/10.3390/smartcities6040080
16. PLEEC-Project (2014). http://www.pleecproject.eu/. Accessed 24 August 2015

Navigating the Digital Landscape: Challenges of BIM and Digital Twin Adaptation

Kamal Jaafar and Mohamed Awais

Introduction

Interaction across Architecture, Engineer, and Construction (AEC) ecosystem is tailored to provide optimal and bespoke solutions to the owners. This collaboration has existed for many years, and due to the advent of the digital era, technology is enabling these sectors to collaborate and exchange data more efficiently [10]. Trends predict an increase in the adoption of these technologies for optimizing the construction process and enabling leaner and sustainable construction methods [2, 7]. This digital transformation is already well adopted in developed nations where almost 75% of establishments in USA are currently employing digital technology in the construction processes [9], and these numbers have only been crawling closer to a 100% over the past few years. Moreover, there has been an increasing number of firms in developing nations aiming to adopt these changes. The impact of this abrupt transformation and its implications need to be understood and evaluated so that the built environment can effectively adapt to these changes.

The primary tools and techniques used for this digital transformation of construction processes are BIM (Building Information Modeling), DT (digital twin), and IoT (Internet of Things). BIM is a set of digital tools employed for producing and managing information and data of construction projects [4]. It comprises detailed digital models, which can be used for several purposes including visualization or executing simulations and tests. Whereas a digital twin is the term affiliated with BIM models if they have real-time data incorporated within—virtual models

---

K. Jaafar (✉)
Faculty of Engineering and Information Science, University of Wollongong in Dubai, Dubai, UAE
e-mail: KamalJaafar@uowdubai.ac.ae

M. Awais
Construction/Project Manager PMP, Comtech Group, Dubai, UAE

© The Author(s), under exclusive license to Springer Nature Switzerland AG 2024
A. Mishra et al. (eds.), *Transforming Industry using Digital Twin Technology*,
https://doi.org/10.1007/978-3-031-58523-4_15

Table 1 Comparison of BIM and DT [7]

Concept	Differentiator				
	Application focus	Users	Supporting technology	Software	Stage of life cycle
BIM	Design visualization and consistency, Clash detection, Lean construction, Time and cost estimation, Stakeholders' interoperability [10]	AEC, Facility manager [14]	Detailed 3D model, Common data environment (CDE), Industry Foundation Class (IFC), Construction Operations Building Information Exchange (COBic) [14]	Revit, MicroStation, ArchiCAD, Open source BIMserver, Grevit [14]	Design, Construction, Use (maintenance), Demolition
Digital Twin of Building	Predictive maintenance, Tenant comfort enhancement, Resource consumption efficiency, What-if analysis, Closed-loop design	Architect, Facility manager	3D model, WSN, Data analytics, Machine learning	Predix, Dasher 360, Ecodomus	Use (operation)

representing the real buildings. This real-time data is collected utilizing the Internet of Things (IoT)—which is a network of smart technologies like sensors. Collected information is then used to replicate actual physical conditions and run simulations or tests in BIM [4]. Table 1 below briefly summarizes the uses and differences between BIM and DTs.

Furthermore, the chapter will also explore the following objectives:

- Is it currently standard practice for design and construction companies to integrate BIM and DT when carrying out their operations?
- What is the disruption process that an organization must account for when implementing these technologies?
- How easily these technologies can be adapted and implemented?
- What are the critical success factors and challenges of implementations?

Literature Review

This chapter explores the impact and process disruption that the implementation of BIM and DT has on design and engineering firms by addressing the progress of emerging technologies in the context of AEC. The study investigates how certain aspects of the construction process have changed in recent years, what are the critical success factors of implementing these technologies, their challenges, benefits, and how current companies and projects cope with the changes.

Domains of BIM/DT Integration Within AEC Ecosystem

When implementing BIM and DT, many aspects of a building and its construction process can benefit from the integration of modern technologies [14]. However, this also means that each sector will need to upgrade to adapt to these modern solutions to utilize their respective advantages; these adaptations could demand significant resources and effort when shifting from traditional norms to modern technologies. Shu Tang divides the applications of BIM into six domains (Table 2) and discusses potential benefits in each domain.

The six categories are:

- Communication and collaboration.
- Performance checks and monitoring.
- Health and safety (H&S) monitoring.
- Leaner construction.
- Facility management.
- Operation and maintenance.

Furthermore, the implementation can have an array of additional benefits including added benefits of technical efficiency, exchange of information across platforms,

Table 2 Application of BIM in various domains of AEC

Domains	Application
Communication and collaboration	Integration of BIM and IoT enhances construction operations by enabling real-time communication and collaboration
Performance checks	BIM tools and onsite sensors can help evaluate the performance of activities and construction status and give real-time data on onsite conditions during different phases of a building's lifecycle
H&S monitoring	The real-time data collected and analyzed from IoT can help identify potential risk, visualize, and notify hazards over the BIM model (e.g., when working in a confined zone)
Leaner construction	By employing reliable data and information. Multiple alternative design solutions can be studied to implement the most suitable ones. Moreover, automating allows for optimized workflow, eliminating waste of resources
Facility management	Automation of operations, maintenance, emergency response, performance, and energy management can promote facility management and produce efficient task achievement
Operation and maintenance	BIM and IoT incorporates crease efficient platforms which give access to real-time data and enable elaborate maintainability by protecting assets of digital data banks and management tools

enhanced building life cycle, data standardization, procurement, planning and scheduling, decreased expenses, conflict analysis, and decision making [5].

Data from reputable and relevant publications are sourced to infer and analyze the influence of digital transformation when employing BIM in different phases of construction and maintenance processes. The literature provided an elaborated assessment of how BIM can affect the construction process positively. However, there is a lack of implications on how a company's operations are affected when shifting from its traditional norms. Further investigation is required to understand this gap and how the transformation can impact the company's resources, financial attributes, people transformation, and process disruption.

BIM Applications

BIM, digital twins, and IoT have been piercing their way around the design and construction processes [13]. To understand the impact, it has on company's operational norms, and it is vital to first develop a solid understanding of its applications and utilization. The emergence of these new technologies offers many insights into the construction process and aids with making better decisions throughout the life cycle of a project [4]. The author conducts a systematic review to identify the expansion of these technologies and how they are being implemented. The knowledge is developed by extensively reviewing reliable publications from the past decade only and then filtering the relevant information qualitatively. Deng M. classified his findings into various classes of which BIM application, simulations, and digital twins are acknowledged for the purposes of this study.

General Applications

BIM, in general, can be utilized for risk management and analysis throughout the construction life cycle. It supports overcoming technical limitations and provides a wider scope of operations potential by creating a multidisciplinary system; it aids the development process by uniting different platforms [4]. Deng M. elaborates and states that it also creates a sustainable build environment by eliminating waste products, in terms of time, efficiency, and other resources.

Design Phase

The author states that during the design process, BIM provides flexibility and enhances the scope of work which otherwise would not be possible without the means of modern technology. It can review design themes and help analyze designed elements and features to optimize energy and thermal impact, cost reduction,

HVAC, acoustics, and water management. Data on these aspects can help produce a functional and effective design [4].

Construction Phase

Constructability takes advantage of BIM implementation by improving information and its exchange, reducing and eliminating designs errors, and enhancing management efficiency, which helps reduce the cost and duration of construction [4]. Moreover, it also increases buildings sustainability by improving the environmental, economic, and social impact of the process [4]. The energy-reducing solutions attained from implementing BIM play a vital role in improving the lifecycle management of a building.

Building Operation Phase

The literature explains that BIM and digital twins can represent physical data and information as digital copies, these data can be used to evaluate building performance which can aid in the operation, maintenance, and facility management. However, these operations are still in their development phase and face many challenges like information interoperability and accurate data acquisition.

Simulations and Testing

Besides prevailing applications, BIM also supports simulations of projects and data which according to the authors are useful in the analysis and management of attributes throughout the lifecycle [4]. Deng M. explains that the major areas of influence for digital simulations are in the construction process; energy performance, and thermal environment; moreover, other simulating techniques including carbon emissions calculations, lighting, and acoustic are also under development.

Integration Digital Twins

To further elaborate the applications of BIM and IoT to form digital twins of buildings can be of existential importance to utilize the technology at its maximum potential. Deng M narrates that besides running simulations, this real-world data and its digitization can help with monitoring the process of construction, aid in energy management, and control indoor and outdoor environment, thermal comforts, space management, and H&S management [4]. All these aspects are crucial in a user's comfort in and around a building and can dictate the successful construction of a lean and sustainable building.

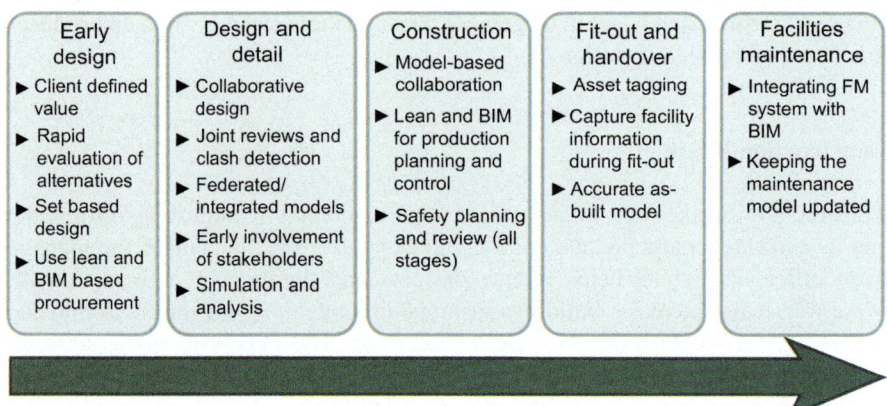

Fig. 1 Building information management (BIM) workflow

Determined from qualifying methods, it is deduced that BIM provides geometric and semantic data of the construction process, which improves information trade between different groups and individuals involved in a project (summarized in Fig. 1), thereby promoting the design, construction, and operations of a building. Moreover, simulation and DTs are very effective and enable monitoring, deduction of trends in the future, and optimization of the process. The literature depicts a further rise in demand and implementation of these technologies in the built environment around the globe. However, the challenges posed by this shift of operational means are not explained. Further analysis of how these implementations disrupt the current operation of design and engineering firms needs to be made. Moreover, the author's claims focus on raw and theoretical learning, yet it is necessary to understand the deviations from expectation when other aspects are in play such as construction schedules and other real-world implications.

Adoption of Digital Tools

Yeliz Tulubas studied the factors that affect the acceptance of digital technology within the AEC industry and states that the use of BIM is increasing. He narrowed his study in three major factors of application: design tasks, organizational competence, and designer competence. The increase in the adoption rate of BIM is deemed to be the normal trend in the coming times; there has been an increase in adoption rates, where over 50% increase was observed in the first 5 years of this decade alone; this indicates the rise in demand of BIM within the industry [15]. The incline is mostly due to the nature of construction projects getting more complicated over time and the need for accurate and easy documentation, which BIM facilitates. Yeliz states that the organizations who are aware of BIM's potential are well adapted to integrate it within their process; moreover, they are actively promoting the use of

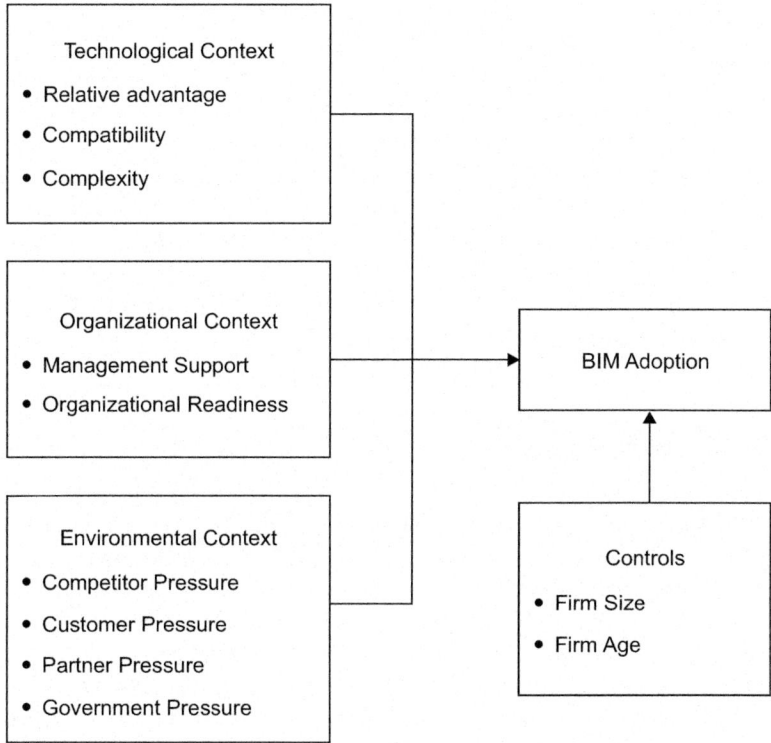

Fig. 2 BIM adoption model [3]

tech within their organization. The willingness of organizations and support are a major influencing factor in the successful adoption of technology as listed below (Fig. 2):

The study proves and signifies the positive impacts on the processes throughout the design and construction lifecycle; however, it is empirical to also understand the negative aspects. As stated, firms believe that they have the resource required to adopt and integrate BIM [3]; however, there is a lack of clarity in terms of what these resources are. A better understanding needs to be developed in terms of how other sectors like engineering and contraction deal with adoption and how competent are they to do so.

Product, People, and Process

A constant increase in the adoption of BIM has been noticed in past few years; this is due to multiple factors such as inclination towards green and sustainable construction, facility management, and complex multi-departmental collaboration [6]. Ning speaks about acceptance of digital tools and technology and what organizations expect them to deliver in terms of the product, users, and processes. Design teams tend to view BIM as a supplementary addition to CAD drawings and models, whereas managers and engineers expect it to be a source of the data management system (DMS) that can aid in other processes including simulation, cost analysis, and audits [6]. He further elaborates that both these sectors are yet to become coherent in terms of tools and technologies they utilize; however, the vendors and manufacturers of BIM and DT tools are working on developing these areas of potential growth. These results could be the product of narrow research as the chapter unambiguously focused on researching design departments alone; further, understanding is required to validate the authenticity of the stated results concerning other domains of construction industry.

Competitive Advantage

BIM enables users to operate and produce more sustainable construction processes. Manzoor reports that BIM is extensively being utilized across Malaysia throughout the construction process of complex projects like high-rise towers; this theme is in line with studies conducted by Mai Nguyen [11]. A harmony of information across the subject that remains to be clarified is current awareness and usage of BIM and DT in the current AEC ecosystem; Bilal states that currently the awareness level and usage are very limited in Malaysia. BIM helps improve the process by giving the operators a competitive advantage in four domains, i.e., productivity, collaboration, lean construction, and improved safety [11]. The study further elaborates the future of construction processes to be more efficient and waste-free and how BIM can enable a firm to achieve required sustainability goals.

Mai Nguyen explores the perspective of apparent advantages of adopting BIM by carrying out a comparative analysis of two focus groups where one group had adopted the BIM and the other one did not. His results state that there is a list of common driving factors that would encourage a design firm to adopt and integrate BIM within their operations.

- It increases the scope of consulting capacity that a design firm can handle.
- It gives them a competitive edge over the firms that do not utilize BIM.
- It enables them to provide revolutionary services in AEC.
- It allows them to adapt, adjust, and entertain design alteration more easily.
- It accelerates the design and construction process.
- It improves accuracy and reduces errors.

Following (Fig. 3) are the sources of competitive advantages that adoption of BIM enables:

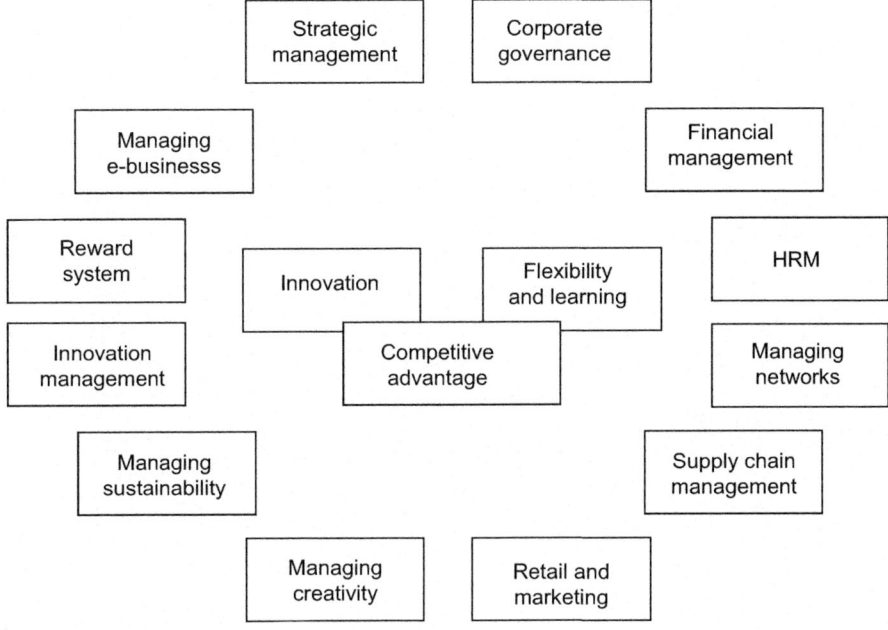

Fig. 3 Sources of competitive advantage [12]

Challenges of Digital Transformation

Despite its substantial advantages, implementing BIM and DT in the construction process has numerous challenges [16]. Rebekka Volk studies and outlines these challenges and how they need to be considered when employing these technologies. These challenges are the obstacles that the design and engineering firms need to address and evaluate to ensure their impact is well understood and accounted for. The information is collected by acquiring relevant data concerning BIM and sustainable buildings from relevant publications which address associated challenges with regards to both existing and new buildings. It is established that many property owners are reluctant to integrate BIM technology due to the added costs [16]. However, when assessed from a broader perspective, it has proven to be more cost-effective in long term, as it can enhance sustainable maintenance and construction. Furthermore, there seems to be a trend of industry transformation. Rebekka Volk states that utilization of these technologies would be the new norm of this industry in the near future. The authors outline the following challenges.

Functional and Technical Problems

Acquiring and linking accurate information of buildings, objects, and other relevant data for an optimal BIM operation remains to be an unreliable process [16]. Lack of required data when modeling new or existing buildings can result in inaccurate results.

Accuracy and Capability

Inconsistent data can reap devastating outcomes from BIM systems. Rebekka Volk states that current technology allows very little flexibility in terms of discrepancies within data.

Information Exchange

Different sectors of the construction ecosystem use different BIM technologies and software depending on what their area of expertise and needs are. It is highly ineffective to exchange these data across sectors [16]. However, this idea is in contrast with much of other literature where authors state that information exchange is very adequate in today's time. For example, Ali G. states that among other benefits of BIM is its ability to interoperability with other platforms.

Effectiveness

BIM is developing technology and overall efficacy is yet to be fully developed to deliver as per its potential [5]. Though the challenges are outlined in detail, the literature still lacks the impact of overcoming these challenges on design and construction firms. Further investigation is required to understand how current construction companies are affected by these challenges and how well suited they are to overcome them.

BIM Collaboration and Digital Transformation

BIM and DT remodeled the construction process and are revolutionizing the built environment; this transformation is changing how design and engineering firms operate [2]. BIM affects three dimensions of sustainable development: environment, society, and economy; this fact was studied by Carvalho J.P. where he states that digital transformation has been a key step to optimize the construction process and aid sustainable construction. Through conducting experiments and engaging in research on the evolution of business structures during this transformation (Fig. 4), we have compiled an assembly that outlines the variances between the traditional and the new processes. This assembly highlights the simplification of collaboration among various sectors of AEC through the utilization of BIM (Building Information Modeling) as a modular platform for all activities. This information can provide a comprehensive understanding of how businesses in AEC are changing and the process disruption the technology is causing.

The study concludes that the transformation is very beneficial, aids sustainable construction, and results in cleaner future environments positively. It enables

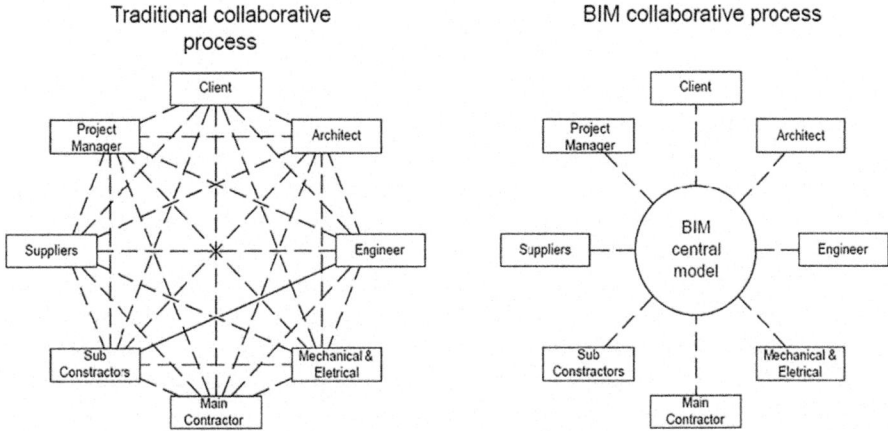

Fig. 4 Difference between the traditional approach and BIM approach [2]

designers to evaluate proposals and aid in decision-making [2]. However, Carvalho J.P also states the limitations of the statements where there is a lack of accuracy in data acquisition and the technology is still evolving and improvements would be expected. This means the process disruption is not at its peak, and the companies will have to keep enduring the impact of transformation for an unforeseeable future. The work also lacks sufficient information on the negative impacts of these transformations and what measures companies had to take to facilitate adjust for the change.

Critical Success Factors (CSF) of Implementation

With the construction process transforming in recent years, many factors have emerged which dictate how successfully a company integrates BIM and DT in the AEC industry [1]. The author states that achieving the goals by addressing the CSFs has proved to aid efficient integration of BIM and digital transformation.

The list of success factors was generated by conducting elaborate research on major design and construction firms in leading nations that have already implemented the technologies successfully. These factors were categorized under the following domains:

- Collaboration among different AEC domains.
- Accuracy and details in 3d modeling.
- Management of construction work.
- Exchange of data and information.
- Information management.
- Improve planning and safety.
- Top level management commitment and willingness to promote and integrate BIM.

Future of Digital Transformation

With modern technology revolutionizing the construction process, it is crucial to predict future trends and what further changes are expected that the design and engineering companies have to adapt to. Build environments requirements of interexchangeable data are being partly fulfilled by the platforms, which promises seamless syntheses of data across AEC sectors; however, this can become a challenging aspect as the number of BIM tools and their complexity increases over time [8]. On the other hand, Koseoglu et al. further states that these tools are still developing and are expected to improve user interface, workability, and ease of implementation. This suggests that the design and engineer firm's operations will also be changing and evolving with time; however the impact of adapting to these constant changes is lacking sufficient studies and is yet to be measured.

However, the demand for monitoring and regulating assets (manufacture and constructed components of buildings) has been inspiring researchers to study their impact further [8]. Koseoglu et al. simplifies and summarizes the complex system of DT into three major segments: physical elements, digital counterparts/models of those elements, and a web of data flow (Fig. 5). However, most of the tech. Integration are being examined under the limited scope of digitization alone, and the matter requires investigation from a wider scope where other elements like implementation impact, manufacturing, building, and planning should also be considered.

Research Design

Research

The research aims to assess the impact of digital transformation on the construction industries and how it may affect the operational norms of design and engineering firms. As digital technology continues to evolve, these companies face the dilemma of either sticking to the traditional construction processes or adapting to a new paradigm shift. Digital transformation substantially improves and optimizes the construction process throughout the project life cycle [4], but accessing disruption caused due to this transformation is crucial; so, we can develop an understanding of the risks and challenges involved and be equipped with the knowledge to overcome them. A systematic literature review was conducted which helped develop the significance of finding the answers to the research questions and their implications. Furthermore, the review aided in a comprehensive understanding of the obstacles, applications, and key factors of successfully implementing BIM and DT.

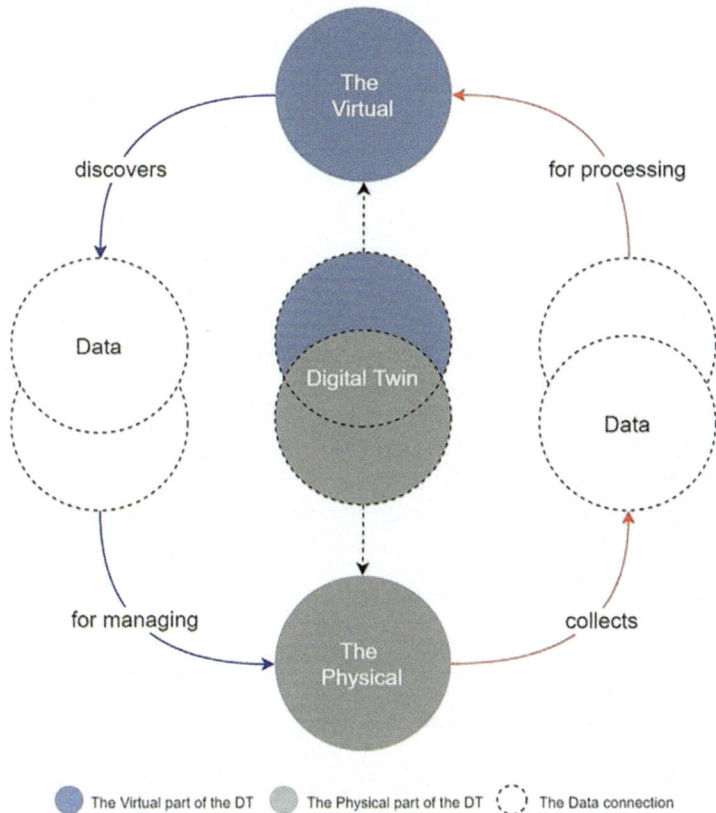

Fig. 5 Digital twin model [8]

Research Question and Objectives

One research question and four additional objectives are posed to promote understanding of BIM implementation in AEC.

What is the impact of implementing BIM and DT on companies operating within the AEC ecosystems?

- Is it currently standard practice for design and construction companies to integrate BIM and DT when carrying out their operations?
- What is the disruption process that an organization must account for when implementing these technologies?
- How easily these technologies can be adapted and implemented?
- What are the critical success factors and challenges of implementations?

Thereafter, elaborate qualitative research was conducted which included focus group discussions, surveying individuals operating in the AEC ecosystem, and interviewing key players operating at high positions in construction industries in UAE and other nations. This aided in filling in the gaps in the literature review and helped furbish first-hand information of how digital transformation may affect a company's operations, discover the challenges of implementation, and outline the resources required to overcome these challenges.

Methodologies

Systematic Literature Review

Over 45 journals and articles with high impact and good reputation were selected to extract information on BIM, DT, and IoT integration within the architectural and engineering domain of AEC. These publications were selected based on their keywords and commitment to exploring the implications of digital transformation in construction industries around the world. Any unrelated chapters were discarded based on their relevance to the research question of this chapter.

Following steps were taken to ensure the acquisition of good quality information only:

(i) Authentic sources: Science direct, Scopus, and IEEE were selected to acquire related articles and journals.
(ii) Selection criteria of only high-impact publications from the past 8 years were set.
(iii) Chapters were searched in reputable individual databases and libraries.
(iv) Irrelevant and similar data was eliminated.
(v) Data categorization and content analysis were conducted based on discussions, reviews, and chapter type.

In the first step, high-impact publications (journal, articles, and previously reviewed conference chapters) in the digital transformation of AEC were distinguished and acquired through the above-mentioned sources. The selection was narrowed down by eliminating any publications with a low cite score. Thereafter, advanced searches were carried out to develop a sound understanding of relevant concepts and applications of BIM and digital transformation.

Qualitative Analysis

To develop a sound understanding about the impact of adopting BIM in an AEC company, the research utilized qualitative analysis where surveys and interviews were conducted to gather information which was then used to confirm assumptions and literature. Later the chapter suggests the implication of the furbished data and how this information could be used for relevant personnel.

Survey

After establishing a comprehensive understanding of the topic by evaluating the literature review and gathering information from the available publications, the literature gaps were identified and the missing information was outlined. This information was deemed necessary for the completion of this study and for acquiring satisfactory answers to the research questions. Based on these research gaps and additional required data, a questionnaire was prepared that comprised of queries that would help extract the aforementioned information and establish a deeper understanding of the topic. The survey consisting of 16 questions was aimed to explore the perspective of individuals employed in the construction industry and how these engineers and designers relate to the digital transformation of construction processes.

A total of 234 candidates were selected for the questionnaire, which was a diverse group of people based in different countries around the globe, including the USA, Canada, UK, France, S. Korea, Australia, Kuwait, UAE, and Pakistan among other nations. These individuals were of different backgrounds, fields, and experience levels. However, to ensure relevance, only suitable candidates were selected who had a professional background in construction engineering, designing, project management, or real estate development.

After concluding the survey, the data was further refined before evaluation. Only unbiased and impartial responses were considered for further evaluation, and inconsistent data from candidates not meeting the selection criteria was discarded. This narrowed the final number of responses to a total of 225 and ensured that only error-free and high-quality data was considered to develop accurate conclusions.

Interviews

Focus group discussions were held to gather information from technical operators, business owners, and other stakeholders working in the construction industry of the United Arab Emirates and other nations. These discussions aimed to gather information on what companies have been struggling with to integrate BIM and DT in their operations and what are the challenges and limiting factors they must overcome. Furthermore, eight interviews were conducted, where only the individuals employed at key positions in construction industries were selected. These participants were practicing engineers and architects from different nations around the world including Canada, South Korea, and UAE. Five out of the eight interviewees were based in UAE. The interview aimed at extracting answers to six core questions and further information was extracted based on the progression of the interview and the professional background of the interviewee. The gathered information gave clear insights into how the shift from traditional norms of using 2d drawings and CAD tools have evolved to digitized workflow with the aid of BIM and DT is affecting the workability of the organizations. Table 3 summarizes how the data collection and experiments were conducted.

Table 3 Synthesis of qualitative analysis steps

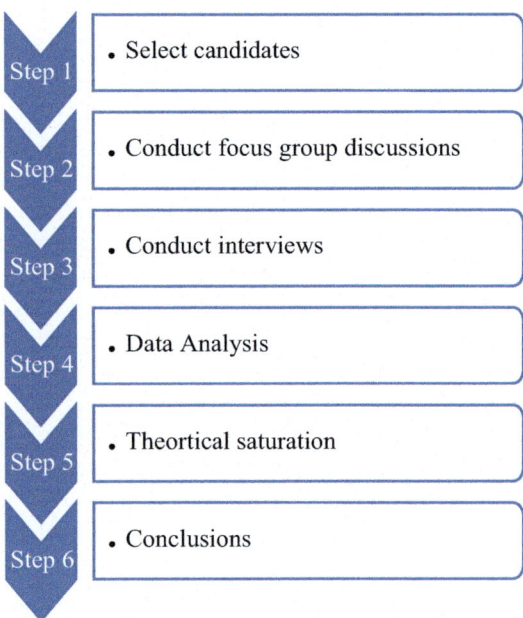

- Step 1: Select candidates
- Step 2: Conduct focus group discussions
- Step 3: Conduct interviews
- Step 4: Data Analysis
- Step 5: Theortical saturation
- Step 6: Conclusions

These methodologies are selected because they are valid when limited information is available to analyze a complex topic. A diverse range of candidates is a definite strong aspect of this method. The diverse background and location of sources will eliminate vagueness and regional biases which can be influenced by cultural values. A similar design was followed by Liy Y in 2017, where he focused on researching and acquiring information on very comparable research questions to this study.

Findings and Discussions

Following an extensive literature review, the chapter conducted an elaborate qualitative analysis to fill in the gaps and further develop the understanding of the impacts of implementing BIM and digital twin across AEC ecosystems. The findings reveal and underline the key challenges that a design or engineering firms need to overcome and what may or may not cause process disruption; in contrast, these also explore and further clarify the benefits that can be evident once the technology has been adapted. However, the analysis found that the challenges and disruptions are not a one-time process; as the technologies are constantly evolving, the companies need to stay up to date and regularly keep updating their platforms and skills.

Relevance of Candidates and Results

Only relevant candidates were narrowed down for this study where all of them were operating in architecture or engineering departments across the globe. All the candidates were highly qualified with 47% of them holding a post-graduate or a doctorate and the rest were at least graduates. Figure 6 shows how the majority of the responses were received from candidates with Architectural, Civil, and Mechanical backgrounds where they collectively accounted for 60% of the total; and the rest were of relevant fields.

Furthermore, the findings conclusions are developed from collecting data from around the globe (Fig. 7), where candidates were selected from a wide range of countries. However, it must be noted that most of the responses received were from

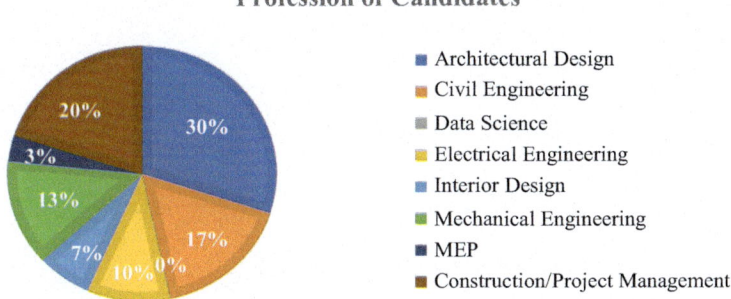

Fig. 6 Profession of candidates

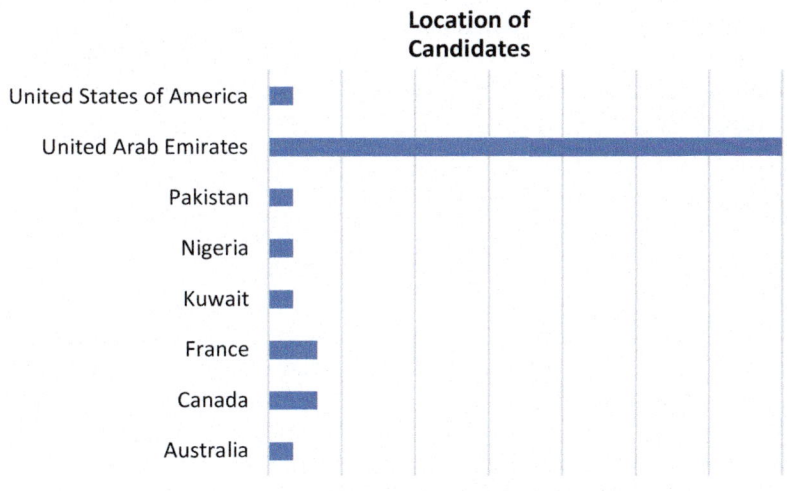

Fig. 7 Location of candidates

Table 4 Experience of candidates

Experience (years)	No. of candidates (%)
0–3	10
3–8	30
8–12	20
12–18	20
18+	10

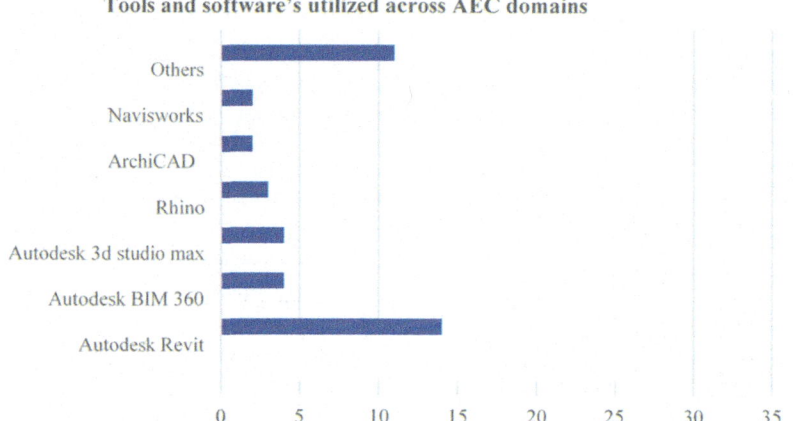

Fig. 8 Tools and software utilized across AEC domains

individuals operating in the United Arab Emirates AEC industry. Nevertheless, diversity accounts for a general allocation of different cultures and environments.

The experience levels of all candidates varied widely (Table 4), contributing to a broader range of results. It appeared that the use of technology diminished with age, as older individuals found it harder to cope with ever-changing technology. However, each individual still possessed a basic understanding of its operations.

Findings

Questionnaire

After establishing the professional background of each candidate, further analyses were conducted based on their background; this helped create and display a bigger picture of the impact of BIM. The first step was to identify the basic tools and software that the candidates utilized (Fig. 8). As the strata consisted of a diverse range of professionals with various backgrounds, the answer to what software they utilize enabled understanding of the fact that whether there is a unifying platform where different departments of the construction industry could share their digital data. The results indicated that each individual would operate and utilized their preferred tool depending on the department depending on what domain of AEC they operate in; however, CAD tools like Autodesk AutoCAD seem to be the only platform that each individual from every department and background is familiar with. AutoCAD is not a BIM tool, rather is used for drafting and drawing; however, with it being the unifying element, it can be presumed that there is at least some scope of coherence where every domain can unify and fall back to if needed.

Upon further analysis and the data from interviews, it is acknowledged that interexchange ability among the different platforms remains to be an issue. Nevertheless, the literature reveals that this is an area of improvement where the vendors are working on creating a seamless integration of tools for the exchange of data across the platform. However, this is only noticed among the platforms that are manufactured by the same vendor, and little effort is being made to standardize software and data created by platforms that are managed by different companies.

To answer the research question, it was important to realize how effective a tool BIM and DT are for its user. Though getting expert training and a good hold over these tools require training, time, and resources, it is yet deemed to be worth the results that modern digital technologies offer. About 73% of the user believe that it is a "very effective" tool with the rest still stating that it is at least "moderately effective" in construction operations. None of the responses received would state that it is ineffective, and the majority agrees that at the very least BIM is more effective than traditional CAD tools.

The majority of the people also acknowledge that BIM and DT can model and represent very accurate data if knowledgeable operators are available. This is a theme that seems to contrast with views from the literature review, as some authors have stated that accuracy in models is recurring issue and an area in need of improvement. Moreover, everyone agrees that implementing BIM and DT enhances the construction process at least to some extent and significantly increases productivity by optimizing the process as shown in Fig. 9. The optimization that digital

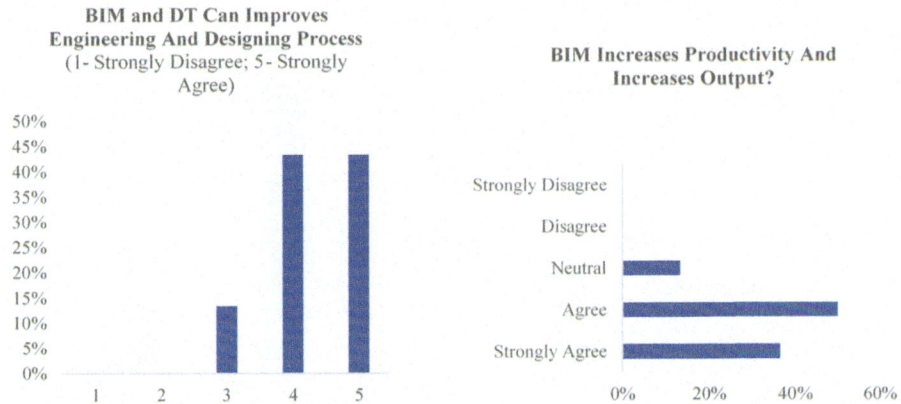

Fig. 9 Impact of BIM and DT in term of productivity and efficiency

transformation enables seems to be the main driving force of why it is being so widely accepted and adopted. The majority of the responses state that introduction of BIM and DT does enable individuals and companies to indulge in practices that lead to leaner and more sustainable construction. Under 13% of respondents had different views where they do believe that it enhances the process but does not facilitate leaner construction.

Another notable finding is that despite its enormous advantages, the majority of people do believe that the technology has yet to advance further. About 77% of the people state that as of currently, the digital transformation is most effective in the design phase and requires further development in the construction and operational phases. This indicates that digital twin that is more prominent in construction and operational phases has not yet become a norm in the industry; many are not aware of its existence and the technology has yet to advance further till it becomes standard practice across AEC sectors.

The analysis proceeds to investigate the opinion of how challenging was it for the interviewees or their organization to acquire the required recourse to implement BIM and DT. These resources included skilled operators, time, financial investment to procure suitable equipment among other aspects. Most believe that it is at least a little challenging and cumbersome to attain these resources (Fig. 10). Symbolizing the initial efforts and investment required from the management to facilitate attainable implementation. The most dominant factor seemed to be the time required to train staff; as discussed earlier, each individual would prefer to use their software and tool of choice; however, it is empirical for every department to agree on the same or similar tools as every team needs to be working on the same platform due to discrepancies and inaccuracies when exchanging data over various platforms.

The results narrate that there is a significant impact in terms of recourse procurement when implementing digital transformation and is a hurdle new companies need to overcome. However, a significant number of people do believe that these

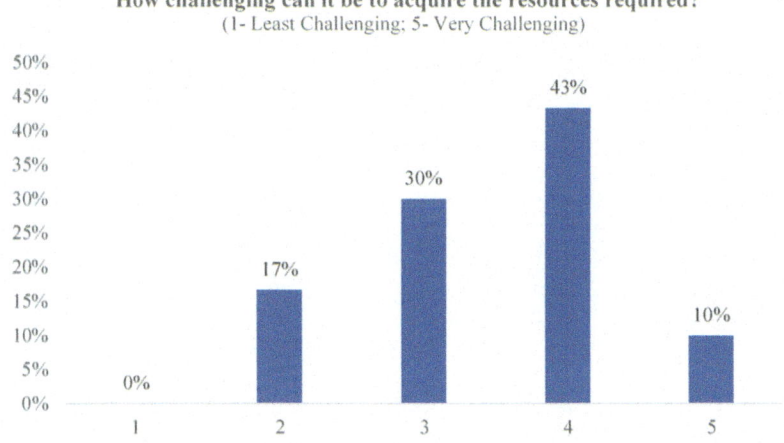

Fig. 10 Resource required and its acquisition

challenges will become easier to overcome as the recourses will be easily available over time. Just under 25% of the population agrees that it will still be just as difficult in the coming times as well. Most of the people also seem to relate these recourses with a significant investment. Where companies will have to acquire adequate talent and pay for suitable equipment, this would not be an issue for companies who do not want to implement digitization. Moving on towards process disruption caused by the implementation of BIM and DT, when asked how it could alter the ongoing practices in a firm, everyone had very polarized reviews on if they considered it to be disruptive or not (Fig. 11).

Where some perceive it as being disruptive to the process, many take this disturbance as a necessary improvement and prefer considering it evolution instead of disruption. Either way, it is established that implementation would alter the ongoing practices and operational norms in some way. Awareness of future implications is of existential importance and the fact of process disruption is an impact that BIM and DT implementation will have in the organization. Management needs to address and overcome it, so that they can attain flawless integration, as it can dictate a successful or failed integration. Moreover, once the integration has been attained, it should also be noted that the companies will continually need to keep their system updated as the technology is bound to keep changing over time. If not kept up with the digital evolution, then companies may find themselves losing their competitive edge as their technology is bound to become obsolete over time.

With the advent of digital transformation, traditional practices seem to be diminishing with the passage of time and new solutions are being discovered as the technology evolves. Many believe that it has become a prerequisite for an architect or engineer to be well aware of BIM tools in order to be a practicing individual in the construction industry.

When asked if the traditional construction norms of CAD designing and other tools are to become obsolete and unified BIM tools would take over, most agree that this would be the case. A notable 94% of the population believes that new advances would make traditional practices obsolete. However, there is still a remaining 6% of the people who strongly believe that the practices will evolve, but older norms are the fundamental bases of evolution and will remain relevant in one way or another. Regardless of the impact it has on the AEC ecosystem, the majority of the people do favor this transformation and are looking forward to the time when current

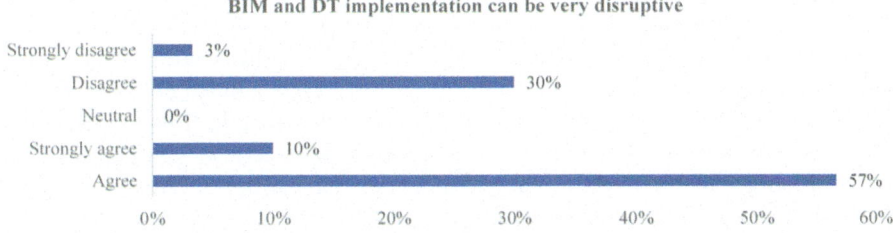

Fig. 11 Process disruption due to BIM and DT implementation

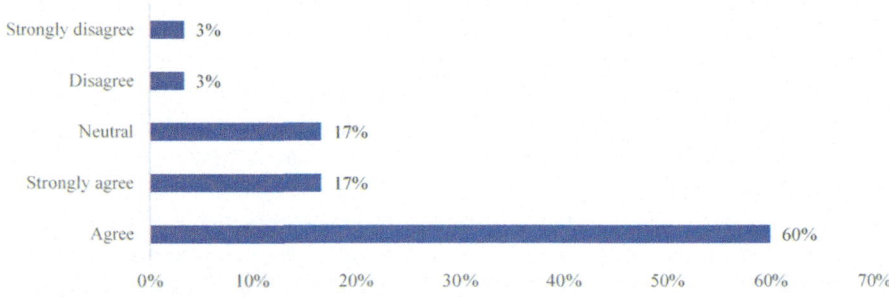

Fig. 12 Evolution of construction norms and practices

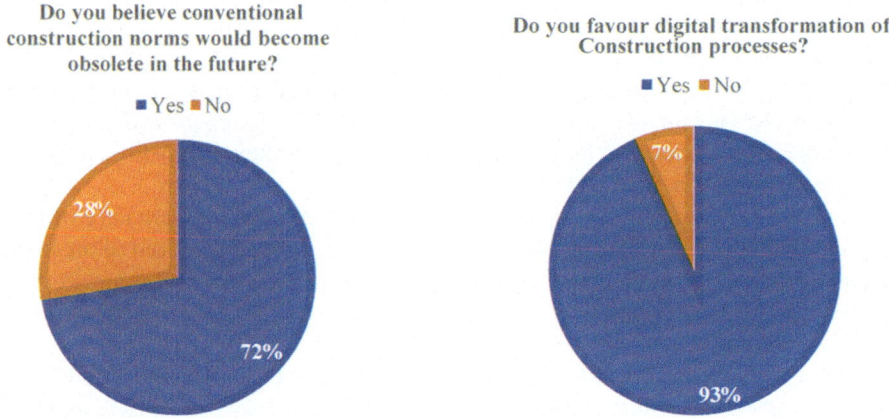

Fig. 13 Conventional norms and digital transformation

challenges are easy to overcome in the future The charts below (Figs. 12 and 13) summarize the reviews of current users in the industry.

Interviews

After analyzing the questionnaire, a more tailored interview session was held where only the key players with extensive experience in the market were considered for gathering information. These individuals were working at notable positions or had significant experience in the market. This ensured only meaningful and high-quality data was collected. Six questions were discussed in the interview, which would lead to developing sound answers to the research question and more cohesive development of information gathered from literature reviews and questionaries.

Usage and Adoption

The first question aimed to understand if the candidate were currently utilizing BIM and DT in their organization, what tools were they using, and to what extent? As evident from the questionnaire, each BIM tool has its strengths and weaknesses; hence, most individuals use specific tools that are tailored to their distinct domain within the AEC ecosystem. Every individual or team chooses the software and technology which can give them the most efficient results. One common theme that was discovered is that there is a huge variation in the tools that designers and engineers use. Designers currently are not inclined to use BIM as it is beyond their scope of work and mostly just utilize CAD software like Rhino 3D or Autodesk 3d studio for purposes of their job description. However, most of the designers operating CAD software are also aware of at least the basic operation of the common BIM software like Revit and BIM 360. This is because they often need to interact with engineering departments and need to transfer their data and models in formats that the engineers can relate to.

Moreover, another major piece of information that was found is that most small companies operating in AEC are not well acquainted with the operation of BIM tools and their process is limited to CAD tools at best. On the other hand, bigger companies dealing with the complex project do consider BIM tools to be a norm and a necessary part of their operations. Some smaller design firms find it easier to outsource the BIM modeling to experts as it is easier for them to hand over the responsibility to experience professional instead of going through the process of adoption and generating a new internal team of the workforce that can handle and operate the unfamiliar BIM tools.

Impact of Implementing BIM and DT

Following this discussion, the next step was aimed to explore how the integration of digital tools may impact the operation of interviewees' teams and what were the positive or negative impacts of utilizing these tools. Along with this, the challenges that individuals and companies have to overcome to attain meaningful utilization of BIM and DT tools and technology were also addressed. Most agree that BIM has enormous advantages and enable the organization to enhance their capacities and potential by

- Increasing productivity.
- Increasing accuracy.
- Increasing workflow.
- Enabling firms to efficiently work on complex projects.
- Providing time-efficient operations.
- Making it easier to entertain design updates and alterations.
- Allowing to create and analyze detailed construction drawings.
- Facilitating construction management processes.

- Improving communications across domains.

However, many also argue that the technology is yet to be optimized and many issues are yet to be addressed. A common factor of disturbance was the theme of inadequacy of BIM and DT tools to communicate or exchange data and models among themselves. Often operators of different backgrounds, for example, an architect and a civil engineer, will rely on third-party software to export their models to where both the parties could interact and use the model for communicating and exchanging information. This transfer of data either across software or through third-party tools presents its challenges as it is often not easy to transfer data across mediums; even when possible, there is a high probability that the models will be reliable for further analysis as the data may get corrupted. However, others state that data exchange is currently seamless, and they do not face any issues when transferring across different platforms.

The ambiguity in the findings could be due to the lack of operational awareness of BIM and CAD tools among individual operators; or alternatively, it could be the result of particular tools, which may not allow the exchange of models across different software. Nevertheless, a common theme of knowledge is that this is a field of development and companies are working to improve data transfer over time, where tools and software are constantly being updated. Among the top negative impacts of implementation was the steep learning curve that the teams and individuals must go through when implementing digital technology in firms. Everyone agrees that the benefits far out-weigh the negative impacts and the transformation is inevitable for any firm looking to stay relevant in the market. Nevertheless, organizations and individuals yet affirm the existence of negative impacts and consider them to be the challenges that companies need to overcome in order to successfully implement BIM and DT. These aspects include:

- Steep learning curves.
- The complexity of BIM tools.
- The time required to train individuals in BIM tools.
- A certain level of skill set is required from the operators.
- Even advanced users can at times find themselves stuck in an abnormal situation and would require the assistance of technical personnel to resolve the issues.
- Technology is constantly evolving which means the operators need to stay up to date with new variants and techniques required to operate these tools.
- It requires financial investment.
- Critical success factors and future of digital transformation.

The next line of questions aimed to understand the critical success factors, process disruption, and what individuals and companies imagine the future of digital transformation to be like.

It is imperial to address the CSFs in order to attain successful implementation of BIM and DT. Besides overcoming the challenges and accounting for the negative impacts of BIM integration, companies must also be mindful of the following CSFs:

- The complexity of digitally transforming operations and processes.

- Willingness to adapt and support top management.
- Hiring and acquiring talented employees.
- Training employees.
- Staying up to date with current market situations and norms.
- Having a growth mindset that encourages innovation and adaptability.
- Acquiring required resources before implementation.
- Risk management.
- Time to adapt and inherit the technology.

Despite the challenges, everyone agrees that BIM, DT, and digitization are the way forward as it enables easier and more convenient project delivery systems. Every interviewee unanimously agreed that they encourage this transformation and they acknowledge the challenges but also state that the industry is swiftly moving towards adopting these trends and it is just a matter of time till all the stated challenges are relatively easy to overcome, even by smaller firms.

Conclusion

The purpose of this research chapter was to analyze the current AEC market norms and how digitization is affecting the constructions processes. Through comprehensive literature review and extensive qualitative analysis, the study confirms that there are enormous benefits of implementing building information modeling systems and digital twins throughout the construction process. However, it also explores the challenges that need to be overcome by companies and the process disruption that implementations cause.

Based on acquired data and analyzing the current AEC ecosystem across the world, the study found that computer-aided operations have become a norm in the construction industry. The reason for this is the optimization and efficiency that digitization enables. It has been established that digital twins is an extension of BIM tools which intern has evolved from CAD tools. Integrating BIM allows simplifying designing and constructing processes of complex projects, which otherwise may not be possible. Although BIM is perceived to be a norm in the industry, it is only being utilized just by big firms, as smaller firms lack the resources or mindset to implement it within their daily operations. They are comfortable with traditional CAD tools as it is proving to be sufficient when working on smaller-scale projects. However, it is also denoted that it is just a matter of time until all sectors and domains adopt BIM and DT and integrate it within the organizations.

Digital twin originates from BIM models but accounts for real-time data from the construction site during the construction and operation phases. DT is a relatively new concept, and few companies are aware of it. However, just like BIM and CAD, it is also deemed to become a part of regular practice in the construction industry as it facilitates leaner and sustainable construction processes, even more so than what BIM and CAD offer.

When implementing these technologies, companies will have to constantly adapt and evolve with continuous development, to stay up to date with the ever-changing technologies. They will have to ensure that they are well equipped to handle the challenges, including the time required for adaptation, financial investment, acquisition of experienced operators, and training new employees to handle the technology. Despite a wide list of challenges that need to be overcome, it is not hard to implement digitization in the construction process for most companies, small and big alike. They, however, need to be mindful of all the critical success factors, which would dictate how successfully they can implement the technology. These CSF mostly relate to the skill set of staff, the mindset of the organization, support from management, available resources, and a growth mindset.

Implications and Limitations

The data and information gathered in this study can be used across various dimensions that are related to the digital transformation of the construction process. It gives an extensive explanation of what construction processes were like and what they are right now and depicts how these practices are evolving over time. It denotes and signifies how individuals and companies operating in the AEC ecosystem around the world need to be accepting of change and evolution and that adapting to the innovative technology can optimize and improve the construction process. The implication promotes a process transformation that encourages more sustainable construction practices in the future that present optimal solutions to traditional problems.

This chapter conducted an extensive analysis that concluded and fulfilled the research objectives; however, it needs to be apprehended that the qualitative analysis was limited to a relatively few numbers candidates. And although the data was collected from various countries across the world, the results may vary when specific regions and areas are considered. This is because of external factors like local norms and the cultural background of the specific regions. However, the conclusion and finding from this chapter can be applied as general standards across the AEC ecosystem around the globe and the information reflects and accounts for variation caused due to unforeseen factors.

References

1. Antwi-Afari, M. F., Li, H., Pärn, E. A., & Edwards, D. J. (2018). Critical success factors for implementing building information modelling (BIM): A longitudinal review. *Automation in construction, 91*, 100–110.
2. Carvalho, J. P., Bragança, L., & Mateus, R. (2019). Optimising building sustainability assessment using BIM. *Automation in Construction, 102*, 170–182. https://doi.org/10.1016/j.autcon.2019.02.021

3. Chen, Y., Yin, Y., Browne, G. J., & Li, D. (2019). Adoption of building information modeling in Chinese construction industry: The technology-organization-environment framework. *Engineering, Construction and Architectural Management, 26*(9), 1878–1898. https://doi.org/10.1108/ECAM-11-2017-0246
4. Deng, M., Menassa, C., & Kamat, V. (2021). From BIM to digital twins: A systematic review of the evolution of intelligent building representations in the AEC-FM industry. *Journal of Information Technology in Construction, 26*, 58–83. https://doi.org/10.36680/j.itcon.2021.005
5. Ghaffarianhoseini, A., Tookey, J., Ghaffarianhoseini, A., Naismith, N., Azhar, S., Efimova, O., & Raahemifar, K. (2017). Building Information Modelling (BIM) uptake: Clear benefits, understanding its implementation, risks and challenges. *Renewable and Sustainable Energy Reviews, 75*(C), 1046–1053.
6. Gu, N., & London, K. (2010). Understanding and facilitating BIM adoption in the AEC industry. *Automation in Construction, 19*, 988–999. https://doi.org/10.1016/j.autcon.2010.09.002
7. Khajavi, S., Hossein Motlagh, N., Jaribion, A., Werner, L., & Holmström, J. (2019). Digital twin: Vision, benefits, boundaries, and creation for buildings. *IEEE Access, 7*, 147406–147419. https://doi.org/10.1109/ACCESS.2019.2946515
8. Koseoglu, O., Sakin, M., & Arayici, Y. (2018). Exploring the BIM and lean synergies in the Istanbul Grand Airport construction project. *Engineering, Construction and Architectural Management, 25*(10), 1339–1354.
9. Liu, H., & Ran, Y. (2013). Study on BIM technology application prospect in engineering management industry. In *2013 6th international conference on information management, innovation management and industrial engineering* (Vol. 3, pp. 278–280).
10. Liu, Y., van Nederveen, S., & Hertogh, M. (2016). Understanding effects of BIM on collaborative design and construction: An empirical study in China. *International Journal of Project Management, 35*, 686. https://doi.org/10.1016/j.ijproman.2016.06.007
11. Manzoor, B., Othman, I., Kang, J. M., & Geem, Z. W. (2021). Influence of Building Information Modeling (BIM) implementation in high-rise buildings towards sustainability. *Applied Sciences, 11*, 7626. https://doi.org/10.3390/app11167626
12. Nguyen, Q., Thúy, T., & Vu, N.-N. (2021). BIM-based competitive advantages and competitive strategies for construction consultancy SMEs: A case study in Vietnam. *International Journal of Sustainable Construction Engineering Technology, 12*, 1–11. https://doi.org/10.30880/ijscet.2021.12.03.001
13. Ohueri, C. C., Bamgbade, J. A., Chuin, A. L. S., Hing, M. W. N., & Enegbuma, W. I. (2022). Best practices in building information modelling process implementation in green building design: The architects' insights. *Journal of Construction in Developing Countries, 27*(1), 79–93. https://doi.org/10.21315/jcdc2022.27.1.5
14. Tang, S., Shelden, D., Eastman, C., Pishdad-Bozorgi, P., & Gao, X. (2019). A review of building information modeling (BIM) and the internet of things (IoT) devices integration: Present status and future trends. *Automation in Construction, 101*, 127–139. https://doi.org/10.1016/j.autcon.2019.01.020
15. Tulubas Gokuc, Y., & Arditi, D. (2017). Adoption of BIM in architectural design firms. *Architectural Science Review, 60*, 483–492. https://doi.org/10.1080/00038628.2017.1383228
16. Volk, R., Stengel, J., & Schultmann, F. (2014). Building Information Modeling (BIM) for existing buildings – Literature review and future needs. *Automation in Construction, 38*, 109–127. https://doi.org/10.1016/j.autcon.2013.10.023

Index

A

Aerospace, 6–7, 71, 72, 75, 167, 187, 210, 211, 228–229, 239, 241, 243, 252–253, 260

Agriculture, 11–12, 40, 45, 74, 239, 241, 243, 250–251

AlphaPose, 97, 99

Analytics in manufacturing performance management, 121

Application performance, 90

Applications, 3, 22, 45, 59, 69, 96, 118, 151, 166, 187, 221, 239, 259, 284, 299

Arbitrary Extra-Large You Only Look Once Version 5 (AXYOLOV5), 95–111

Architecture, 2, 15, 22, 51, 82, 96, 97, 99–101, 121, 122, 126, 132, 134, 137, 139–141, 146, 156, 220, 230–231, 234, 235, 255, 286, 287, 313

Artificial intelligence (AI), 2, 3, 15, 16, 19, 22, 23, 26–38, 41, 42, 51, 72, 74, 142, 176, 177, 182, 190, 234, 242, 260, 261, 290

B

Blockchain, 15, 20, 21, 38, 39, 95–111, 145–147, 150–163

Building information modeling (BIM), 243, 251, 297–322

C

Construction, 8, 9, 21, 26, 50, 51, 73, 148, 150, 162, 167, 231, 235, 239, 240, 243, 251–252, 271, 278, 297–311, 314–319, 321, 322

Cyber security, 59–66, 121–123

D

Data integration, 46, 51, 79, 199, 235, 242, 245, 253–255, 284

Data security, 36–39, 64, 98, 150, 187, 192, 241, 243, 245, 248, 249, 252–255

Deep learning (DL), 19, 22, 27, 35, 74, 97, 104, 108, 110, 190, 197–199, 243

Design and development, 75, 165, 166, 220, 228, 284

Digital transformation, 170, 188, 252, 255, 261, 297, 300, 305–308, 310, 311, 315–318, 320, 322

Digital trust, 59–66

Digital twin (DT), 1, 19, 45, 71, 96, 115, 147, 165, 187, 219, 239, 259, 282, 297

Digital twin technology, 26, 30, 32, 36, 39, 72–74, 86, 96, 111, 117, 120, 141, 148–151, 155, 158, 161–163, 167, 176, 179, 180, 203, 209, 234, 235, 240–242, 255–257, 272, 278, 281–295

E

Education, 14, 41, 45–55, 63, 165–184, 192, 201, 212, 225, 226, 246, 247, 251, 256, 288

Energy, 8–10, 21, 26, 28, 31, 32, 40, 41, 45–55, 74, 75, 155, 158, 160, 161, 166, 220, 222, 224, 225, 227–228, 230, 231, 233–235, 239–244, 247–248, 252, 260, 263–265, 268, 270, 281, 285, 286, 293, 299–301

Energy and education application, 45–55

F
Firewalls, 37, 63

H
Healthcare, 7–8, 15, 20–26, 31, 32, 38–41, 47, 48, 51, 71, 73–76, 95–111, 160–161, 166, 187–192, 195, 196, 198, 201–205, 207–216, 219, 220, 225–226, 239, 240, 242, 243, 245–247, 260, 288, 293
Human digital twin (HDT), 187–216
Human-centered, 165–184

I
Internet of Everything (IoE), 286–290
Internet of Things (IoT), 10, 15, 23, 47, 48, 51, 72, 73, 79, 95, 96, 115, 142, 165, 190, 201, 222, 224–225, 252, 256, 260, 261, 267, 270, 271, 278, 282–286, 297, 298
Internet of Things (IoT) in manufacturing performance management, 122

L
Literature reviews, 150–151, 162, 170, 241–243, 298–312, 315, 318, 321

M
Machine learning (ML), 2, 15, 19, 26–29, 33, 35, 41, 48–51, 54, 74, 80, 95, 96, 124, 142, 177, 190, 194, 196–197, 211, 214, 224, 234, 255, 290
Manufacturing, 3–5, 15, 24, 25, 28, 32, 41, 45, 46, 50, 51, 72–75, 95, 96, 115–121, 123, 124, 126–129, 131–134, 136–139, 141, 142, 148, 150, 151, 159–160, 166, 170, 181, 187, 192, 197, 210, 219–224, 239, 242–245, 250, 259, 285, 293, 308
Manufacturing digital twin, 119
Manufacturing digital twin solution architecture, 115, 116, 122, 140
Modelling software, 170

N
Network security, 9, 63–64

O
Overall equipment effectiveness (OEE) calculation, 121, 128, 130, 132

P
Practical Byzantine Fault Tolerance (pBFT), 102–104
Predictive maintenance, 20, 31, 41, 50, 120, 138, 139, 167, 201, 209, 220, 222–224, 227–230, 232, 235, 239, 242, 244–245, 247, 252–253, 293, 294

R
Real-time monitoring, 11, 20, 28, 32, 34, 39, 41, 75, 77, 80, 159, 199, 205–206, 210, 222–224, 227, 230, 231, 234, 241, 243, 249, 251–254, 270, 282, 284–286, 289, 293, 294
Real time simulation, 79
Requirements, 4, 10, 48, 69, 72, 79–81, 115–118, 121, 122, 125, 126, 138, 141, 146, 155, 162, 169, 173, 221, 233, 241, 244, 246, 255, 268, 270, 283, 286, 308

S
Simulations, 1–3, 6, 7, 11, 14, 16, 20, 28, 45–47, 49, 54, 72, 73, 76, 79, 83, 84, 89, 117, 121, 124, 138, 139, 141, 159, 160, 168–170, 176, 179, 187, 191, 195, 196, 199, 201, 203, 204, 208, 209, 220, 239, 242–244, 247, 249, 251, 253, 254, 256, 259, 282, 284, 285, 294, 295, 297, 298, 300–302, 304
Smart cities, 9–10, 15, 28, 32, 41, 48, 75, 150, 160, 166, 181, 219, 220, 222, 224, 225, 239, 241, 243, 249–250, 259–278, 281, 282, 286–293
Smart contracts, 147, 150, 153, 154, 156, 157, 161, 162
Software architecture, 87
Software testing, 89
Students learning, 174, 184

T
Technology, 2, 19, 45, 71, 96, 115, 145, 167, 187, 225, 239, 261, 282, 297

Traffic flows, 32, 167, 224, 230, 231, 240, 241, 249, 254, 265, 268, 270, 282, 294, 295

Transportations, 10, 26, 31, 32, 40, 41, 48, 73–75, 160, 166–167, 219, 220, 222–225, 230–233, 239, 240, 243, 249–250, 254, 261, 268–270, 281–295

V

Virtual replica, 1, 28, 31, 39, 72, 86, 89, 159, 165, 166, 169, 187, 188, 190–192, 204, 209, 210, 212, 221–224, 226–232, 234, 239, 240, 244, 246–248, 252, 254, 285, 293

Virtual sensors, 126, 129, 139

Virtual technologies, 24

SPRINGER NATURE

GPSR Compliance

The European Union's (EU) General Product Safety Regulation (GPSR) is a set of rules that requires consumer products to be safe and our obligations to ensure this.

If you have any concerns about our products, you can contact us on ProductSafety@springernature.com

In case Publisher is established outside the EU, the EU authorized representative is:

Springer Nature Customer Service Center GmbH
Europaplatz 3
69115 Heidelberg, Germany

The manufacturer's authorised representative in the EU is Springer Nature Customer Service Centre GmbH, Europaplatz 3, 69115 Heidelberg, Germany. If you have any concerns regarding our products, please contact ProductSafety@springernature.com

Printed and bound by CPI Group (UK) Ltd, Croydon, CR0 4YY

25/03/2026

02078169-0006